CONTROL AND DYNAMIC SYSTEMS

Advances in Theory and Applications

Volume 18

CONTRIBUTORS TO THIS VOLUME

MARK J. BALAS
THURBER R. HARPER
C. A. HARVEY
C. D. JOHNSON
KAPRIEL V. KRIKORIAN
C. T. LEONDES
R. E. POPE
C. JOSEPH RUBIS
CRAIG S. SIMS
ROBERT E. SKELTON
GEORGE R. SPALDING
AJMAL YOUSUFF

CONTROL AND DYNAMIC SYSTEMS

ADVANCES IN THEORY AND APPLICATIONS

Edited by
C. T. LEONDES

School of Engineering and Applied Science
University of California
Los Angeles, California

VOLUME 18 1982

ACADEMIC PRESS
A Subsidiary of Harcourt Brace Jovanovich, Publishers

New York London
Paris San Diego San Francisco São Paulo Sydney Tokyo Toronto

ACADEMIC PRESS RAPID MANUSCRIPT REPRODUCTION

ACADEMIC PRESS, INC.
111 Fifth Avenue, New York, New York 10003

United Kingdom Edition published by
ACADEMIC PRESS, INC. (LONDON) LTD.
24/28 Oval Road, London NW1 7DX

LIBRARY OF CONGRESS CATALOG CARD NUMBER: 64-8027

ISBN: 0-12-012718-0

PRINTED IN THE UNITED STATES OF AMERICA

82 83 84 85 9 8 7 6 5 4 3 2 1

CONTENTS

CONTRIBUTORS

Numbers in parentheses indicate the pages on which the authors' contributions begin.

Mark J. Balas *(361), Electrical, Computer, and Systems Engineering Department, Rensselaer Polytechnic Institute, Troy, New York 12181*

Thurber R. Harper *(317), Propulsion Dynamics, Inc., Annapolis, Maryland 21401*

C. A. Harvey *(161), Aerospace and Defense Group, Honeywell Systems and Research Center, Minneapolis, Minnesota 55413*

C. D. Johnson *(224), Department of Electrical Engineering, The University of Alabama in Huntsville, Huntsville, Alabama 35899*

Kapriel V. Krikorian* *(131), Aerospace Groups, Hughes Aircraft Co., Culver City, California 90230*

C. T. Leondes *(131), School of Engineering and Applied Science, University of California, Los Angeles 90024*

R. E. Pope *(161), Aerospace and Defense Group, Honeywell Systems and Research Center, Minneapolis, Minnesota 55413*

C. Joseph Rubis *(317), Propulsion Dynamics, Inc., Annapolis, Maryland 21401*

Craig S. Sims† *(55), School of Electrical Engineering, Oklahoma State University, Stillwater, Oklahoma 74074*

Robert E. Skelton *(1), School of Aeronautics and Astronautics, Purdue University, West Lafayette, Indiana 47907*

George R. Spalding *(105), Department of Engineering, Wright State University, Dayton, Ohio 45433*

Ajmal Yousuff *(1), School of Aeronautics and Astronautics, Purdue University, West Lafayette, Indiana 47907*

*Present address: Hughes Aircraft Co., Los Angeles, California 90009.
†Present address: Department of Electrical Engineering, West Virginia University, Morgantown, West Virginia 26506.

PREFACE

As noted in earlier volumes, this annual international series went to the format of theme volumes beginning with Volume 12. The theme for this volume is techniques for the analysis and synthesis of large-scale complex systems. In the first chapter, Skelton and Yousuff present a comprehensive treatment of component cost analysis of large-scale systems. The applied implications include cost balancing methods for system design, failure mode analysis, model reduction techniques, the design of lower-order controllers that meet on-line controller software limitations, and other significant issues.

The second chapter, by Sims, deals with the increasingly important problem of reduced-order modeling and filtering. An important application of this material is to deal more effectively with the on-line, real-time computer control problem for large-scale complex systems. One of the most important classes of large-scale complex systems is that of distributed parameter systems. Techniques for the development of mathematical descriptions of such systems in order to implement effect optimal controllers are in a constant state of evolution and development. The third chapter, by Spalding, presents some powerful practical techniques for modeling distributed parameter systems.

Techniques for representing and controlling complex systems through singular perturbation techniques have been examined in recent literature. Krikorian and Leondes (Chapter 4) present these important and powerful techniques, and show how they can be applied to the optimal design of complex systems. For this application the system is partitioned into a fast-modes subsystem and a slow-modes subsystem, and these are amalgamated in a singular optimal system design.

The next chapter, by Harvey and Pope, is a comprehensive compilation of linear multivariable systems synthesis techniques with a collection of significant illustrative applications. This chapter may indicate the direction of new research pursuits in this rapidly changing field.

C. D. Johnson is identified with a number of significant advances in modern control theory. Not the least of these is his significant pioneering research in disturbance-accommodating controllers, which was first presented in an earlier volume in this series. In Chapter 6, Johnson extends his important work in this area to digital control of dynamical systems.

In many cases, system simulation for complex systems (e.g., the development of the system model) is very much an art relying on experience with principles or techniques. Therefore, Chapter 7, "Ship Propulsion Dynamics Simulation" by Rubis and Harper, both of whom are preeminent in this field, is of special significance because these issues are best understood by the development and presentation of significant examples.

In the final chapter, by Balas, "Toward a More Practical Control Theory for Distributed Parameter Systems," the practitioner will find many powerful techniques for the analysis and synthesis of complex distributed parameter systems.

The authors of this volume have made splendid contributions to what will undoubtedly be an excellent reference source for years to come.

CONTENTS OF PREVIOUS VOLUMES

Volume 4

Volume 5

Volume 6

Volume 7

Volume 8

Volume 9

Volume 10

Volume 11

Volume 12

Volume 13

Volume 16

Volume 17

Component Cost Analysis of Large-Scale Systems

ROBERT E. SKELTON

AJMAL YOUSUFF

School of Aeronautics and Astronautics
Purdue University
West Lafayette, Indiana

I. INTRODUCTION

One of the most fundamental concepts in systems theory is the basic definition of a dynamic system. A dynamic system may be defined as an interconnection of entities (which we

ISBN 0-12-012718-0

shall call "components") causally related in time. It seems equally natural and basic, therefore, to characterize the system's behavior in terms of contributions from each of the system's building blocks--"components." The performance of the dynamic system is quite often evaluated in terms of some performance metric we choose to call the "cost function" V. The cost function might represent the system energy or a norm of the output errors over some interval of time. Concerning the physical or mathematical components of the system, it is only natural then to ask question CC: *What fraction of the overall system cost V is due to each component of the system?*

This chapter is devoted to a precise answer to question CC and to several applications of the mathematical machinery developed for answering the question. Such an analysis will be called *Component Cost Analysis* (CCA). Conceptually, it is easy to imagine several uses for CCA.

(a) Knowledge of the magnitude of each component's contribution to the system performance can be used to suggest which components might be *redesigned* if better performance is needed. By redesigning so as to reduce the cost associated with these "critical" components (those with larger contributions to system performance), one is following a "cost-balancing" strategy for system design. *Thus, CCA can be useful in system design strategies.*

(b) Knowledge of the magnitude of each component's contribution to the system performance can be used to predict the performance degradation in the event of a *failure* of any component. *Thus, CCA can be useful in failure mode analysis.*

(c) Knowledge of the magnitude of each component's contribution to the performance of a higher order model of the system can be used to decide which components to delete from the model to produce lower order models. *Thus, CCA can be useful in model reduction.*

(d) Alternately, if one defines the components to include each dynamical element of a linear feedback controller, the knowledge of the magnitude of each component's contribution to the closed-loop system performance can be used to determine which dynamical elements of the controller to delete so as to cause the smallest change in performance which respect to the performance of the high-order controller. *Thus, CCA can be useful in the design of low-order controllers* that meet on-line controller software limitations.

This chapter will focus on possibility (c) in some detail.

This notion of using a performance metric is basic in the most well-developed and simplest problem of optimal control: the Linear Quadratic problem. However, one of the fundamental deficiencies of modern control theory is its absolute reliance on the fidelity of the mathematical model of the underlying physical system, which is essentially infinte dimensional. Many "failures" of modern control applications are due to modeling errors. Thus, theories that can more systematically relate the modeling problem and the control problem are sorely needed since these two problems are not truly separable, although most practice and theory presently treat them as separable. This chapter presents one such unifying theory and can be viewed as an *application* chapter in the sense that it is concerned with making the linear quadratic theory more *practical*. Thus, the proposed insights into the behavior of dynamic systems are available within the standard mathematical tools of linear quadratic and Linear Quadratic Gaussian (LQG) theories [6]. Hence, the contributions of CCA lie not in the development of new mathematical theories, but in the presentation of cost decomposition procedures that readily reveal the "price" of system components. A similar notion of "pricing" of system components is a common strategy in operations research and mathematical programming problems such as Dantzig-Wolfe decomposition and the dual algorithm by Benders [1,2]. However, such useful notions of pricing seem not to have found their way into common control practice. This chapter calls attention to the manner in which such notions can be used in dynamic systems. The mathematical details are quite different from the pricing of the static models of operations research, but the concepts are similar.

The concepts of CCA evolved in a series of presentations [3-5]. However, these introductory papers left unanswered the most important questions of stability, the best choice of coordinates, and development of the theory of minimal realizations with respect to quadratic performance metrics. This chapter, therefore, presents a complete theory for CCA and, in addition, develops the theory of minimal realizations with respect to quadratic performance metrics.

II. COMPONENT DESCRIPTIONS

The entities that compose dynamic systems are herein labeled "components." To illustrate the flexibility in the definition of components consider example 1.

Example 1. Let the vertical motion of a throttlable rocket be described by

$$m\dot{v} = f - mg, \tag{1a}$$

where m is the assumed constant mass, g is the gravitational constant, and f is the rocket thrust that is regulated by a fuel valve with the dynamics

$$\dot{f} = af + u \tag{1b}$$

for a given command u. Thus, for the system

$$\begin{pmatrix} \dot{v} \\ \dot{f} \end{pmatrix} = \begin{bmatrix} 0 & 1/m \\ 0 & a \end{bmatrix} \begin{pmatrix} v \\ f \end{pmatrix} + \begin{bmatrix} -1 & 0 \\ 0 & 1 \end{bmatrix} \begin{pmatrix} g \\ u \end{pmatrix}, \tag{2}$$

the *vehicle dynamics* (1a) with state v might be chosen as one system component, and the *valve dynamics* (1b) with state f might be chosen as another component. In this case one might wish to ascertain the relative contribution of the dynamics of the vehicle and the dynamics of the valve in the overall system performance metric

$$V = \frac{1}{T}\left\{\int_0^T f^2(t)\,dt + [v(T) - \bar{v}]^2\right\}, \tag{3}$$

where T is the terminal time at which the velocity $v(T) = \bar{v}$ is desired.

Alternatively, one may define components of (2) in *any* transformed set of coordinates of (2). For example, one might wish to know the relative contribution in (3) of the *modal* coordinates of (2), in which case the system components are q_1 and q_2 described by

$$\begin{pmatrix} \dot{q}_1 \\ \dot{q}_2 \end{pmatrix} = \begin{bmatrix} 0 & 0 \\ 0 & a \end{bmatrix} \begin{pmatrix} q_1 \\ q_2 \end{pmatrix} + \begin{bmatrix} -1 & -1/ma \\ 0 & 1/ma \end{bmatrix} \begin{pmatrix} g \\ u \end{pmatrix}. \tag{4}$$

As described in the Introduction, "component cost analysis" (CCA), is the procedure developed for answering question CC for *any* choice of component definitions. In the special case where the components are *modal* coordinates, the procedure is called "modal cost analysis" (MCA) [4]. It is possible to use CCA with any choice of component definitions including the "balanced" coordinates of Moore [7], the "output-decoupled" coordinates used in Tse *et al.* [8], etc. For any choice of

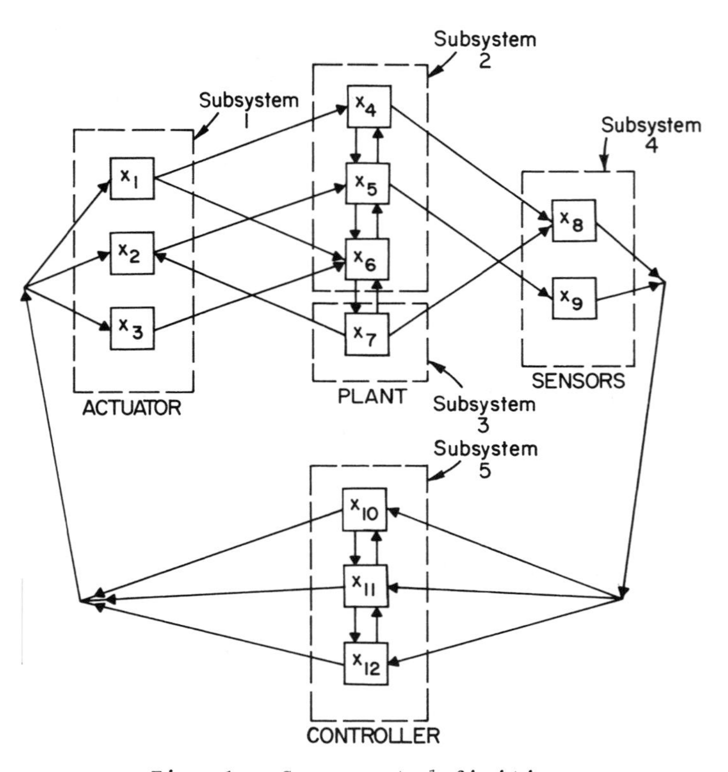

Fig. 1. Component definitions.

coordinates the n components may be described in the form

$$\dot{x}_i = \sum_{j=1}^{n} A_{ij}x_j + D_i w, \qquad x_i \in R^{n_i}, \qquad y = \sum_{j=1}^{n} C_j x_j. \tag{5}$$

For notational convenience, we shall later need to differentiate between the definitions of coordinates, components, and subsystems. These distinctions can best be introduced via example. A certain system has state x. Let the symbols x_i, χ_i, X_i all represent partitions of the state vector to various levels of detail. The scalars χ_i, $i = 1,\ldots, N$, will be called coordinates. The vectors $x_i \in R^{n_i}$, $i = 1,\ldots, n$, will be called the states of the components and $X_i \in R^{N_i}$, $i = 1,\ldots, s$, will be called the states of the subsystems. Then for $n = 3$, $s = 2$,

$$x = \underset{\substack{N\\\text{coordi-}\\\text{nates}}}{\begin{bmatrix} \chi_1 \\ \chi_2 \\ \vdots \\ \chi_{n_1} \\ \hline \vdots \\ \hline \cdot \\ \chi_N \end{bmatrix}} = \underset{\substack{n\\\text{compo-}\\\text{nents}}}{\begin{bmatrix} x_1 \\ \hline x_2 \\ \hline x_3 \end{bmatrix}} = \underset{\substack{s\\\text{sub-}\\\text{systems}}}{\begin{bmatrix} X_1 \\ \hline X_2 \end{bmatrix}}, \qquad \left(N = \sum_{i=1}^{n} n_i = \sum_{i=1}^{s} N_i\right). \tag{6}$$

As an example of component definitions, consider Fig. 1, where dynamic elements x_i, $i = 1,\ldots, 12$, and their interconnections are described. Each of these dynamic elements (selected *a priori* by the analyst) with state x_i, $i = 1, \ldots, 12$, is called a component of the system. However, each component may have additional dynamical variables $\chi_1, \chi_2, \ldots,$ called coordinates. This coordinate view of the system is the *microscopic* view of the system, whereas, the view of certain collections of components, called subsystems, is the *macroscopic* view of the system. For the example in Fig. 1, see from (6) that

$$X_1 = \begin{pmatrix} x_1 \\ x_2 \\ x_3 \end{pmatrix}, \quad X_2 = \begin{pmatrix} x_4 \\ x_5 \\ x_6 \end{pmatrix}, \quad X_3 = (x_7),$$
$$X_4 = \begin{pmatrix} x_8 \\ x_9 \end{pmatrix}, \quad X_5 = \begin{pmatrix} x_{10} \\ x_{11} \\ x_{12} \end{pmatrix}. \tag{7}$$

Of course, when the analyst chooses $n_i = 1$ and $N_i = 1$, there is no distinction between coordinate, component, and subsystem.

III. CONCEPTS OF COST DECOMPOSITION

In our preliminary discussions, we presume that *the linear system model*

$$\dot{x} = A(t)x + D(t)w, \quad x \in R^n \tag{8a}$$

$$y = C(t)x, \quad y \in R^k, \ w \in R^d \tag{8b}$$

having components $x_i \in R^{n_i}$ *exists for the purpose of accurately modeling the outputs* $y(t)$ *over the interval* $0 \leq t \leq T$. To make this notion more precise, we construct the performance metric

$$V(T) = \frac{1}{T} E\left\{\int_0^T Y(t)\,dt + Y(T)\right\},$$

$$Y(t) \triangleq \|y(t)\|^2_{Q(t)} \triangleq y^T(t)Q(t)y(t), \tag{9}$$

where $Q(t)$ and $Q(T)$ are positive definite and symmetric, and where the expected value operator E is needed if either the initial condition $x(0)$ or the disturbance w is random. The basic idea of component cost decomposition is illustrated by the following example.

Example 2. The quadratic function of $x \in R^2$ given by

$$V \triangleq x^T Q x = x_1^2 Q_{11} + x_1 x_2 Q_{12} + x_2 x_1 Q_{12} + x_2^2 Q_{22} \tag{10}$$

may be decomposed into costs due to components x_1 and x_2 by defining the component cost by

$$V_1 \triangleq \frac{1}{2} \frac{\partial V}{\partial x_1} x_1 = x_1^2 Q_{11} + x_1 x_2 Q_{12}, \tag{11a}$$

$$V_2 \triangleq \frac{1}{2} \frac{\partial V}{\partial x_2} x_2 = x_2^2 Q_{22} + x_2 x_1 Q_{12}. \tag{11b}$$

Hence, the total cost is the sum of the component costs

$$V = \sum_{i=1}^{n} V_i, \qquad V_i = \frac{1}{2} \frac{\partial V}{\partial x_i} x_i \tag{12}$$

where $n = 2$ in this example.

To extend this component cost concept to the systems (8) and (9), we must first specify the character of the excitations of (8). The situation is now separately described for deterministic and stochastic inputs.

A. *COMPONENT COSTS FOR STOCHASTIC SYSTEMS*

Let any inputs $w(t)$ that are correlated with time or state be described by a Gauss-Markov model. We will assume, however, in order to simplify notation, that $w(t)$ in (8) is a zero-mean white noise process with intensity $W(t) > 0$ and that $x(0)$ has covariance $X(0) \geq 0$.

The first definition follows the lead provided by (10)-(12).

Definition 1. The component cost V_i for the ith component of (5) and (8) with respect to the performance metric (9) is defined by

$$V_i(T) \triangleq \frac{1}{2T} E\left\{\int_0^T \frac{\partial Y(t)}{\partial x_i(t)} x_i(t)dt + \frac{\partial Y(T)}{\partial x_i(T)} x_i(T)\right\}. \tag{13}$$

Two important properties of the component costs $V_i(T)$ are

(a) the superposition property of component costs

$$V(T) = \sum_{i=1}^{n} V_i(T), \tag{14}$$

(b) the component cost formula

$$V_i(T) = \frac{1}{T} \operatorname{tr}\left\{\int_0^T XC^TQCdt + X(T)C^T(T)Q(T)C(T)\right\}_{ii}, \tag{15}$$

where X is the state covariance satisfying

$$\dot{X} = AX + XA^T + DWD^T, \quad X(0) = X_0, \tag{16}$$

and $\{\cdot\}_{ii}$ denotes the $n_i \times n_i$ matrix corresponding to the position of x_i in x.

These results allow one to examine individual component contributions in a variety of situations involving (i) *specified times* T, (ii) *specified time intervals* $t \in [0, T]$, and (iii) *time-varying systems*. Examples of situation (i) includes circumstances in which the system goes through "critical" times T, and at this time it is required to have more precise knowledge about the dynamical interactions of the system components than at other times. Some critical times in engineering problems include

(i1) spacecraft reentry time T,

(i2) time of rendezvous T of two spacecraft,

(i3) critical times T in a nuclear reactor,

(i4) switch-in or switch-out time T of a power substation in a larger power network, and

(i5) time T of maximum dynamic pressure of an aircraft or rocket.

Examples of situation (ii) include finite-time control problems:

(ii1) air-to-air missile intercept; quidance of a rocket to orbital insertion,

(ii2) rapid repositioning of a flexible space vehicle,

(ii3) a finite time industrial process.

Examples of the time-varying situation (iii) are common and will not be enumerated.

For time-invariant systems with $T \to \infty$, (9) and (15) simplify to

$$V(\infty) = \lim_{t\to\infty} EY(t) = \operatorname{tr} X(\infty)C^T QC, \tag{17}$$

$$V_i(\infty) = \frac{1}{2} \lim_{t\to\infty} E \frac{\partial Y(t)}{\partial x_i(t)} x_i(t) \tag{18a}$$

$$= \operatorname{tr}[X(\infty)C^T QC]_{ii}, \tag{18b}$$

where $X(\infty)$ exists if and only if the disturbable (controllable) modes of (A, D) are stable, and $X(\infty)$ is the positive definite solution of

$$0 = AX(\infty) + X(\infty)A^T + DWD^T \tag{19}$$

if (A, D) is a disturbable pair [6].

B. COMPONENT COSTS FOR DETERMINISTIC SYSTEMS

If all disturbances are written in differential equation form (8) *without* the noise w and with specified initial conditions, then we may simplify the form of (8) and (9) to

$$\dot{x} = Ax, \quad x(0) = x_0, \quad y = Cx, \tag{20}$$

$$V(T) = \frac{1}{T}\left\{\int_0^T Y(t)\,dt + Y(T)\right\}, \tag{21}$$

and (13) becomes

$$V_i(T) = \frac{1}{2T}\left\{\int_0^T \frac{\partial Y(t)}{\partial x_i(t)} x_i(t)\,dt + \frac{\partial Y(T)}{\partial x_i(T)} x_i(T)\right\}, \tag{22}$$

where (15) still holds except that (16) is now replaced by

$$\dot{X} = AX + XA^T, \quad X(0) \triangleq x(0)x^T(0), \tag{23}$$

which has nontrivial solutions $X(t)$, $t \in [0, T]$, for finite T.

Example 3. A finite-time deterministic problem.

For the example (2), find the component costs for vehicle and actuator components $v(t)$ and $f(t)$, if $T = 1000$, $m = 1$, $a = -1$, $\bar{v} = 100$, $v(0) = 0$, $f(0) = 0$, $g = 9.8$, and $u = 1$ is a step input. The deterministic model of the inputs augmented to (2) yields (20), where

$$A = \begin{bmatrix} 0 & 1/m & -1 \\ 0 & a & 1/9.8 \\ 0 & 0 & 0 \end{bmatrix}, \quad x = \begin{pmatrix} v - \bar{v} \\ f \\ g \end{pmatrix}, \quad x(0) = \begin{pmatrix} -100 \\ 0 \\ 9.8 \end{pmatrix}.$$

Putting (3) into the form (20) leads to

$$C(t) = [0 \quad 1 \quad 0], \quad Q(t) = 1, \quad 0 \le t < T$$

$$C(T) = [1 \quad 0 \quad 0], \quad Q(T) = 1.$$

Solving (15), subject to (23), yields for (22),

$$V_1(T) = 7.92 \times 10^4, \quad V_2(T) = 1.00, \quad V_3(T) = 0,$$

where $V_1(T)/V(T) = 0.9999$ is the fraction of the cost associated with vehicle dynamics, $V_2(T)/V(T) = 1.27 \times 10^{-5}$ is the fraction of the cost associated with actuator dynamics, and

$V_3(T)/V(T) = 0$ is the fraction of the cost due to the biases in the system (gravity). Clearly the vehicle dynamics dominate the performance.

C. *INFINITE-TIME DETERMINISTIC PROBLEMS*

In the limit $T \to \infty$, (21) yields $V(T) = 0$ if A is asymptotically stable. Hence, a *different* performance metric and component cost definition is required for the special case of *infinite-time deterministic problems*. An appropriate cost function for this case is

$$V_D(\infty) \triangleq \int_0^\infty Y(t)\,dt, \tag{24}$$

which leads to

$$V_D(\infty) = x_0^T K x_0, \qquad 0 = KA + A^T K + C^T QC, \tag{25}$$

where K exists if and only if the observable modes of (A, C) are stable, and K is positive definite if (A, C) is observable. Here the component cost V_{Di} associated with component i is defined as the net effect of the excitation of the ith component state x_i. Hence, in this case the excitation is $x_i(0)$ and the component cost is defined by

$$\begin{aligned} V_{Di}(\infty) &= \frac{1}{2}\,\frac{\partial V_D(\infty)}{\partial x_i(0)}\,x_i(0) = \sum_{j=1}^{n} x_i^T(0) K_{ij} x_j(0) \\ &= \mathrm{tr}[Kx(0)x^T(0)]_{ii}, \end{aligned} \tag{26}$$

where K satisfies (25).

The remainder of this chapter will focus on the *stochastic* problem rather than the deterministic problem of Section III.C. This means that the "output-induced" component costs (18) will be of interest, rather than the "input-induced" definitions of (26). The reader can find details of an input-induced component cost for stochastic systems in [5], which, for deterministic problems, is based upon (26), and for the stochastic problem

$$V = E\int_0^\infty Y(t)\,dt, \quad \dot{x} = Ax, \quad Ex(0)x^T(0) = X(0). \tag{27}$$

[5] utilizes the stochastic version of the component cost definition (26) whose calculation is

$$V_i = \mathrm{tr}[KX(0)]_{ii}. \tag{28}$$

For the stochastic problem

$$V = \lim_{t\to\infty} EY(t), \quad \dot{x} = Ax + Dw, \quad Ew(t)w^T(\tau) = W\delta(t - \tau)$$

$$y = Cx, \quad Y \triangleq \|y\|_Q^2, \tag{29}$$

[5] utilizes the input-induced component cost definition

$$V_i \triangleq \frac{1}{2} \lim_{t\to\infty} E\,\frac{\partial Y}{\partial \omega_i}\,\omega_i, \quad \omega_i \triangleq D_i w \tag{30}$$

whose calculation is

$$V_i = \mathrm{tr}[KDWD^T]_{ii}. \tag{31}$$

The *input-induced* definitions (30) of component costs V_i represent the effect in V of excitations of component i, whereas the *output-induced* definitions (13) and (18) represent the total contributions of x_i in V in the presence of all excitations. The latter and more recent definition seems to be a much more complete notion of the contribution of component x_i in the system cost while the system is subject to all its natural environmental disturbances. Hence, this chapter will present a theory only for output-induced definitions of component cost, although the same procedures could be used to work out a corresponding theory for the input-induced case. To further simplify the presentation, only time-invariant systems with infinite terminal time T will be treated in detail. The application of the concepts to the finite-time case will be straightforward.

IV. BASIC THEOREMS OF COMPONENT COST ANALYSIS FOR MODEL REDUCTION

Given the time-invariant linear system

$$\dot{x} = Ax + Dw, \quad y = Cx, \tag{32}$$

$$V = \lim_{t\to\infty} E\|y(t)\|_Q^2, \quad Q > 0 \tag{33}$$

with components described by (5), and with zero-mean white noise disturbances w(t) with intensity W, then the value of component i whose state is $x_i \in R^{n_i}$ has been shown in previous sections to be

$$V_i = \mathrm{tr}[XC^TQC]_{ii}, \quad 0 = AX + XA^T + DWD^T \tag{34}$$

and V_i is called the ith component cost. The fractional part of the ith component's contribution to V is V_i/V, where $V = \Sigma_{j=1}^n V_i$. This component cost information (34) might be useful to guide system redesigns, failure mode analysis, and model reductions as mentioned in the Introduction. In the context of model reduction there may be considerable freedom in the selection of coordinates before model reduction begins. That is to say the definition of components is up to the analyst. For any selected component definition, the model reduction scheme proposed is simply to discard (truncate) some of the component equations (5). Suppose the component index i belongs to the set $R(i \in R)$ corresponding to the *retained* components x_i, and $i \in T$ denotes the set of indices associated with the truncated (deleted) component equations. The reduced model is

$$\dot{\hat{x}}_R = A_R \hat{x}_R + D_R w, \quad \hat{x}_R \in R^{n_r}$$
$$\hat{y} = C_R \hat{x}_R, \quad n_r = \sum_{i=1}^{r} n_i \tag{35}$$

where A_R is composed of the set $\{i \in \mathcal{R}\}$ of columns and rows of A, D_R is composed of the set $\{i \in \mathcal{R}\}$ of rows of D, and C_R is composed of the set $\{i \in \mathcal{R}\}$ of columns of C. The set $\mathcal{R}$ is determined by those r integers (here denoted generically by 1, 2,..., r) associated with the r largest component costs

$$\mathcal{V}_1 \geq \mathcal{V}_2 \geq \mathcal{V}_3 \geq \cdots \geq \mathcal{V}_r \geq \mathcal{V}_{r+1} \geq \cdots \geq \mathcal{V}_n. \tag{36}$$

The CCA algorithm for model reduction is therefore characterized by these two basic steps:

The Basic CCA model reduction algorithm

I. Compute component costs $\mathcal{V}_i$ by (34) and rank according to (36).

II. Delete the n - r components associated with the n - r smallest component costs. The resulting model is (35).

The remainder of the chapter seeks to characterize various mathematical properties of this CCA algorithm. This is clearly necessary since it is not apparent at this point whether the CCA algorithm produces "good" reduced models. To address this question of model error, we shall define a model error index $\mathcal{Q}$ in Section VI. But first a brief review of modal coordinates is in order.

V. MODAL COST ANALYSIS (MCA)

There is an important case in which the input- and output-induced definition of component costs yield the same result, and this case is summarized below.

Proposition 1. Consider system (29) where x_i is the ith modal coordinate and assume for convenience that A has distinct eigenvalues. Hence, A is diagonal. Then the CCA algorithm will produce the same reduced model, whether the output-induced or the input-induced modal cost definitions, (18a) or (30), respectively, are used.

Proof. To prove this result we must show that the component costs as computed by (18b) and (31) are identical if A_{ij} in (5) has the property $A_{ij} = \lambda_i \delta_{ij}$. First we shall show that for all real λ_i,

$$V_i = [XC^*QC]_{ii} = [KDWD^*]_{ii} \quad \text{for all } i = 1, \ldots, n, \tag{37}$$

where X and K satisfy

$$0 = XA^* + AX + DWD^*, \quad D^* = \text{complex conjugate transpose} = (\bar{D})^T \tag{38}$$

$$0 = KA + A^*K + C^*QC, \tag{39}$$

when $A_{ij} = \lambda_i \delta_{ij}$. The complex notation * is required due to the complex matrices A, D, and C. It is well known [6] that the total cost is the same by either calculation $V = \operatorname{tr} XC^*QC$ or $V = \operatorname{tr} KDWD^*$, but the issue here is whether *each* modal cost (37) is the same. Denoting the ith row of D by d_i^* and the ith column of C by c_i, the solutions of (38) and (39), respectively, are

$$X_{ij} = -d_i^* W d_j / (\lambda_i + \bar{\lambda}_j), \tag{40}$$

$$K_{ij} = -c_i^* Q c_j / (\bar{\lambda}_i + \lambda_j). \tag{41}$$

We also remark that mode i is observable (disturbable) if and only if $c_i (d_i)$ is not zero. Use (40) and (41) to obtain, respectively,

$$[XC^* QC]_{ii} = -d_i^* W \sum_{j=1}^{n} \left[\frac{d_j c_j^*}{\lambda_i + \bar{\lambda}_j} \right] Q c_i \tag{42}$$

$$[KDWD^*]_{ii} = -c_i^* Q \sum_{j=1}^{n} \left[\frac{c_j d_j^*}{\bar{\lambda}_i + \lambda_j} \right] W d_i. \tag{43}$$

Since the complex number on the right-hand side of (42) is the conjugate of the complex number on the right-hand side of (43), (37) is therefore verified for the special case of real eigenvalues of A. For a particular complex eigenvalue λ_i, let $\lambda_{i+1} = \bar{\lambda}_i$. Equations (42) and (43) show that the component cost of any x_i associated with a complex eigenvalue λ_i will be a complex number and that the component cost of x_{i+1} corresponding to the eigenvalue $\lambda_{i+1} = \bar{\lambda}_i$ will be the complex conjugate of V_i. That is $V_{i+1} = \bar{V}_i$ and V_i have the same norm. Hence, replacing V_i by $|V_i|$ in the CCA (presently MCA) truncation rule (36), the MCA model reduction algorithm would always truncate modal components so that complex conjugate *pairs* of eigenvalues are truncated. Note also that in the case of proposition 1, $n_i = 1$ and for a complex conjugate pair $\lambda_{i+1} = \bar{\lambda}_i$, it is true that

$$V_{i+1} + V_i = 2 \mathrm{Re} V_i. \tag{44}$$

Hence, the total cost V is real, and the sum of the modal cost of any two modal components associated with complex conjugate pairs of eigenvalues will be real. The proof is concluded by noting that complex conjugates are truncated in pairs and from the fact

$$[XC^*QC]_{i+1,i+1} = \overline{[XC^*QC]_{ii}} = \overline{[KDWD^*]_{ii}}, \tag{45}$$

we conclude that for a complex pair

$$\begin{aligned} V_i + V_{i+1} &= [XC^*QC]_{ii} + [XC^*QC]_{i+1,i+1} \\ &= [KDWD^*]_{ii} + [KDWD^*]_{i+1,i+1}. \end{aligned} \tag{46}$$

Hence the same modes will be truncated by either definition of modal cost. #

Under special conditions the modal costs (42) and (43) simplify greatly.

Proposition 2. If either (a), (b), or (c) holds:

(a) $d_i^* W d_j = 0, \quad i \neq j$ (disturbance decoupled modes);

(b) $c_i^* Q c_j = 0, \quad i \neq j$ (output decoupled modes);

(c) $(\mathrm{Re}\lambda_i / \mathrm{Im}\lambda_i)$ arbitrarily small, and $\mathrm{Im}\lambda_i \neq \mathrm{Im}\lambda_j$ (lightly damped modes);

then the modal costs of a linear system are given by

$$V_{ci} \triangleq V_i + \overline{V}_i = -\frac{c_i^* Q c_i d_i^* W d_i}{\mathrm{Re}\lambda_i} = \tau_i \|c_i\|_Q^2 \|d_i\|_W^2, \tag{47}$$

which holds for either the input-induced definition (30) or the output-induced definition of modal cost and where V is defined by (29). If λ_i is real, then the ith modal cost is $V_{ci} \triangleq V_i$.

The proof of parts (a) and (b) follow immediately from (42) and (43). The proof of part (c) is given in [5]. #

Examples of case (c) in proposition 2 appear in [5] and [17] where MCA is applied to systems of order up to 200. It should be noted that since the MCA formulas (47) are explicit [hence, the linear matrix Eq. (38) does not have to be solved numerically], the MCA algorithm may be applied to any system for which modal data are available. It will subsequently be shown that modal coordinates might *not* be the best coordinates in which to perform model truncation. However, much insight is available from (47) indicating that modal costs are composed of the product of three properties of a mode: (1) time constant, (2) observability norm, and (3) disturbability norm.

VI. MODEL ERROR INDICES

Having a reduced model (35), we now turn our attention to the definition and calculation of a convenient measure of "model error" when comparing the reduced model (35) with the evaluation model (32).

Definition 2. The errors associated with model (35) produced by the CCA algorithm are measured by the model error index

$$Q \triangleq \frac{1}{V}|(V - V_R)|, \tag{48}$$

where V_R is the performance metric associated with (35). If the disturbable modes of (A_R, D_R) are stable, then

$$V_R = \text{tr}\, X_R C_R^T Q C_R, \qquad 0 = A_R X_R + X_R A_R^T + D_R W D_R^T, \tag{49}$$

and V is the performance metric associated with the "evaluation" model (32), as given by (33).

Of course, V_R can be computed only after model reduction. The information available *a priori* will be called the *predicted model error index* $\hat{Q}$.

Definition 3. The predicted model error index is defined by

$$\hat{Q} \triangleq \frac{1}{V}|(V - \hat{V}_R)|, \tag{50}$$

where

$$\hat{V}_R \triangleq \Sigma V_i, \quad i \in R. \tag{51}$$

From (36), (50), and (51), it follows also that

$$\hat{Q} = \frac{1}{V}|\hat{V}_T|, \quad \hat{V}_T \triangleq \Sigma V_i, \quad i \in T, \tag{52}$$

and

$$V = \hat{V}_R + \hat{V}_T. \tag{53}$$

When applying the model error index Q to the reduction of the *closed*-loop system (to reduce controllers rather than models), Q plays a role similar to the "suboptimality index" of Siljak [14]. Note also that the Q chosen here (48) is the difference in the norms of y and $\hat{y}$, whereas the model error index chosen in [5] is the norm of the difference $y - \hat{y}$. This choice (48) is primarily motivated by the controller reduction problem where V_R represents the performance using the reduced controller. In *that* problem $V_R \geq V$ since V_R represents the suboptimal controller perfromance. Since V_R is minimized if Q is minimized, the difference of norms represented by (48) is a more logical choice for controller reduction. This chapter now focuses on the prerequisite problem of model reduction where all the essential mathematical results are derived for subsequent application to controller reduction.

For the *model reduction* problem, the model error index (48) would be a meaningless index if the parameters of the reduced model (A_R, D_R, C_R) were arbitrary, since in this case parameters can always be found to make $Q = 0$. Reasonableness is added to the problem, however, by the fact that the search for small Q is subjected to the parameters (A_R, D_R, C_R), which are *constrained* to be a transformed subset of the original system parameters (A, D, C). We now continue with this model reduction problem.

The questions that naturally arise and are to be answered in the sequel are

(QI) Under what conditions is the predicted model error index $\hat{Q}$ exact $(\hat{Q} = Q)$?

(QII) Under what conditions is the model error index Q zero?

(QIII) Under what conditions is the model error index Q minimized by the CCA algorithm?

(QIV) Given that A is stable, under what conditions is the reduced model produced by CCA stable?

VII. COST-EQUIVALENT REALIZATIONS AND MINIMAL REALIZATIONS WITH RESPECT TO COST

Toward the development of the mathematical machinery required to answer questions (QI)-(QIV), we introduce the following definitions.

Definition 4. Cost-equivalent realizations

Let $\{A_R, D_R, C_R, X_R(0), W_R\}$ characterize the partial realization (35) and let $\{A, D, C, X(0), W\}$ characterize the evaluation model ("complete" realization) (32). The partial realization is said to be "cost-equivalent" if $Q = 0$.

Definition 5. Minimal cost-equivalent realizations

With respect to the given components (5), the partial realization (35) is said to be a "minimal cost-equivalent" realization if r is the smallest integer for which $Q = 0$.

To simplify our bookkeeping, let us assume that the components (5) are arranged in order of their component costs and define

$$x_R^T \triangleq (\ldots, x_i^T, \ldots), \quad i \in \mathcal{R}, \tag{54a}$$

$$x_T^T \triangleq (\ldots, x_i^T, \ldots), \quad i \in \mathcal{T}. \tag{54b}$$

Then (32) may be written in the form

$$\begin{aligned} \begin{pmatrix} \dot{x}_R \\ \dot{x}_T \end{pmatrix} &= \begin{bmatrix} A_R & A_{RT} \\ A_{TR} & A_T \end{bmatrix} \begin{pmatrix} x_R \\ x_T \end{pmatrix} + \begin{bmatrix} D_R \\ D_T \end{bmatrix} w, \\ y &= [C_R \quad C_T] \begin{pmatrix} x_R \\ x_T \end{pmatrix}. \end{aligned} \tag{55}$$

Let X as defined by (34) be likewise partitioned in the manner

$$X = \begin{bmatrix} \hat{X}_R & \hat{X}_{RT} \\ \hat{X}_{RT}^T & \hat{X}_T \end{bmatrix}. \tag{56}$$

Due to symmetry of X, the partitioned form of the linear equation (34) using (55) and (56) yields three linear equations of smaller dimensions. Two of these equations are

$$0 = A_R \hat{X}_{RT} + \hat{X}_{RT} A_T^T + A_{RT} \hat{X}_T + \hat{X}_R A_{TR}^T + D_R W D_T^T, \tag{57a}$$

$$0 = A_T \hat{X}_T + \hat{X}_T A_T^T + A_{TR} \hat{X}_{RT} + \hat{X}_{RT}^T A_{TR}^T + D_T W D_T^T. \tag{57b}$$

The remaining equation in $\hat{X}_R$ is subtracted from (49) to yield

$$0 = A_R \tilde{X}_R + \tilde{X}_R A_R^T + A_{RT} \hat{X}_{RT}^T + \hat{X}_{RT} A_{RT}^T, \tag{57c}$$

where $\tilde{X}_R \triangleq \hat{X}_R - X_R$.

A. *ANSWERS TO QUESTIONS QI, QII, AND QIII*

Using the above symbols, question QI can now be answered.

Proposition 3. The predicted model error index $\hat{Q}$ is exact in the sense $\hat{Q} = Q$ under any of the following conditions:

(a) if $\mathrm{tr}\left(\tilde{X}_R C_R^T Q C_R + \hat{X}_{RT} C_T^T Q C_R\right) = 0$ and $V \geq V_R$; (58a)

(b) if $\mathrm{tr}\ \tilde{X}_R C_R^T Q C_R + \mathrm{tr}\ 2\hat{X}_T C_T^T Q C_T + \mathrm{tr}\ 3\hat{X}_{RT} C_T^T Q C_R$

$= 0$ and $V < V_R$; (58b)

(c) if $\hat{X}_{RT} = 0$;

(d) if x_T is unobservable;

(e) if x_T is undisturbable.

Proof. Noting (48) and (50), it follows that the proof requires that $\hat{V}_R = V_R$ if $V \geq V_R$ and requires that $2V = V_R + \hat{V}_R$ if $V < V_R$. To show that $V_R = \hat{V}_R$ when (58a) holds, we first write from (34), using (55) and (56),

$$\hat{V}_R = \sum_i \mathrm{tr}\,[XC^TQC]_{ii} = \mathrm{tr}\left(\hat{X}_R C_R^T Q C_R + \hat{X}_{RT} C_T^T Q C_R\right), \quad i \in R$$

$$= \mathrm{tr}\left\{(\tilde{X}_R + X_R) C_R^T Q C_R + \hat{X}_{RT} C_T^T Q C_R\right\}. \tag{59}$$

Now subtract (49) from (59) to obtain (58a) directly. To prove (58b), write, using (34), (55), and (56),

$$V = \mathrm{tr}\ \hat{X}_R C_R^T Q C_R + 2\ \mathrm{tr}\ \hat{X}_{RT} C_T^T Q C_R + \mathrm{tr}\ \hat{X}_T C_T^T Q C_T. \tag{60}$$

Substitute (60) into $2V = V_R + \hat{V}_R$, using (49) and (59) to get

$$2\left(\mathrm{tr}\ \hat{X}_R C_R^T Q C_R + \mathrm{tr}\ 2\hat{X}_{RT} C_T^T Q C_R + \mathrm{tr}\ \hat{X}_T C_T^T Q C_T\right)$$

$$= \mathrm{tr}\ X_R C_R^T Q C_R + \mathrm{tr}\ \hat{X}_R C_R^T Q C_R + \mathrm{tr}\ \hat{X}_{RT} C_T^T Q C_R, \tag{61}$$

which reduces to (58b). To prove (c), set $\hat{X}_{RT} = 0$ in (57c) to obtain $\tilde{X}_R = 0$. Furthermore since $\hat{X}_{RT} = 0$, we have from (60) and (49)

$$V - V_R = \mathrm{tr}\ \hat{X}_R C_R^T Q C_R + \mathrm{tr}\ \hat{X}_T C_T^T Q C_T - \mathrm{tr}\ X_R C_R^T Q C_R$$

$$= \mathrm{tr}\ \tilde{X}_R C_R^T Q C_R + \mathrm{tr}\ \hat{X}_T C_T^T Q C_T = \mathrm{tr}\ \hat{X}_T C_T^T Q C_T$$

Now, since the state covariance X is at least positive semi-definite [6], $\hat{X}_T \geq 0$ and hence $V \geq V_R$. Hence, (58a) is applicable, and this proves (c). To prove (d), one may without loss of generality assume $A_{RT} = 0$ and $C_T = 0$ since x_T is unobservable. This yields from (57c) $\tilde{X}_R = 0$, which immediately leads to (58), since $C_T = 0$. To prove (e), assume x_T is undisturbable (i.e., set $A_{TR} = 0$, $D_T = 0$). This yields from (57a) and (57b) $\hat{X}_T = 0$, $\hat{X}_{RT} = 0$, and (57c) yields $\tilde{X}_R = 0$. Hence, condition (58) is again satisfied. #

It may be comforting to know that the predicted model error index is accurate, but the initial issue of the "best" choice of coordinates and components is still unresolved. That is, some choice of coordinates may lead to smaller model error indices than other choices, even though the predicted model error index may be exact for each choice. Before we try to resolve the question of the best set of coordinates, we shall define the limiting case where the reduced model is "perfect." Thus, the following result answers question QII.

Proposition 4. The partial realization (35) is a cost-equivalent realization of (32) under either of these conditions:

(a) if and only if

$$\mathrm{tr}\ \tilde{X}_R C_R^T Q C_R + \mathrm{tr}\ 2\hat{X}_{RT} C_T^T Q C_R + \mathrm{tr}\ \hat{X}_T C_T^T Q C_T = 0; \tag{62}$$

(b) if x_T is unobservable;

(c) if x_T is undisturbable.

Proof. From (48), it follows that proof of (a) relies upon a proof that $V = V_R$ if (62) holds. Subtract (49) from (60) to get (62) directly. To prove (b) we rely on the proof of proposition 3, which showed that $\tilde{X}_R = 0$ if x_T is unobservable and that $C_T = 0$ may be assumed. The conditions $\tilde{X}_R = 0$, $C_T = 0$ lead to satisfaction of (62). To prove (c), note from the proof of proposition 3 that $\tilde{X}_R = 0$, $\hat{X}_T = 0$, $\hat{X}_{RT} = 0$ if x_T is undisturbable. These substitutions in (62) conclude the proof. #

Having answered questions QI and QII, it is now possible to provide an answer to QIII. This answer is summarized by proposition 5.

Proposition 5. Given a specified r and the components (5), which satisfy proposition 3 ($\hat{Q} = Q$), the model error index Q is minimized by the CCA algorithm.

Proof. Since $\hat{Q} = Q$ the model error index is given by (52). Among the set of $\{V_i, i = 1, 2, \ldots n\}$, the $\hat{V}_T$ in (52) is composed (*by definition*) of the n - r smallest subset of V_is, according to (36). Hence, $\hat{Q}$ cannot be decreased by any other choice of r components from the given set of n components (5). #

We must not read too much into proposition 5. It only guarantees that there are no better r choices of the *given* n components. The *a priori choices of component definitions* that can be made are infinite. In any model truncation problem these three factors are important:

(a) choice of coordinates,

(b) choice of a truncation criterion, and

(c) choice of an evaluation criterion for the reduced model.

In CCA, the best choice (a) has not yet been determined, choice (b) is given by (36) and (52), and choice (c) is given by (48). One suggestion for choice (a) is introduced in the next section. It should be noted, however, that depending upon the question being addressed, the analyst may *not* have a choice of coordinates. In this case the results of Section VII.A apply, but the coordinate transformation of Section VII.B will not be permitted.

B. *COST-DECOUPLED COMPONENTS*

The previous section describes CCA for any given choice of components, and this flexibility is important for the analysis of component costs using *physical* components. However, in model reduction the analyst may be free to *choose* the reference coordinates and may *not* be restricted to the analysis of physical components. The component costs for some choices of components (i.e., choices associated with the underlying coordinate transformations) are more convenient to interpret than others. As an example of possible confusion, note that even though the *sum* of component costs V_i is positive (14), an individual V_i defined by (15) or (34) can be negative. All propositions of previous sections are still valid, but one might obtain better reduced models by using absolute value signs around each V_i in (36). Clearly, such issues need not be of concern if all V_i are proven to be nonnegative. The cost-decoupled components to be defined in this section will prove to have this property.

It may also be observed from the basic ideas of (11), from the general component cost formula (15), and from the steady-state cases of (18) and (34), that the component costs V_i and V_j are not generally independent. That is, the ith component cost V_i is influenced by component j. This presents no problem for the *in situ* component cost analysis for purposes other than model reduction. But for model reduction such dependence between V_i and V_j leads to errors in the predicted model quality index $\hat{Q}$, since in this case the deletion of component j also modifies the cost of the retained component i. This nuisance can be removed by choosing components that have independent costs. Thus, the motivation for such component choices is to gain the property $\hat{Q} = Q$ of proposition 3. For the purposes of this section, define the *components* $x_i \in R^1$ to be *each coordinate* of the cost-decoupled state x_2. Hence $n_i = 1$ for all i in this case. From part (c) of proposition 3, it is clear that uncorrelated components (i.e., $X_{ij} = 0$, $i \neq j$) yield the property $\hat{Q} = Q$. An additional property is added to obtain the "cost-decoupled" coordinates defined as follows.

Definition 6. The "cost-decoupled" coordinates of a linear system are any coordinates for which the covariance X and the state weighing C^TQC are both diagonal matrices.

A convenient choice of cost-decoupled coordinates may be computed as follows. Let $\{x°, X°, C°, A°, D°\}$ represent an original set of coordinates and data, and let $\{x, X, C, A, D\}$ represent the transformed data according to the transformation

$$x° = \Theta x, \quad 0 = X°A°^T + A°X° + D°WD°^T, \tag{63a}$$

where $\Theta = \Theta_x \Theta_y$ and Θ_x, Θ_y satisfy

$$X^\circ = \Theta_x \Theta_x^T \tag{63b}$$

$$\Theta_y \triangleq E_y \Omega_y, \qquad \Theta_x^T C^{\circ T} Q C^\circ \Theta_x = E_y \Lambda_y^2 E_y^T. \tag{63c}$$

Note that Θ_x is the square root of the covariance matrix X°. The nonsingular diagonal matrix Ω_y is arbitrary. The orthonormal matrix of eigenvectors of $\Theta_x^T C^{\circ T} Q C^\circ \Theta_x$ is E_y and the corresponding eigenvalues (which are also singular values [9] since the matrix is symmetric) are elements of the diagonal matrix Λ_y^2.

In cost-decoupled coordinates, the system (32) is transformed by

$$A = \Theta_y^{-1} \Theta_x^{-1} A^\circ \Theta_x \Theta_y, \tag{64a}$$

$$D = \Theta_y^{-1} \Theta_x^{-1} D^\circ, \tag{64b}$$

$$C = C^\circ \Theta_x \Theta_y. \tag{64c}$$

The calculation of the steady-state covariance matrix of a stable system in cost-decoupled coordinates reveals that

$$X = \Omega_y^{-2} \tag{65}$$

and the *state* weighting matrix $[C^T Q C]$ in the performance metric

$$V = \lim_{t \to \infty} E \| y \|_Q^2 = \text{tr } X C^T Q C$$

is

$$C^T Q C = \Omega_y^2 \Lambda_y^2. \tag{66}$$

Hence from (65) and (66),

$$V = \text{tr } X C^T Q C = \text{tr } \Lambda_y^2 = \sum_{i=1}^{k} \lambda_i \left[\Theta_x^T C^T Q C \Theta_x \right], \tag{67a}$$

where $\lambda_i[\cdot]$ denotes eigenvalue of $[\cdot]$, and the summation is only up to k, since there are only k nonzero eigenvalues of $\Theta_X^T C^T Q C \Theta_X$, since rank C = k. Note that the component costs in cost-decoupled coordinates are

$$V_i = \lambda_i\left[\Theta_x^T C^T Q C \Theta_x\right], \tag{67b}$$

which leads to this simple interpretation of cost-decoupled coordinates and component costs: In coordinates (components) that are uncorrelated ($X_{ij} = 0$) and output decoupled ($[C^T QC]_{ij} = 0$), the component costs are the eigenvalues of the state-weighting matrix. In view of (67a), which holds for any Ω_y, there seems to be no disadvantage in the choice $\Omega_y = I$, although a different choice for Ω_y will be chosen in Section VII.D. for convenient comparisons with the work of others. Temporarily, we choose $\Omega_y = I$.

The useful properties of cost-decoupled coordinates are now summarized in the following proposition.

Proposition 6. In cost-decoupled coordinates, the full-order model (55) has the following properties:

(1) $V_i \geq 0$ (the component costs are all nonnegative);

(2) A_R has no eigenvalue in the open right half plane;

(3) A_R is asymptotically stable if and only if the pair (A_R, D_R) is disturbable.

Proof. Claim (1) follows immediately from (67b) since

$$V_i = \lambda_i\left[\Theta_X^T C^T Q C \Theta_X\right] \geq 0. \tag{68}$$

To prove claim (2) and (3), partition (65) (with $\Omega_y = I$) as

$$X = \begin{bmatrix} \hat{X}_R & \hat{X}_{RT} \\ \hat{X}_{RT}^T & \hat{X}_T \end{bmatrix} = \begin{bmatrix} I & 0 \\ 0 & I \end{bmatrix}. \tag{69a}$$

This reveals that $\hat{X}_{RT} = 0$. Hence, writing the partioned form of (34), partitioned compatibly with (55), yields

$$0 = A_R^T + A_R + D_R W D_R^T, \tag{69b}$$

$$0 = A_{TR}^T + A_{RT} + D_R W D_T^T, \tag{69c}$$

$$0 = A_T^T + A_T + D_T W D_T^T. \tag{69d}$$

Either (A_R, D_R) is disturbable or not. If (A_R, D_R) is disturbable, then the state covariance of (35) from (69b) is

$$X_R = \int_0^\infty e^{A_R^T t} D_R W D_R^T e^{A_R^T t} dt = I \tag{69e}$$

and the finiteness of X_R guarantees asymptotic stability of A_R(X_R would not be bounded for unstable A_R under the disturbability assumption). This proves the "if" part of claim (3). If (A_R, D_R) is not disturbable then there exists an orthonormal transformation

$$\chi = \begin{bmatrix} x_1 \\ x_2 \end{bmatrix} = T^T x_R = \begin{bmatrix} T_1^T \\ T_2^T \end{bmatrix} x_R \tag{70a}$$

to take the system (35) to the controllable canonical form

$$\begin{bmatrix} \dot{x}_1 \\ \dot{x}_2 \end{bmatrix} = \begin{bmatrix} A_{11} & A_{12} \\ 0 & A_{22} \end{bmatrix} \begin{bmatrix} x_1 \\ x_2 \end{bmatrix} + \begin{bmatrix} D_1 \\ 0 \end{bmatrix} w, \tag{70b}$$

$$y_R = [C_1 \quad C_2] \begin{bmatrix} x_1 \\ x_2 \end{bmatrix}.$$

Where (A_{11}, D_1) is completely disturbable. Now (69b) becomes

$$0 = A_{11}^T + A_{11} + D_1 W D_1^T, \tag{71a}$$

$$0 = A_{12}, \tag{71b}$$

$$0 = A_{22}^T + A_{22}. \tag{71c}$$

The eigenvalues of A_R are those of A_{11} and A_{22}. Since (A_{11}, D_1) is disturbable and $\int_0^\infty e^{A_{11}t} D_1 W D_1^T e^{A_{11}^T t}\, dt = I < \infty$, the eigenvalues of A_{11} must lie in the open left-hand half plane by reasons mentioned above. The eigenvalues of A_{22} must lie on the imaginary axis since A_{22} is skew-symmetric $(A_{22} = -A_{22}^T)$. Hence, no eigenvalues of A_R can lie in the right-hand half plane but there are eigenvalues with zero real parts. This proves claim (2). Moreover this proves that A_R is not asymptotically stable if (A_R, D_R) is not disturbable, the "only if" part of claim (3). #

Proposition 7. If the CCA algorithm using *cost-decoupled coordinates* produces a disturbable pair (A_R, D_R) then the following properties hold:

(1) $\hat{Q} = Q$ (the predicted model error index is exact);

(2) Q is minimized for a given r;

(3) $Q = 0$ if $r \geq k$ (the CCA algorithm produces a minimal cost-equivalent realization of order k = rank C).

Proof. Claim (1) is proven by showing that (58a) holds. By virtue of the fact that $\hat{X}_{RT} = 0$ (since X in definition 6 is diagonal) it follows from (57c) that $\tilde{X}_R = 0$. Hence (58a) is satisfied if $V \geq V_R$. To show that $V \geq V_R$ note from (67a) that

$$V_R = \sum_i \lambda_i \left[\Theta_x^T C^T Q C \Theta_x\right], \quad i \in R.$$

Hence, since $\lambda_i \geq 0$ for all i,

$$V = V_R \quad \text{if } r \geq k, \quad k = [\text{rank } C]$$

$$V > V_R \quad \text{if } r < k$$

and (1) is proven. The proof of claim (2) follows from (1) and proposition 5. Claim (3) follows from claim (1) together with (67b) and the fact that $\left[\Theta_x^T C^T Q C \Theta_x\right]$ can have no more than $k \triangleq$ rank C nonzero eigenvalues. #

It may be readily verified that proposition 7 holds for the general cost-decoupled coordinates in definition 6, and proposition 7 is not restricted to the special choice of cost-decoupled coordinates given by (63). Furthermore, claim (3) of proposition 7 shows that the CCA algorithm using cost-decoupled coordinates yields a cost-equivalent realization of (32) if $r \geq k$ and if the reduced-order model is disturbable. These are only sufficient conditions. We shall now present the precise conditions in which such cost-equivalent realizations are obtained.

Proposition 8. The CCA algorithm using *cost-decoupled coordinates* yields *cost-equivalent realizations* of (32) if and only if (a) $r \geq k$ and (b) the undisturbable subspace of (A_R, D_R) is unobservable.

Proof. For any pair (A_R, D_R), the transformation defined in (70) exists. (If (A_R, D_R) is disturbable then $I = T = T_1$, $A_{11} = A_R$, $D_1 = D_R$, and $C_1 = C_R$.) Then from equations (70) and (71), it can be seen that A_R is not asymptotically stable. Those eigenvalues of A_R which are not asymptotically stable are contained in the set of eigenvalues of A_{22}, which corresponds to the undisturbable part of (A_R, D_R). Hence, the undisturbable modes are the only ones that are not asymptotically stable. Since the unstable (and undisturbable) part of A_R does not contribute to the cost V_R [6], the model (35) can be further reduced to yield

$$\dot{x}_1 = A_{11}x_1 + D_1 w, \qquad y_1 = C_1 x_1, \tag{72}$$

such that

$$V_R = \lim_{t\to\infty} E\|y_R(t)\|_Q^2 = \lim_{t\to\infty} E\|y_1(t)\|_Q^2.$$

Now from (71a) and (49) we have

$$\begin{aligned} V_R &= \text{tr } C_1^T Q C_1 = \text{tr } T_1^T C_R^T Q C_R T_1 \\ &= \text{tr } C_R^T Q C_R - \text{tr } T_2^T C_R^T Q C_R T_2, \end{aligned} \tag{73a}$$

where the orthonormal property of $T(T_1T_1^T + T_2T_2^T = I)$ is used. From (66), (67a), and the partitioning of C in (55), it can be seen that

$$\text{tr } C_R^T Q C_R = \sum_{i=1}^{r} \lambda_i \left[\theta_X^T C^T Q C \theta_X\right].$$

Hence,

$$V_R = \sum_{i=1}^{r} \lambda_i \left[\theta_X^T C^T Q C \theta_X\right] - \text{tr } C_2^T Q C_2 \tag{73b}$$

where $C_2 \triangleq C_R T_2$. Now, since $A_{12} = 0$ from (71b), considering (70b) to be in observable canonical form [6], it can be said that $C_2 = 0$ if and only if condition (b) holds. Furthermore, since the columns of T_2 span the undisturbable subspace of

(A_R, D_R) [6], we have the following:

(i) if condition (b) holds (equivalently if $C_2 = 0$) then

$$V_R = \begin{cases} \sum_{i=1}^{r} \lambda_i \left[\theta_X^T C^T Q C \theta_X\right] = V \quad \text{if } r \geq k & (74a) \\ \sum_{i=1}^{r} \lambda_i \left[\theta_X^T C^T Q C \theta_X\right] < V \quad \text{if } r < k & (74b) \end{cases}$$

(ii) if condition (b) does not hold (i.e., $C_2 \neq 0$), then

$$V_R = \sum_{i=1}^{r} \lambda_i \left[\theta_X^T C^T Q C \theta_X\right] - \alpha < V, \tag{74c}$$

where $\alpha \triangleq \text{tr } C_2^T Q C_2 > 0$. Obviously $\alpha = 0$ if (A_R, D_R) is disturbable since $[T_1 \; T_2] \to [T_1] = I$ and $[C_1] \to [C_R]$ implying $C_2 \to 0$. Note, therefore, from (74) that if condition (a) does not hold, then $V_R < V$ and (35) is not a cost-equivalent realization. #

One obvious conclusion from proposition 8 is that the order of the minimal CER is never less than k, the number of independent outputs. It is of interest to classify those systems whose minimal CER is of order greater than k.

Proposition 9. For all systems (32) whose first Markov Parameters is zero (CD = 0) the order of the minimal CER is greater than k.

Proof. Let the system (55) be in cost-decoupled coordinates and let r = k. Hence, assuming $\Omega_Y^2 = I$, from (66) we have

$$C^T Q C = \begin{bmatrix} C_R^T Q C_R & C_R^T Q C_T \\ C_T^T Q C_R & C_T^T Q C_T \end{bmatrix} = \begin{bmatrix} \Lambda^2 & 0 \\ 0 & 0 \end{bmatrix}, \tag{75a}$$

where $\Lambda^2 \triangleq \text{diag}\{\lambda_1[C^TQC], \lambda_2[C^TQC], \ldots, \lambda_k[C^TQC]\}$. Now since rank $C = k$, $\lambda_i[C^TQC] \neq 0$, $i = 1, 2, \ldots, k$. Hence, equating

$$C_R^TQC_R = \Lambda^2, \tag{75b}$$

$$C_T^TQC_T = 0, \tag{75c}$$

and recognizing that $Q > 0$, it can then be claimed that C_R (of dimension $k \times k$) is square and of full rank and that $C_T = 0$. Now, since Markov parameters are invariant under similarity transformation, we have

$$CD = 0 \leftrightarrow [C_R \;\; 0]\begin{bmatrix} D_R \\ D_T \end{bmatrix} = C_RD_R = 0. \tag{76}$$

Equation (76) is satisfied if and only if $D_R = 0$, since C_R is square and of full rank. In this event, the pair (A_R, D_R) is obviously undisturbable. Furthermore, due to full rank of C_R, the pair (C_R, A_R) is completely observable. Therefore, the undisturbable subspace of (A_R, D_R) cannot be also unobservable. This violates condition (b) of proposition 8. Hence the minimal CER cannot be of order k. #

Nevertheless, a minimal CER of order $r > k$, can be constructed for the systems defined in proposition 9, by increasing r until condition (b) of proposition 8 is satisfied.

C. *THE ALGORITHM FOR COST-EQUIVALENT REALIZATIONS (CER)*

Cost-equivalent realizations (CERs) are provided by the CCA algorithm using cost-decoupled coordinates and the CERs have all the properties of propositions 6 and 7. The two steps of the basic CCA algorithm are described in Section IV and the cost-decoupled coordinates are described in Section VII.B. Combining these two ingredients leads to the following CER algorithm.

The CER algorithm

Step 1. Given the model and performance objectives of (32) and (33):

(A, D, C, Q, W) where Q > 0, W > 0, A stable.

(Choose $\Omega_y = I$.)

Step 2. Compute covariance X from[1]

$$0 = XA^T + AX + DWD^T. \tag{77}$$

Step 3. Compute Θ_x the square root of X[2]

$$X = \Theta_x \Theta_x^T. \tag{78}$$

Step 4. Compute Θ_y, the orthonormal modal matrix of $\Theta_x^T C^T Q C \Theta_x$.[3] The component costs are

$$\Theta_y^T \Theta_x^T C^T Q C \Theta_x \Theta_y = \text{diag}\{V_1, V_2, \ldots, V_k, 0, \ldots, 0\}, \tag{79}$$

where the number of nonzero component costs are k = rank[C].

Step 5. Rearrange the columns of Θ_y so that the V_i appear in order

$$V_1 \geq V_2 \geq \cdots \geq V_k. \quad \text{Set } r = k = \text{rank } C. \tag{80}$$

Step 6. Then define Θ_R by

$$\Theta_y = [\Theta_R, \Theta_T], \quad \Theta_R \in R^{n\times r}. \tag{81}$$

[1] *For efficient solution of the linear Liapunov equation, use the algorithm in [11].*

[2] *For efficient calculation of Θ_x, see the computer codes in [12].*

[3] *For this task use singular value decomposition [9] or use an eigenvalue/eigenvector program specialized for* symmetric *matrices.*

Step 7. Compute

$$\left.\begin{aligned} A_R &= \theta_R^T \theta_X^{-1} A \theta_X \theta_R \\ D_R &= \theta_R^T \theta_X^{-1} D \\ C_R &= C \theta_X \theta_R \end{aligned}\right\} \text{CER.} \tag{82}$$

Step 8. Compute modal data for A_R:

$$A_R e_i = \lambda_i e_i, \quad i = 1, 2, \ldots, r$$

If $\|C_R e_i\| > 0$ for any i such that $R_e \lambda_i = 0$, where $R_e(\cdot)$ denotes "real part" of $(\cdot)$, set $r = r + 1$ and go to step 6. Otherwise stop.

Remark: The product $C_R e_i$ is defined as the *observability vector* associated with mode i [5] (mode i in a nondefective system is unobservable if and only if its observability vector is zero). Hence, the purpose of step 8 is to check if the unstable mode ($R_e \lambda_i = 0$) is observable. Since in cost-decoupled coordinates the unstable modes of A_R are also undisturbable, step 8 amounts to checking if the condition (b) of proposition 8 holds.

This algorithm guarantees the construction of a CER. However, the construction of a *minimal* CER is guaranteed only if the algorithm converges within the first two iterations, in which case the CER is of order $r = k$ or $r = k + 1$. For the minimal CER of order k the triple (A_R, D_R, C_R) is both disturbable and observable and asymptotically stable. For any other CER produced by the algorithm, the disturbable, observable spectrum of (A_R, D_R, C_R) is asymptotically stable. *After* the first iteration of the algorithm, the selection of the best sequence of eigenvector calculations in step 5 has not been determined and is under investigation.

If the CER algorithm yields a CER of order r with an unstable (and undisturbable and unobservable) spectrum, then the CER can be further reduced, as shown in (70) and (72), to yield a realization of order less than r. This smaller realization is still a CER as is proved following proposition 8.

D. *RELATIONSHIPS BETWEEN THE COST-DECOUPLED COORDINATES AND THE BALANCED COORDINATES OF MOORE [7]*

The balanced coordinates of Moore [7] are defined by the transformation that diagonalizes the controllability and observability matrices (X and K in this chapter). Singular value analysis provides the efficient tools to compute the balanced coordinates. As mentioned in the introduction, CCA can be applied to any choice of coordinates, including balanced coordinates. The most powerful results from CCA are obtained with the use of the cost-decoupled coordinates defined in the last section using the CER algorithm. Moore [7] introduced balanced coordinates to reduce numerical ill-conditioning, thereby making data more manageable in the computer. On the other hand, the primary goal of CCA is specifically to tailor the reduced model to the control or output response objectives (33). It would be of interest to know whether there are circumstances under which balanced coordinates of Moore are cost-decoupled.

To obtain *balanced* coordinates, a coordinate transformation is selected so that the new (balanced) coordinates have the properties [see Eqs. (5.1) and (5.2) in [7]],

$$X = K = \Sigma^2 = \text{diag}, \tag{83}$$

where X is the disturbability matrix

$$X \triangleq \int_0^\infty e^{At} DWD^T e^{A^T t}\, dt, \qquad W > 0 \tag{84}$$

satisfying

$$0 = XA^T + AX + DWD^T, \tag{85}$$

and K is the observability matrix

$$K \triangleq \int_0^\infty e^{A^T t} C^T Q C e^{At}\, dt, \qquad Q > 0 \tag{86}$$

satisfying

$$0 = KA + A^T K + C^T QC. \tag{87}$$

To obtain the *cost-decoupled* coordinates of Section VII.B, a coordinate transformation is selected so that the new (cost-decoupled) coordinates have the properties from definition 6,

$$X = \text{diag}, \qquad C^T QC = \text{diag}, \tag{88}$$

To summarize these results from (83) and (88), proposition 10 specifies the condition under which cost-equivalent realizations can be obtained from balanced coordinates.

Proposition 10. If in balanced coordinates the state weighting $C^T QC$ happens to be diagonal, then balanced coordinates are cost-decoupled and hence have all the properties of proposition 6.

Proof. Cost-decoupled coordinates are defined by (88) and balanced coordinates satisfy (83). The comparison of (83) and (88) concludes the proof. #

VIII. SHOULD REDUCED MODELS DEPEND UPON THE WEIGHTS IN THE QUADRATIC COST?

The reader should be reminded of the fact that the state weighting $[C^TQC]$ in the performance metric

$$V = \lim_{t\to\infty} E\|y\|_Q^2 = \lim_{t\to\infty} Ex^T[C^TQC]x$$

often contains parameters chosen in an *ad hoc* fashion. Why then, one might ask, should one adopt a model reduction strategy in which the reduced models depend upon the weight C^TQC? This question is briefly answered as follows. The selection of a performance metric V reflects, to the best of one's ability, the objective of the model analysis (to describe accurately specific outputs y). Thus, it is important to keep in mind that there are many problems in which the entire state weighting matrix C^TQC is *not* arbitrary, but only the output weighting Q might be free to be manipulated. This notion of penalizing only specific physical variables represented by y allows the number of free parameters in the $n \times n$ state weighting C^TQC to be reduced from $n(n + 1)$ to $k(k + 1)$, the free parameters in Q. Thus, C^TQC contains important information by its very structure. For example, a certain spacecraft may have a mission to keep optical line-of-sight errors small in a space telescope. These error variables, collected in the vector herein labeled y, make up only a small subset of all the state variables $y = Cx$. Alternatively, the same spacecraft may have a communications mission where one is interested in the RMS deflections over the entire surface of a flexible antenna. These two problems have entirely *different* modeling (and control) objectives and *it is precisely the weights* C^TQC *that distinguish between the two objectives.* That is, the reduced-order model that is best for the analysis (estimation, control) of optical errors is different from the model that is best for analysis of errors in the parabolic shape of the antenna. To ignore these weights C^TQC is to force a complete and artificial *separation* between the control problem and the modeling problem, a state of affairs which the authors believe is not realistic. The authors' opinion is that one's ability to evaluate the quality of *any* reduced model (obtained by any method) is no better and no worse than his ability to choose a precise performance metric. Of course, if one has no physical objective to motivate the choice of specific output variables $y = Cx$ and if he instead arbitrarily chooses an *equal* weighting on all *balanced* coordinates ($C^TQC = I$), *then* the CCA algorithm produces the same partial realization as balanced coordinate methods of model reduction. A primary goal of this chapter is therefore to promote a systematic beginning for the integration of the modeling and the estimation/control problems, to allow modeling decisions to be influenced by specific quadratic control or estimation objectives, without relying upon nonlinear programming methods.

It should also be mentioned that for *scalar* input-output systems the reduced models produced by the CCA algorithm are *independent* of the choices of the output weighting Q and the noise intensity W. This can be readily verified by noting that Q and W in (34) and also in (47) are scalars that are common factors in every component cost V_i, and cannot therefore influence the cost ordering (36).

IX. STABILITY CONSIDERATIONS

Stability may or may not be an important feature of a reduced model. In fact, several schemes for "improving" reduced models upon which state estimators are based include the *intentional destabilization* of the model as a means to reduce or eliminate Kalman filter divergence. This means of improving models is discussed in [15] and its references. Also note that the technique of guaranteeing stability margins in linear regulator problems by multiplying the state weight in the quadratic cost function by $\exp(2\alpha t)$ causes the closed loop eigenvalues to lie to the left of the line $-\alpha$ in the complex plane [16]. This method is also equivalent to *intentionally destabilizing* a stable plant model by replacing A by $A + \alpha I$ in lieu of multiplying the state weight by $\exp(2\alpha t)$. It is not our purpose to recommend necessarily such methods for estimator or control design, but merely to point out that stability is neither a necessary nor sufficient qualification for a reduced model to be a "good" model of a stable system.

The model error index Q is finite if the observable modes of (A, C) and the observable modes of (A_R, C_R) are stable. Hence, stability is a sufficient but not a necessary condition for the existence of Q. If stability is an overriding concern in the selection of a partial realization, then one may choose special coordinates for which the CCA algorithm guarantees stability.

Presently, if the order of the partial realization is fixed *a priori*, the only coordinates for which asymptotic stability of the partial realizations produced by CCA has been guaranteed is modal coordinates. The modal cost analysis (MCA) of Section V produces stable models since the eigenvalues of the reduced model are a subset of the eigenvalues of the original (stable) system. However, since other coordinate choices (such as the cost-equivalent coordinates of Section VII) may produce better models, it is suggested that the CER be found first and examined for stability. If stability of the reduced model is required and not obtained from the CER, then obtain the reduced model by application of MCA, which guarantees stability. Note, however, that if both realizations (from CER and MCA) of order r are stable, the authors have not found a single example in which the CER failed to yield a smaller model error index Q.

Furthermore, if the order of the partial realization is *not* fixed *a priori*, then the CER algorithm always yields a CER that is asymptotically stable.

X. CER EXAMPLES

The concepts are best illustrated with simple problems. We begin with a second-order example.

Example 1. The CER for the system (32) with parameters

$$A = \left[\begin{array}{c|c} -1 & 0 \\ \hline 0 & -10 \end{array}\right], \quad D = \left[\begin{array}{c|c} 1 & 1 \\ \hline 70 & 1 \end{array}\right], \quad Q = 1$$

$$C = [1, -0.2], \quad W = I$$

with transfer function

$$y(s) = G(s)w(s),$$

$$G(s) = [(s + 1)(s + 10)]^{-1}[-13s - 4, 0.8s + 9.8]$$

is

$$A_R = -10.318, \quad C_R = -2.867, \quad D_R = [4.534, -0.279],$$

which has the transfer function

$$\hat{y}(s) = G_R(s)w(s), \quad G_R(s) = [-13,\ 0.8][(s + 10.33)]^{-1}.$$

The reduced-order model has an eigenvalue near the fast mode (-10) of the original system as a consequence of the fact that this mode is highly disturbable from w(t).

Example 2. Several authors on model reduction have cited the fact that there seems to be no simple way to say that

$$G(s) = \frac{(s + 1.1)}{(s + 1)(s + 10)} \approx \frac{1}{s + 10}.$$

We consider a little more general situation: We find that for

$$G(s) = \frac{s + \alpha}{(s + 1)(s + 10)}$$

the minimal CER is

$$G_R(s) = \frac{1}{s + \dfrac{110}{10 + \alpha^2}}.$$

Table I provides the results for a variety of choices of α, and the corresponding CER. The table illustrates (for α = 1, 1.1, 10) the proper use of zero information in a near

Table I.

α	*Example G(s)*	$G_R(s)$ *of CER*
1	$\frac{s + 1}{(s + 1)(s + 10)}$	$\frac{1}{s + 10}$
1.1	$\frac{s + 1.1}{(s + 1)(s + 10)}$	$\frac{1}{s + 9.8}$
10	$\frac{s + 10}{(s + 1)(s + 10)}$	$\frac{1}{s + 1}$
-10	$\frac{s - 10}{(s + 1)(s + 10)}$	$\frac{1}{s + 1}$
- 1	$\frac{s - 1}{(s + 1)(s + 10)}$	$\frac{1}{s + 10}$
0	$\frac{s}{(s + 1)(s + 10)}$	$\frac{1}{s + 11}$

pole-zero cancellation situation--a situation that frustrates many model reduction schemes. The reader is reminded that for scalar input-output systems, the CER parameters (A_R, D_R, C_R) are *independent* of the noise intensity $W > 0$ and output weighting $Q > 0$.

Example 3. Consider the following system whose first Markov parameter is zero (i.e., $CD = 0$)

$$\dot{x} = Ax + Dw, \quad w \sim N(0, 1)$$

$$y = Cx$$

where

$$A = \begin{bmatrix} -10 & 1 & 0 \\ -5 & 0 & 1 \\ -1 & 0 & 0 \end{bmatrix}, \quad D = \begin{bmatrix} 0 \\ 1 \\ 1 \end{bmatrix}, \quad C = [1 \ 0 \ 0]$$

From proposition 9 a minimal CER of order 1 does not exist for this system. However a CER of order 2 exists and is given by

$$\dot{\hat{x}}_R = A_R \hat{x}_R + D_R w$$

$$y_R = C_R \hat{x}_R,$$

where

$$A_R = \begin{bmatrix} 0 & 0.7384 \\ -0.7384 & -8.166 \end{bmatrix}, \quad D_R = \begin{bmatrix} 0 \\ 4.0413 \end{bmatrix}$$

$$C_R = [0.335, 0].$$

This CER is asymptotically stable, disturbable and observable.

XI. APPLICATION OF CCA TO CONTROLLER REDUCTION

Given a model of high-order $n > n_c$, the traditional approach to designing a linear controller of specified order n_c is first to use model reduction methods to reduce the model to order n_c and then design a controller that is perhaps optimal for the reduced model. There are at least two objections to this strategy. The first disadvantage is that most model reduction techniques ignore the effect of the (yet to be determined) control inputs, and it is well known that the inputs (whether they be functions of time or state) can have a drastic effect on the quality of the reduced model. The second disadvantage is that optimal control theory applied to a poor model can certainly yield poor results, often destabilizing the actual system to which the low-order "optimal" controller is applied.

The design strategy suggested for obtaining a controller of order n_c given a model of order $n >> n_c$ is as follows:

A controller-reduction algorithm

1. Apply CCA to reduce the *model* to order $N_R > n_c$, where N_R is the largest dimension of a Riccati equation that can be reliably solved on the local computer.

2. Solve for the optimal controller of order N_R, using the reduced model of order N_R.

3. Apply CCA to reduce the *controller* to order $n_c < N_R$

The purpose of this section is to show how to accomplish step 3. The intended advantage of this algorithm over the traditional approach (which skips step 3 and sets $N_R = n_c$) is that more information about the higher order system and its would-be optimal controller is made available for the design of the reduced-order controller.

The controller reduction can be presented as a restricted model reduction problem as follows: Consider the plant,

$$\dot{x} = Ax + Bu + Dw,$$
$$y = Cx, \qquad (89)$$
$$z = Mx + v,$$

$$x \in R^n, \quad u \in R^m, \quad w \in R^d,$$

$$y \in R^k, \quad z \in R^\ell, \quad rk[B] = m \leq n,$$

where $w(t)$ and $v(t)$ are uncorrelated zero-mean white noise processes with intensities $W > 0$ and $V > 0$, respectively. The measurement is z, y is the output to be controlled, and u is the control chosen to minimize

$$V = \lim_{t\to\infty} E\left(\|y\|_Q^2 + \|u\|_R^2\right), \quad Q > 0,\ R > 0. \tag{90}$$

Under the assumptions that (A, B) and (A, D) are controllable, (A, C) and (A, M) observable, the optimal controller for (89) takes the form

$$\dot{x}_c = A_c x_c + Fz, \quad x_c \in R^n,$$
$$u = Gx_c, \tag{91a}$$

where

$$A_c \triangleq A + BG - FM, \tag{91b}$$

$$G = -R^{-1}B^T K, \quad KA + A^T K - KBR^{-1}B^T K + C^T QC = 0, \tag{92}$$

$$F = PM^T V^{-1}, \quad PA^T + AP - PM^T V^{-1} MP + DWD^T = 0. \tag{93}$$

Augmenting the plant (89) and the controller (91) yields the closed-loop system.

$$\begin{bmatrix} \dot{x} \\ \dot{x}_c \end{bmatrix} = \begin{bmatrix} A & BG \\ FM & A_c \end{bmatrix} \begin{bmatrix} x \\ x_c \end{bmatrix} + \begin{bmatrix} D & 0 \\ 0 & F \end{bmatrix} \begin{bmatrix} w \\ v \end{bmatrix} \tag{94}$$

$$\begin{bmatrix} y \\ u \end{bmatrix} = \begin{bmatrix} C & 0 \\ 0 & G \end{bmatrix} \begin{bmatrix} x \\ x_c \end{bmatrix}.$$

The cost V can be expressed as

$$V = \mathrm{tr}\ \hat{X}_{11} C^T QC + \mathrm{tr}\ \hat{X}_{22} G^T RG, \tag{95}$$

where

$$\begin{bmatrix} \hat{X}_{11} & \hat{X}_{12} \\ \hat{X}_{12}^T & \hat{X}_{22} \end{bmatrix} \begin{bmatrix} A^T & M^T F^T \\ G^T B^T & A_c^T \end{bmatrix} + \begin{bmatrix} A & BG \\ FM & A_c \end{bmatrix} \begin{bmatrix} \hat{X}_{11} & \hat{X}_{12} \\ \hat{X}_{12}^T & \hat{X}_{22} \end{bmatrix} + \begin{bmatrix} DWD^T & 0 \\ 0 & FVF^T \end{bmatrix} = 0. \tag{96}$$

Now if the two "components" of (94) are defined as the plant (with state x) and the controller (with state x_c), then the component costs for x and x_c are denoted V° and V^c, respectively, where

$$V = V^\circ + V^c \tag{97}$$

and

$$V^\circ \triangleq \operatorname{tr} \hat{X}_{11} C^T Q C, \tag{98}$$

$$V^c \triangleq \operatorname{tr} \hat{X}_{22} G^T R G. \tag{99}$$

Since we desire to reduce the dimension of the controller and not the plant, we further decompose V^c into individual component costs associated with controller states.

$$V = V^\circ + \sum_{i=1}^{n} V_i^c, \tag{100}$$

where

$$V_i^c \triangleq \left(\hat{X}_{22} G^T R G\right)_{ii}. \tag{101}$$

Having defined the controller components, the controller reduction can be shown to be a special "model reduction" problem by simply interpreting (94) in the form of (32). That is, substitute

$$x^T \to \begin{bmatrix} x^T & x_c^T \end{bmatrix}, \quad y^T \to \begin{bmatrix} y^T & u^T \end{bmatrix}$$

$$A \to \begin{bmatrix} A & BG \\ FM & A_c \end{bmatrix}, \quad D \to \begin{bmatrix} D & 0 \\ 0 & F \end{bmatrix}, \quad W \to \begin{bmatrix} W & 0 \\ 0 & V \end{bmatrix}$$

$$Q \to \begin{bmatrix} Q & 0 \\ 0 & R \end{bmatrix}, \quad C \to \begin{bmatrix} C & 0 \\ 0 & G \end{bmatrix}$$

Now with a very minor modification the standard CCA algorithm can be applied to obtain the reduced-order model of dimension $N_c = n + n_c$, where n_c is the dimension of the reduced-order controller desired. The minor restriction is that the plant component of dimension n is not to be truncated, regardless of the value of $V°$.

Motivated by the theory of cost-decoupled coordinates CERs and definition 6, we desire to transform the controller coordinates so that both $\hat{X}_{22}$ and G^TRG in (101) are diagonal.

The cost-decoupled controller (CDC) algorithm

Step 1. Given the model and performance objectives (A, B, D, C, M, W, V, Q, R).

Step 2. Compute the optimal controller (A_c, F, G) from (91)-(93) and the covariances $\hat{X}_{11}$ and $\hat{X}_{22}$ satisfying (96).

Step 3. Compute Θ_1 the square root of $\hat{X}_{22}$

$$\hat{X}_{22} = \Theta_1 \Theta_1^T$$

Step 4. Compute Θ_2 the orthonormal modal matrix of $\Theta_1^T G^T R G \Theta_1$. The controller component costs are

$$\Theta_2^T \Theta_1^T G^T R G \Theta_1 \Theta_2 = \text{diag}\left\{V_1^c, V_2^c, \ldots, V_m^c, 0, \ldots 0\right\}$$

where the number of nonzero controller component costs are m = rank B.

Step 5. Rearrange the columns of Θ_2 so that the V_i^c appear in order

$$V_1^c \geq V_2^c \geq \cdots \geq V_m^c.$$

Then define Θ_R by

$$\Theta_2 = [\Theta_R, \Theta_T], \quad \Theta_R \in R^{n\times m}.$$

Step 6. The reduced CDC is

$$\dot{\hat{x}}_R = A_R\hat{x}_R + F_R z, \quad \hat{x}_R \in R^m,$$

$$u = G_R\hat{x}_R,$$

where

$$A_R \triangleq \Theta_R^T\Theta_1^{-1}A_c\Theta_1\Theta_R,$$

$$F_R \triangleq \Theta_R^T\Theta_1^{-1}F,$$

$$G_R \triangleq G\Theta_1\Theta_R.$$

Additional properties of the CDC must be explored in future investigations. Space limitations suggest this convenient stopping point in the presentation of the CER theory and its application to both model and controller reduction.

XII. CONCLUSIONS

A summary of the ideas of cost decomposition is given to aid in the determination of the relative cost (or "price") of each component of a linear dynamic system using quadratic performance criteria. In addition to the insights into system behavior that are afforded by such a component cost analysis (CCA), these CCA ideas naturally lead to a theory for cost-equivalent realizations.

Cost-equivalent realizations (CERs) of linear systems are defined, and an algorithm for their construction is given. The partial realizations of order r produced by this algorithm have these properties:

1. a minimized model error index;

2. the model error index is zero (i.e., the original system and the partial realization have the same value of the quadratic performance metric), if $r \geq k$, where k is the number of independent outputs;

3. the algorithm does not require the computation of modal data of the plant matrix A;

4. the method is applicable to large-scale systems, limited only by the necessity to solve a linear Liapunov-type algebraic equation;

5. the CER algorithm produces stable realizations of a stable system.

The algorithm is based upon component cost analysis (CCA), which is described for time-varying systems, for time-invariant systems, and for systems for which accurate modeling is of concern only over a finite interval of time. These component costs are shown herein to be useful in obtaining the above cost-equivalent realizations, but they are also useful in closed-loop applications where controllers, rather than models, are to be simplified.

Property 2 above reveals that cost-equivalent realizations can be *smaller* than Kalman's minimal realization, which is always of the dimension of the controllable, observable subspace.

Section VI is a point of departure for further research using different model error criteria. Instead of using the difference of norms as in (48), an error criterion using the norm of the differences can be studied much more extensively than done in [5], where only input-induced component costs were used. The model error criterion utilized herein, (48), is chosen for its appropriateness to the reduction of optimal controllers. Other uses of component cost analysis (CCA), which warrant further research include decentralized control, failure analysis, and system redesign strategies based upon "cost-balancing."

ACKNOWLEDGMENTS

Portions of this work have been supported by the National Science Foundation under Grant ENG-7711703, and portions were supported under consulting work for the Naval Research Lab.

REFERENCES

1. G. B. DANTZIG and P. WOLFE, "The Decomposition Algorithm for Linear Programming," *Econometrica 29*, No. 4 (1961); *Operations Research 8*, (1960).

2. J. F. BENDERS, "Partitioning Procedures for Solving Mixed Variables Programming Problems," *Numer. Math. 4*, 238-252 (1962).

3. R. E. SKELTON, "Cost-Sensitive Model Reduction for Control Design," *Proc. AIAA Guidance Control Conf.*, Palo Alto, California (1978).

4. R. E. SKELTON and C. Z. GREGORY, "Measurement Feedback and Model Reduction by Modal Cost Analysis," *Proc. Joint Automatic Control Conf.*, Denver, Colorado (1979).

5. R. E. SKELTON, "Cost Decomposition of Linear Systems with Application to Model Reduction," *Int. J. Control 32*, No. 6, 1031-1055 (1980).

6. H. KWAKERNAAK and R. SIVAN, "Linear Optimal Control Systems," Wiley, New York, 1972.

7. B. C. MOORE, "Principal Component Analysis in Linear Systems, Controllability, Observability, and Model Reduction," to appear, *IEEE Trans. Auto. Control.*

8. E. E. Y. TSE, J. V. MEDANIC, and W. R. PERKINS, "Generalized Hessenberg Transformations for Reduced-Order Modeling of Large Scale Systems," *Int. J. Control 27*, No. 4, 493-512 (1978).

9. G. FORSYTHE, M. A. MALCOLM, and C. B. MOLER, "Computer Methods for Mathematical Computation," Prentice-Hall, Englewood Cliffs, New Jersey, 1977.

10. B. C. MOORE, "Singular Value Analysis of Linear Systems," *IEEE CDC Proc.*, pp. 66-73 (1978).

11. G. H. GOLUB, S. NASH, and C. VAN LOAN, "A Hessenberg-Schur Method for the Problem AX + XB = C," *IEEE Trans. Auto. Control AC-24*, No. 6, 909-912 (1979).

12. G. J. BIERMAN, "Factorization Methods for Discrete Sequential Estimation," p. 54, Academic Press, New York, 1977.

13. D. A. WILSON, "Model Reduction for Multivariable Systems," *Int. J. Control 20*, No. 1, 57-64 (1974).

14. D. D. SILJAK, "Large-Scale Dynamic Systems: Stability and Structure," North-Holland, New York, 1978.

15. A. E. BRYSON, JR., "Kalman Filter Divergence and Aircraft Motion Estimators," *J. Guidance and Control 1*, No. 1, 71-80 (1978).

16. B. D. O. ANDERSON and B. C. MOORE, "Linear Optimal Control," Prentice-Hall, Englewood Cliffs, New Jersey, 1971.

17. R. E. SKELTON and P. C. HUGHES, "Modal Cost Analysis for Linear Matrix-Second-Order Systems," *J. Dyn. Sys. Meas. Control 102*, 151-158 (1980).

Reduced-Order Modeling and Filtering

CRAIG S. SIMS
School of Electrical Engineering
Oklahoma State University
Stillwater, Oklahoma

I. INTRODUCTION

A. BACKGROUND

The complexity of a control system or filter is often a primary consideration. An optimal solution designed without regard for the difficulties of implementation may be

ISBN 0-12-012718-0

unsatisfactory because the required calculations cannot be made in the time available. For many years researchers have addressed this difficulty by fixing the structure of the controller or filter, and then optimizing the free parameters with respect to a performance measure. The resulting performance is suboptimal, but the structure of the design is such that implementation is feasible. Such an approach may be used for applications where the computational facilities are limited and the state vector is so large that it is not feasible to perform all the on-line computations required for a Kalman filter or an optimal stochastic controller. One might argue that computer capabilities are increasing so rapidly that the problem of computer throughput is becoming less significant. In fact the reverse is true, since the temptation is now to address problems that would have been out of the question several years ago. An example is near real-time image processing. Even with advanced distributed computer architecture and performing most multiplies in parallel, throughput may be limited by power considerations. Achieving real-time processing for complicated estimation problems is currently, and will continue to be, an important issue.

Historically, the works of Wiener [1], and Newton *et al.* [2] represent the pioneering contributions in the area of parameter optimization. These efforts were directed toward stationary stochastic systems and were not set in a state-space format. Later, when state-space techniques were adopted, the work in the area of limited complexity controllers was referred to as specific optimal control or fixed configuration control [3-5]. These techniques, aimed at the optimization of constant parameters, were then applied to estimation problems [6]. With regard to implementation, order reduction is of primary concern. One may specify limited complexity of a controller by choosing a reduced-order configuration, and selecting optimal time variable parameters. Johansen [7] was the first to take this approach in a linear stochastic control problem; others also took this approach [8].

More recently many researchers have been concerned with the design of filters having less complexity than a Kalman filter [9-16]. Such filters are most often referred to as reduced-order filters. It has been shown that optimization of the parameters of a reduced-order filter leads naturally to a matrix control problem. In certain cases [17], depending on the performance measure, one can obtain a complete solution by solving a linear two-point boundary-value problem. In general, however, certain aspects of the reduced-order filtering problem remain unsolved.

The general topic of model reduction is closely related to the reduced-order filtering problem. If one can obtain a good reduced-order model for a system that is actually of high dimension, then perhaps a Kalman filter designed for the reduced-order model will work adequately. Singular perturbation methods [18] may be used to obtain such lower order filters. Alternatively, optimization theory may be used to obtain the optimal reduced-order model. Several researchers have taken this approach [19-21]. It is interesting to note that researchers have tended to investigate reduced-order stochastic control problems prior to reduced-order filtering and modeling problems. The reduced-order modeling problem is fundamental and one would think that it would have been the first to attract attention. As is often the case, certain aspects of reduced-order modeling problems are handled with little difficulty while others remain open issues.

In this chapter the purpose is to show how one can find a reduced-order model or a reduced-order filter with a reasonable amount of design effort. It is the author's current feeling that any result such as in [8], which requires the solution of a nonlinear matrix two-point boundary-value problem of high order, is not practical. The design procedure in such cases will be prohibitively difficult. It is reasonable to accept a rather large amount of off-line design calculations to achieve on-line computational savings, but clearly if the design is overly burdensome another method will be selected. In this chapter only those aspects of reduced-order design optimization that are amenable to solutions with well-known mathematical techniques are considered.

B. *PROBLEM STATEMENTS*

1. *Reduced-Order Modeling*

One may construct a reduced-order model using several approaches. Here it is assumed that a model of rather high dimension is available, and that it is to be approximated by a low-order model. Optimization theory is used to obtain the gain of the reduced-order model. The known system dynamics are described by the state equation

$$\dot{x}(t) = A(t)x(t) + B(t)w(t), \tag{1}$$

where x is the state vector of dimension n. The matrices A and B are known, and w is zero-mean white noise with covariance

$$E\{w(t)w^T(\tau)\} = Q(t)\delta(t - \tau), \tag{2}$$

and the initial statistics of the state are given as

$$E\{x(t_0)\} = \mu_0, \qquad \text{var}\{x(t_0)\} = V_0. \tag{3}$$

The output equation is

$$y(t) = C(t)x(t), \tag{4}$$

where y is the output of dimension $\ell \leq n$. In the reduced-order modeling problem, y is to be approximated by y_m, which is generated by the model

$$\dot{y}_m(t) = F_m(t)y_m(t) + K_m(t)w(t). \tag{5}$$

The performance measure

$$J_m = E\left\{\int_{t_0}^{t_f} e_m^T(t)U_m(t)e_m(t)\,dt + e_m^T(t_f)\,S_m e_m(t_f)\right\} \tag{6}$$

is to be minimized by properly selecting $K_m(t)$. In Eq. (6), e_m is the error defined as

$$e_m(t) \triangleq y(t) - y_m(t), \tag{7}$$

and it is required that $U_m \geq 0$ and $S_m > 0$. The selection of an an optimal $F_m(t)$ is a difficult matter. With the exception of certain unlikely circumstances [21], this problem remains an open issue and is not considered in detail here. Singular perturbation is the only method for choosing F_m given any consideration in this article, although certain other viewpoints exist [20] that provide insight in the selection of F_m. Optimization with respect to $K_m(t)$ is seen to lead to linear mathematics, so that the solution is rather easily obtained.

2. Reduced-Order Filtering

The purpose of obtaining a reduced-order model may be to design a reduced-order filter based on the model, and thus have a simpler design. Suppose there is a noisy measurement

$$m(t) = y(t) + v(t) \tag{8}$$

with v zero-mean white noise having covariance

$$E\{v(t)v^T(\tau)\} = R(t)\delta(t - \tau). \tag{9}$$

It is assumed throughout that w and v are uncorrelated with each other and with $x(t_0)$. A Kalman filter, based on the assumption that $y = y_m$, would be of the form

$$\dot{\hat{y}}_m(t) = F_m(t)\hat{y}_m(t) + K_{KF}(t)[m(t) - \hat{y}_m(t)]. \tag{10}$$

The total estimation error e_T is composed of two components

$$e_T \triangleq y - \hat{y}_m = e_m + (y_m - \hat{y}_m), \tag{11}$$

so clearly the modeling error influences the accuracy of the estimate. Consequently it is of interest to consider a more direct approach to reduced-order filtering.

It is assumed that the vector of interest is the output indicated by Eq. (4). (It may be of general interest to assume that the vector of elements to be estimated is any linear transformation of the state [17].) It is desired to estimate y with a filter of the form

$$\dot{\hat{y}}(t) = F(t)\hat{y}(t) + K(t)m(t). \tag{12}$$

The matrix K is chosen to minimize the performance measure

$$J_f = E\left\{\int_{t_0}^{t_f} \left[e_f^T(t)U_f(t)e_f(t)\right]dt + e_f^T(t_f)S_f e_f(t_f)\right\}, \tag{13}$$

where

$$e_f \triangleq y - \hat{y}, \tag{14}$$

and $U_f \geq 0$, $S_f > 0$.

The filter initial condition may also be optimized. For some cases, the selection of F may be shown to be arbitrary [17]. In general, however, there is not an easy method of obtaining the optimum value. In this article, use is made of the reduced-order model, and the Kalman filter designed for that model determines F. This is just one approach and not necessarily the best.

C. ORGANIZATION

In Section II the problem of reduced-order modeling and its relation to reduced-order filtering is considered. It is seen that there is a systematic way of modifying a design based on singular perturbation, or other procedures, so that performance is improved during certain portions of the interval of interest. For example, the model might be better near the end of the interval than at the beginning. The matrix minimum principle is used to lead to a result that only involves the solution of a linear single-point boundary-value problem.

In Section III the reduced-order filtering problem is considered. There is no guarantee that a Kalman filter designed for the optimum reduced-order model is the best filter to use. This fact offers sufficient reason to approach the reduced-order filtering problem directly. The problem is formulated as an optimal control problem, and the result requires that the designer solve a linear two-point boundary-value problem. Both biased and unbiased estimators are considered.

Computational procedures are considered in Section IV. The difficulties inherent in solving a linear matrix two-point boundary-value problem are explored. Approaches for dealing with these difficulties are suggested. Certain classes of problems are seen to be much easier to solve than others.

In Section V an example is presented which illustrates the theory presented in the previous sections. Section VI indicates how one may solve reduced-order modeling and filtering problems stated in a discrete format.

II. REDUCED-ORDER MODELING

A. *INTRODUCTION*

It is useful to clarify the reduced-order modeling problem by means of a specific example. Suppose that the system of interest is of the form

$$\dot{y}(t) = -y(t) + \xi(t), \tag{15}$$

where ξ is not white noise but is a colored noise. The model for ξ is

$$\dot{\xi}(t) = -\alpha\xi(t) + \alpha w(t), \tag{16}$$

where w is zero-mean white noise. This is a model that may have been developed based on an experimental determination of the autocorrelation function of ξ. If $\alpha \gg 1$, then a wide-band assumption is reasonable, and a designer might justifiably decide to use the reduced-order scalar model

$$\dot{y}_m(t) = -y_m(t) + w(t), \tag{17}$$

where y_m is used to approximate y. If estimation is the goal, and there is a scalar observation available,

$$m(t) = y(t) + v(t), \tag{18}$$

where v is white noise, the designer might then build a first-order Kalman filter based on using Eq. (17) as a model. This would be called a reduced-order filter because the optimal Kalman filter is second order, based on the true model indicated by Eqs. (15) and (16). The reason for choosing a first-order filter is that the structure is simple, and easier to implement than the second-order filter. The performance loss should be minimal if α is sufficiently greater than one and if stationary operation is considered. It is difficult to predict what performance to expect when α is close to 1, and a finite time interval is of importance such that transients due to random initial conditions are significant. It is interesting to note that the approach discussed here is similar to the singular perturbation approach [18]. This is easily seen if Eq. (16) is rewritten as

$$\mu\dot{\xi}(t) = -\xi(t) + w(t), \tag{19}$$

where μ is a small parameter ($\mu = 1/\alpha$), if ξ is wide-band noise.

The method to be considered here involves replacing Eq. (17) with

$$\dot{y}_m(t) = -y_m(t) + K_m(t)w(t), \tag{20}$$

where $K_m(t)$ is selected to be optimal with respect to a quadratic performance measure of the form indicated by Eq. (6). Equation (20) is a particular example of the general case indicated by Eq. (5).

It is relatively easy to generalize the procedure that has been discussed using singular perturbation. Suppose that there is a known model of order p,

$$\dot{y}(t) = F_{11}(t)y(t) + F_{12}(t)\xi(t), \tag{21}$$

driven by a vector of inputs ξ described by

$$\mu\dot{\xi}(t) = F_{21}(t)\xi(t) + F_{22}(t)w(t), \tag{22}$$

where μ is a small parameter and w is white, then under certain rather broad circumstances [18] the limiting behavior of ξ as $\mu \to 0$ is that of a white noise having covariance matrix equivalent to that of the process

$$\xi^*(t) = -F_{21}^{-1}(t)F_{22}(t)w(t).$$

For small μ it is then appealing to replace Eq. (21) by an approximate model

$$\dot{y}(t) \cong F_{11}(t)y(t) - F_{12}(t)F_{21}^{-1}(t)F_{22}(t)w(t). \tag{23}$$

Equations (21) and (22) are of the form Eq. (1) and it is seen that Eq. (5) is a generalization of Eq. (23). Thus the model

$$\dot{y}_m(t) = F_{11}(t)y_m(t) + K_m(t)w(t) \tag{24}$$

might provide a better approximation than Eq. (23) for suitably selected K_m. Although performance could also be improved by selecting some F_m other than F_{11}, the associated optimization problem is difficult in general. It is therefore assumed that F_m is a given matrix, having been selected *a priori* by any reasonable approach such as using singular perturbation arguments. It is emphasized that the procedure discussed herein is for gain improvement only.

B. *THE OPTIMIZATION PROBLEM*

In this section the optimization of the model gain K_m is considered. The true model is as indicated by Eqs. (1)-(4), and the approximation is indicated by Eq. (5). The performance measure in Eq. (6) may be written as

$$J_m = \operatorname{tr}\left\{\int_{t_0}^{t_f} U_m(t)P_{mm}(t)\,dt + S_mP_{mm}(t_f)\right\}, \tag{25}$$

where

$$P_{mm}(t) \triangleq E\left\{e_m(t)e_m^T(t)\right\}, \tag{26}$$

and U_m and S_m are appropriately chosen weighting matrices.

The error e_m satisfies the differential equation

$$\dot{e}_m(t) = B_m(t)x(t) + F_m(t)e_m(t) + \beta_m w(t), \tag{27}$$

where

$$B_m(t) \triangleq \dot{C}(t) + C(t)A(t) - F_m(t)C(t) \tag{28}$$

and

$$\beta_m(t) \triangleq C(t)B(t) - K_m(t). \tag{29}$$

Note that if it were possible to select F_m and K_m so that both B_m and β_m were zero, then the problem would be trivially solved. In such a case, $e_m(t)$ is identically zero for all $t \geq t_0$ provided that $e_m(t_0)$ is zero. It is assumed in what follows that no value of $F_m(t)$ is available, which results in a value of zero for $B_m(t)$.

From Eqs. (1) and (27) it may be seen that P_{mm} satisfies

$$\dot{P}_{mm} = B_m P_{xm} + P_{mx} B_m^T + \beta_m Q \beta_m^T + F_m P_{mm} + P_{mm} F_m^T, \tag{30}$$

where

$$P_{xm}(t) \triangleq E\left\{x(t)e_m^T(t)\right\} = P_{mx}^T(t). \tag{31}$$

The matrix P_{xm} satisfies

$$\dot{P}_{xm} = AP_{xm} + P_{xx} B_m^T + P_{xm} F_m^T + BQ\beta_m^T, \tag{32}$$

with P_{xx} the known second moment of the state, which one may obtain by solving the equation

$$\dot{P}_{xx} = AP_{xx} + P_{xx}A^T + BQB^T \tag{33}$$

from initial condition

$$P_{xx}(t_0) = E\left\{x(t_0)x^T(t_0)\right\} = V_0 + \mu_0 \mu_0^T. \tag{34}$$

The problem is ideally suited to solution using the matrix

form of the minimum principle [22]. The Hamiltonian is formed as

$$H_m = \operatorname{tr}\left\{U_m P_{mm} + \dot{P}_{mm}\Lambda_{mm}^T + \dot{P}_{xm}\Lambda_{xm}^T + \dot{P}_{mx}\Lambda_{mx}^T\right\}, \tag{35}$$

where Λ_{xm} and Λ_{mm} are matrices of Lagrange multipliers. These must satisfy the equations

$$\dot{\Lambda}_{xm} = -\partial H_m/\partial P_{xm} = -A^T\Lambda_{xm} - \Lambda_{xm}F_m - B_m^T\Lambda_{mm} \tag{36}$$

and

$$\dot{\Lambda}_{mm} = -\partial H_m/\partial P_{mm} = -U_m - F_m^T\Lambda_{mm} - \Lambda_{mm}F_m \tag{37}$$

subject to the boundary conditions

$$\Lambda_{xm}(t_f) = 0, \tag{38}$$

$$\Lambda_{mm}(t_f) = S_m. \tag{39}$$

It is clear that Λ_{mm} may be regarded as known, having the solution

$$\Lambda_{mm}(t) = \phi_F(t, t_f)S_m\phi_F^T(t, t_f) + \int_t^{t_f} \phi_F(t, \tau)U_m(\tau)\phi_F^T(t, \tau)\,d\tau, \tag{40}$$

where ϕ_F is the transition matrix satisfying

$$\dot{\phi}_F(t, \tau) = -F_m^T(t)\phi_F(t, \tau) \tag{41}$$

$$\phi_F(t, t) = I. \tag{42}$$

Once Λ_{mm} is known, Eq. (36) can be solved for Λ_{xm}:

$$\Lambda_{xm}(t) = \int_t^{t_f} \phi_A(t, \tau)B_m^T(\tau)\Lambda_{mm}(\tau)\phi_F^T(t, \tau)\,d\tau, \tag{43}$$

where

$$\dot{\phi}_A(t, \tau) = -A^T(t)\phi_A(t, \tau) \tag{44}$$

and

$$\phi_A(t, t) = I. \tag{45}$$

The gain K_m is found by requiring that

$$\partial H_m/\partial K_m = 0 \tag{46}$$

or equivalently, the necessary condition for optimality is

$$\Lambda_{mm}(t)K_m(t)Q(t) = [\Lambda_{mx}(t) + \Lambda_{mm}(t)C(t)]B(t)Q(t). \tag{47}$$

If $\Lambda_{mm}(t)$ and $Q(t)$ both are nonsingular, the gain may be expressed as

$$K_m(t) = \left[\Lambda_{mm}^{-1}(t)\Lambda_{mx}(t) + C(t)\right]B(t), \tag{48}$$

where Λ_{mm} and Λ_{mx} are evaluated using Eqs. (40) and (43).

It is of interest to note that the moment Eqs. (30), (32), and (33) do not need to be solved unless one is interested in evaluating performance. The optimal gain is independent of Q if it is positive definite. It is also independent of the initial statistics of x. The gain will be significantly influenced by S_m and U_m, however. This means that one can select a model that is more accurate at some times than it is at other times. Since one is going to suffer losses of accuracy with a reduced-order model, it is desirable to be able to pick the times of relatively accurate or inaccurate performance. The reduced-order model, among other uses, may form the basis for designing a reduced-order filter.

C. THE ASSOCIATED KALMAN FILTER

Once having picked K_m by optimization and F_m as indicated by Eq. (24), or by some other approach, one may design a Kalman filter of the form indicated by Eq. (10). The gain K_{KF}

is given as

$$K_{KF}(t) = P_{KF}(t)R^{-1}(t), \tag{49}$$

where P_{KF} satisfies the Riccati equation

$$\dot{P}_{KF} = F_m P_{KF} + P_{KF}F_m^T - P_{KF}R^{-1}P_{KF} + K_m Q K_m^T. \tag{50}$$

The initial condition for Eq. (50) is

$$P_{KF}(t_0) = C(t_0)V_0 C^T(t_0), \tag{51}$$

and the initial condition for Eq. (10) is

$$\hat{y}_m(t_0) = C(t_0)\mu_0. \tag{52}$$

At this point some remarks are in order about the design procedure that has been discussed. While one can predict that the model Eq. (24) is superior to Eq. (23), in the sense of minimizing the performance measure Eq. (6), it does not logically follow that a Kalman filter derived for Eq. (24) will have better performance than one designed for Eq. (23). In fact the concept of designing a Kalman filter for any reduced-order model, although reasonable, is certainly not optimal. The filter so designed will in general be biased [15] and suboptimal. In Section III, methods for improving the performance of a reduced-order filter are presented. In particular, the gain of the reduced-order filter is improved. The Kalman filter for the reduced-order model may be used as a guide for selecting the filter matrix F indicated in Eq. (12). It should be noted that

$$P_{KF} \neq E\left\{(y - \hat{y}_m)(y - \hat{y}_m)^T\right\}, \tag{53}$$

and thus is not a good measure of performance. The correct equations for evaluating performance are presented in the following section.

III. REDUCED-ORDER FILTERING

A. *INTRODUCTION*

In this section the direct optimization of a reduced-order filter is considered. It is clear that Eq. (10) may be expressed as

$$\dot{\hat{y}}_m(t) = F(t)\hat{y}_m(t) + K_{KF}(t)m(t), \tag{54}$$

where

$$F(t) = F_m(t) - K_{KF}(t). \tag{55}$$

That leads to the consideration of filtering equations of the form Eq. (12). It has been found that for a given F(t), one may generally improve the performance of a reduced-order filter by selection of an optimum gain K. In fact the solution that minimized the quadratic performance measure indicated by Eq. (13) requires only linear mathematics. The resulting estimate is biased, but the bias may be removed as will be discussed in Section III.C.

B. *REDUCED-ORDER FILTER OPTIMIZATION*

The performance measure Eq. (13) may be rewritten in terms of the second moment of the error

$$J_f = \operatorname{tr}\left\{\int_{t_0}^{t_F} U_f(t)P_{ff}(t)\,dt + S_fP_{ff}(t_F)\right\}, \tag{56}$$

where

$$P_{ff}(t) \triangleq E\left\{e_f(t)e_f^T(t)\right\}, \tag{57}$$

and e_f is the filtering error. The differential equation that describes the error is

$$\dot{e}_f(t) = B_f(t)x(t) + F(t)e_f(t) + C(t)B(t)w(t) - K(t)v(t), \tag{58}$$

where

$$B_f(t) \triangleq [\dot{C}(t) + C(t)A(t) - F(t)C(t) - K(t)C(t)]. \tag{59}$$

Equations (1), (4), (8), and (12) are used to obtain Eq. (58). Using Eqs. (1) and (58), the second moment equation for the error is obtained,

$$\dot{P}_{ff} = B_f P_{xf} + P_{fx} B_f^T + FP_{ff} + P_{ff} F^T + CBQB^T C^T + KRK^T, \tag{60}$$

where

$$P_{xf}(t) \triangleq E\left\{x(t) e_f^T(t)\right\} = P_{fx}^T(t). \tag{61}$$

The matrix P_{xf} satisfies the equation

$$\dot{P}_{xf} = AP_{xf} + P_{xx} B_f^T + P_{xf} F^T + BQB^T C^T, \tag{62}$$

where P_{xx} satisfies Eqs. (33) and (34) and is a known matrix. It is seen that the problem is cast in a deterministic framework and is quite similar to the reduced-order modeling problem. It may be solved using the same technique. Equation (56) is to be minimized subject to the constraints imposed by Eqs. (60) and (62).

The Hamiltonian is formed,

$$H_f = \text{tr}\left\{U_f P_{ff} + \dot{P}_{ff} \Lambda_{ff}^T + \dot{P}_{xf} \Lambda_{xf}^T + \dot{P}_{fx} \Lambda_{fx}^T\right\}, \tag{63}$$

and optimization is again by means of the matrix minimum principle. The Lagrange multiplier matrices satisfy the equations

$$\dot{\Lambda}_{ff} = -\partial H_f / \partial P_{ff} = -U_f - F^T \Lambda_{ff} - \Lambda_{ff} F \tag{64}$$

and

$$\dot{\Lambda}_{xf} = -\partial H_f / \partial P_{xf} = -A^T \Lambda_{xf} - \Lambda_{xf} F - B_f^T \Lambda_{ff}, \tag{65}$$

subject to the terminal conditions

$$\Lambda_{ff}(t_F) = S_f, \tag{66}$$

$$\Lambda_{xf}(t_f) = 0. \tag{67}$$

Comparing Eqs. (64)-(67) with Eqs. (36)-(39), it appears that this problem differs from the reduced-order modeling problem only symbolically and not in any important way. This apparent similarity is in fact deceptive, since the difference is substantial, in that this problem is a true two-point boundary-value problem. The difference occurs because B_f is a function of K. The result is that a greater amount of computation is required for the reduced-order filtering problem.

Minimizing H_f with respect to K leads to the resulting necessary condition for the optimum gain

$$K(t) = \left[P_{fx}(t) + \Lambda_{ff}^{-1}(t)\Lambda_{fx}(t)P_{xx}(t)\right]C^T(t)R^{-1}(t), \tag{68}$$

where it has been assumed that the required inverses exist. Since P_{xx} and Λ_{ff} may be precomputed, and therefore regarded as known, obtaining K only requires that a linear two-point boundary-value problem be solved. Upon substitution for the gain, using Eq. (68) in Eqs. (62) and (65), one obtains

$$\dot{P}_{xf} = \bar{A}P_{xf} + P_{xf}F^T - P_{xx}C^TR^{-1}CP_{xx}\Lambda_{xf}\Lambda_{ff}^{-1} + D_1 \tag{69}$$

and

$$\dot{\Lambda}_{xf} = -\bar{A}^T\Lambda_{xf} - \Lambda_{xf}F + C^TR^{-1}CP_{xf}\Lambda_{ff} + D_2, \tag{70}$$

where

$$\bar{A} \triangleq A - P_{xx}C^TR^{-1}C, \tag{71}$$

and D_1 and D_2 are defined as

$$D_1 \triangleq P_{xx}(\dot{C} + CA - FC)^T + BQB^TC^T, \tag{72}$$

$$D_2 \triangleq -(\dot{C} + CA - FC)^T \Lambda_{ff}. \tag{73}$$

The terminal condition for Eq. (70) is as indicated by Eq. (67) and the initial condition for P_{xf} depends on how the filter is initialized. If the estimate is to be initially unbiased so that

$$E\{e_f(t_0)\} = 0, \tag{74}$$

then $\hat{y}(t_0)$ must be selected as

$$\hat{y}(t_0) = C(t_0)\mu_0, \tag{75}$$

and it is easily shown that

$$P_{xf}(t_0) = V_0C^T(t_0). \tag{76}$$

The values of P_{xf} and Λ_{xf} are obtained by solving Eqs. (69) and (70) subject to the boundary conditions indicated by Eqs. (67) and (76). These may be substituted in Eq. (68) to provide the optimum gain for the filter having Eq. (75) as an initial condition. Equation (75) appears to provide a realistic initial condition for the filter, since it results in an unbiased initial estimate. Unless μ_0 is zero, however, the filter will give a biased estimate for $t > t_0$. To see that this is true, one only needs to take the expected value of Eqs. (1) and (58). The resulting equations for the mean values are

$$\dot{\mu}_x(t) = A(t)\mu_x(t) \tag{77}$$

and

$$\dot{\mu}_f(t) = B_f(t)\mu_x(t) + F(t)\mu_f(t), \tag{78}$$

where

$$\mu_x(t) \underline{\Delta} E\{x(t)\}, \tag{79}$$

$$\mu_f(t) \underline{\Delta} E\{e_f(t)\}. \tag{80}$$

The reduced-order filter will be unbiased if Eq. (74) is satisfied and if

$$B_f(t)\mu_x(t) = 0. \tag{81}$$

Since Eq. (81) will not always be satisfied, it is clear that the designer will have to settle for a biased estimate in general. In view of this fact, it appears that there is really no reason to assume that Eq. (75) is the best choice for filter initial condition. In fact, it is not difficult to show that there may be a better choice.

The generalized boundary condition applied at t_0 is

$$\operatorname{tr}\Big\{\delta P_{ff}(t_0)\Lambda_{ff}^T(t_0) + \delta P_{xf}(t_0)\Lambda_{xf}^T(t_0) + \delta P_{fx}(t_0)\Lambda_{fx}^T(t_0)\Big\} = 0. \tag{82}$$

The second moment terms in Eq. (82) are

$$P_{ff}(t_0) = C(t_0)P_{xx}(t_0)C^T(t_0) - \hat{y}(t_0)\mu_0^T C^T(t_0) - C(t_0)\mu_0\hat{y}^T(t_0) + \hat{y}(t_0)\hat{y}^T(t_0) \tag{83}$$

and

$$P_{xf}(t_0) = P_{xx}(t_0)C^T(t_0) - \mu_0\hat{y}^T(t_0). \tag{84}$$

Noting that the only free term in Eqs. (83) and (84) is $\hat{y}(t_0)$, one obtains the relationship

$$\operatorname{tr}\Big\{[\Lambda_{ff}(t_0)[\hat{y}(t_0) - C(t_0)\mu_0] - \Lambda_{fx}(t_0)\mu_0]\delta\hat{y}^T(t_0) + \delta y(t_0)[\Lambda_{ff}(t_0)[\hat{y}(t_0) - C(t_0)\mu_0] - \Lambda_{fx}(t_0)\mu_0]^T\Big\} = 0. \tag{85}$$

The above is satisfied if

$$\Lambda_{ff}(t_0)\hat{y}(t_0) = \Lambda_{ff}(t_0)C(t_0)\mu_0 + \Lambda_{fx}(t_0)\mu_0. \tag{86}$$

When $\Lambda_{ff}^{-1}(t_0)$ exists, the optimum initial condition is evaluated as

$$\hat{y}(t_0) = \left[C(t_0) + \Lambda_{ff}^{-1}(t_0)\Lambda_{fx}(t_0)\right]\mu_0. \tag{87}$$

Hence Eq. (75) does not, in general, provide the best choice, even though it is intuitively appealing. It will be the best when μ_0 is zero, or when $\Lambda_{fx}(t_0)$ is zero. From Eqs. (65) and (67) it can be seen that the latter condition will hold if $B_f(t)$ is zero. The case where $B_f(t)$ is zero has been treated elsewhere [17] and is not considered here. It is a condition that simplifies things greatly, but there are few reduced-order problems for which this will be the case. Hence Eq. (87) represents a nonintuitive result that has application to many reduced-order filtering problems for which the state is not zero mean. The similarity of Eq. (87) to a result obtained in [8] is noteworthy.

It should be clear that Eq. (87) changes the nature of the boundary conditions that must be satisfied. Equation (76) is no longer applicable when Eq. (87) is used. Instead one must use the initial value

$$P_{xf}(t_0) = P_{xx}(t_0)C^T(t_0) - \mu_0\mu_0^T\Psi, \tag{88}$$

where

$$\Psi = \left[C(t_0) + \Lambda_{ff}^{-1}(t_0)\Lambda_{fx}(t_0)\right]. \tag{89}$$

Alternatively, one may replace Eq. (88) with the expression

$$P_{xf}(t_0) = V_0C^T(t_0) - \mu_0\mu_0^T\Lambda_{xf}(t_0)\Lambda_{ff}^{-1}(t_0). \tag{90}$$

There are other ways of treating the bias error in the reduced-order filtering problem. One obvious approach is to include the bias error in the performance measure as a part of the problem statement. One could formulate and solve the optimization problem of selecting K to minimize

$$J' = J_f + E\left\{\int_{t_0}^{t_f} \mu_f^T(t)\Omega(t)\mu_f(t)\,dt + \mu_f^T(t_f)\;\Xi\;\mu_f(t_f)\right\}. \tag{91}$$

This is a reasonable approach that allows one to choose a value for K which strikes a perfect balance between the first and second moments of the error. The problem is somewhat more involved and will not be treated here. It is probably questionable whether a designer ever would have enough insight to choose good values for the weighting matrices in Eqs. (56) and (91). The extra effort involved in solving the problem might not be worthwhile. In the next section it is seen that one can get rid of the problem of bias error at the cost of increasing filter complexity slightly.

C. *BIAS REMOVAL*

In order to eliminate the *a priori* bias present in the reduced-order filter, an alternative filter structure may be used. Instead of Eq. (12) one may use the structure

$$\dot{\hat{y}}(t) = F(t)\hat{y}(t) + K(t)m(t) + b(t). \tag{92}$$

The mean value of the filtering error then satisfies the equation

$$\dot{\mu}_f(t) = B_f(t)\mu_x(t) + F(t)\mu_f(t) - b(t). \tag{93}$$

Hence if $b(t)$ is selected as

$$b(t) = B_f(t)\mu_x(t), \tag{94}$$

where $\mu_x(t)$ satisfies Eq. (77), the result is that the mean value of the error has the homogeneous equation

$$\dot{\mu}_f(t) = F(t)\mu_f(t). \tag{95}$$

It follows that $\mu_f(t)$ is zero provided that the filter is initialized according to Eq. (75) so that Eq. (74) is true. This technique was pointed out in [17]. The error equation is not as indicated by Eq. (58) if bias removal is used. Instead the error equation is

$$\dot{e}_f = B_f\tilde{x} + Fe_f + CBw - Kv, \tag{96}$$

where

$$\tilde{x}(t) = x(t) - \mu_x(t), \tag{97}$$

and satisfies the differential equation

$$\dot{\tilde{x}}(t) = A\tilde{x}(t) + B(t)w(t). \tag{98}$$

Note that $\tilde{x}(t)$ is a zero-mean process. One may then go throu through mathematics similar to that presented here to obtain the optimum gain. The details are presented in [17]. The difference is that the problem is based on Eqs. (96) and (98) instead of Eqs. (1) and (58).

IV. SOLUTION PROCEDURE

A. *THE STATE TRANSITION MATRIX APPROACH*

In this section, solution procedures are discussed for the two-point boundary-value problem described by Eqs. (69) and (70) with boundary conditions indicated by Eqs. (67) and (90). Motivated by the form of Eqs. (68) and (90), one is inclined to consider the matrix Λ defined as

$$\Lambda \triangleq \Lambda_{xf}\Lambda_{ff}^{-1}. \tag{99}$$

The time derivative of Λ is given by

$$\dot{\Lambda} = \dot{\Lambda}_{xf}\Lambda_{ff}^{-1} - \Lambda_{xf}\Lambda_{ff}^{-1}\dot{\Lambda}_{ff}\Lambda_{ff}^{-1}. \tag{100}$$

Substituting for $\dot{\Lambda}_{xf}$ and $\dot{\Lambda}_{ff}$ from Eqs. (64) and (70) leads to the expression

$$\dot{\Lambda} = -\bar{A}^T\Lambda + \Lambda\bar{B} + C^TR^{-1}CP_{xf} + D_2\Lambda_{ff}^{-1}, \tag{101}$$

where

$$\bar{B} \triangleq U_f\Lambda_{ff}^{-1} + F^T. \tag{102}$$

Equation (69) can be written as

$$\dot{P}_{xf} = \bar{A}P_{xf} + P_{xf}F^T - P_{xx}C^TR^{-1}CP_{xx}\Lambda + D_1. \tag{103}$$

The boundary conditions for Eqs. (101) and (103) are

$$P_{xf}(t_0) = V_0C^T(t_0) - \mu_0\mu_0^T\Lambda(t_0) \tag{104}$$

and

$$\Lambda(t_f) = 0. \tag{105}$$

Of course one could always stack the columns of P_{xf} and Λ in a large vector and use linear systems theory or a Riccati transformation to solve the linear two-point boundary-value problem. The Kronecker algebra [23] is useful in this regard. This makes for a computationally burdensome design procedure however. There are situations where that approach is unnecessary, and one can solve the problem in its natural matrix format. The situation where U_f is zero is of particular interest for several reasons [17], including the existence of a singular arc that provides a great deal of freedom in selecting F. The matter of interest here is computational however. When U_f and

and Λ_{ff} are such that

$$U_f(t)\Lambda_{ff}^{-1}(t) = \alpha(t)I, \tag{106}$$

where α is a scalar, it follows that Eqs. (101) and (103) can be written in the form

$$\left[\begin{array}{c}\dot{P}_{xf}\\ \hline \dot{\Lambda}\end{array}\right] = \left[\begin{array}{c|c}G_{11} & G_{12}\\ \hline G_{21} & G_{22}\end{array}\right]\left[\begin{array}{c}P_{xf}\\ \hline \Lambda\end{array}\right] + \left[\begin{array}{c}P_{xf}\\ \hline \Lambda\end{array}\right]F^T + \left[\begin{array}{c}D_1\\ \hline \tilde{D}_2\end{array}\right], \tag{107}$$

where

$$G_{11} = \bar{A}, \quad G_{12} = -P_{xx}C^TR^{-1}CP_{xx}, \quad G_{21} = C^TR^{-1}C,$$
$$G_{22} = -\bar{A}^T + \alpha I, \quad \tilde{D}_2 = D_2\Lambda_{ff}^{-1}. \tag{108}$$

The solution to Eq. (107) is of the form

$$P_{xf}(t) = \phi_{11}(t, t_0)P_{xf}(t_0)\psi(t, t_0) + \phi_{12}(t, t_0)\Lambda(t_0)\psi(t, t_0) + \Gamma_1(t), \tag{109}$$

$$\Lambda(t) = \phi_{21}(t, t_0)P_{xf}(t_0)\psi(t, t_0) + \phi_{22}(t, t_0)\Lambda(t_0)\psi(t, t_0) + \Gamma_2(t), \tag{110}$$

where

$$\dot{\phi} = \left[\begin{array}{c|c}\dot{\phi}_{11} & \dot{\phi}_{12}\\ \hline \dot{\phi}_{21} & \dot{\phi}_{22}\end{array}\right] = \left[\begin{array}{c|c}G_{11} & G_{12}\\ \hline G_{21} & G_{22}\end{array}\right]\left[\begin{array}{c|c}\phi_{11} & \phi_{12}\\ \hline \phi_{21} & \phi_{22}\end{array}\right] \tag{111}$$

and

$$\dot{\psi} = \psi F^T \tag{112}$$

with

$$\Phi(t, t) = I, \quad \psi(t, t) = I. \tag{113}$$

The terms $\Gamma_1(t)$ and $\Gamma_2(t)$ are evaluated as

$$\Gamma_1(t) = \int_{t_0}^{t} [\phi_{11}(t, \tau)D_1(\tau) + \phi_{12}(t, \tau)\tilde{D}_2(\tau)]\psi(t, \tau)d\tau, \tag{114}$$

$$\Gamma_2(t) = \int_{t_0}^{t} [\phi_{21}(t, \tau)D_1(\tau) + \phi_{22}(t, \tau)\tilde{D}_2(\tau)]\psi(t, \tau)d\tau. \tag{115}$$

Applying Eq. (110) at $t = t_f$ results in the expression

$$\Lambda(t_f) = 0 = \phi_{21}(t_f, t_0)P_{xf}(t_0)\psi(t_f, t_0) + \phi_{22}(t_f, t_0)\Lambda(t_0)\psi(t_f, t_0) + \Gamma_2(t_f). \tag{116}$$

Substituting the boundary condition Eq. (104) in Eq. (116) gives an equation for $\Lambda(t_0)$:

$$\phi_{21}(t_f, t_0)\left[V_0C^T(t_0) - \mu_0\mu_0^T\Lambda(t_0)\right]\psi(t_f, t_0) + \phi_{22}(t_f, t_0)\Lambda(t_0)\psi(t_f, t_0) + \Gamma_2(t_f) = 0. \tag{117}$$

After solving Eq. (117) for $\Lambda(t_0)$, $P_{xf}(t_0)$ is evaluated using Eq. (104). Equations (109) and (110) are then used to evaluate $P_{xf}(t)$ and $\Lambda(t)$. The optimal gain $K(t)$ may then be expressed as

$$K(t) = [P_{xf}(t) + P_{xx}(t)\Lambda(t)]^TC^T(t)R^{-1}(t). \tag{118}$$

The optimum filter initial condition is

$$\hat{y}(t_0) = \left[C(t_0) + \Lambda^T(t_0)\right]\mu_0. \tag{119}$$

B. *THE RICCATI TRANSFORMATION*

The problem may also be solved using approaches other than the state transition matrix method. One other approach is often referred to as a Riccati transformation. In this approach it is assumed that P_{xf} and Λ are linearly related

$$P_{xf}(t) = M(t)\Lambda(t) + \beta(t). \tag{120}$$

Then differentiation gives

$$\dot{P}_{xf} = \dot{M}\Lambda + M[G_{21}(M\Lambda + \beta) + G_{22}\Lambda + \tilde{D}_2] + \dot{\beta} + M\Lambda F^T, \tag{121}$$

while directly from Eq. (107), one obtains

$$\dot{P}_{xf} = G_{11}(M\Lambda + \beta) + G_{12}\Lambda + (M\Lambda + \beta)F^T + D_1. \tag{122}$$

It follows that M must satisfy the Riccati equation

$$\dot{M} + MG_{21}M + MG_{22} = G_{11}M + G_{12}, \tag{123}$$

which is seen to be symmetric upon inspecting Eq. (108). This offers some computational savings. The equation for β is

$$\dot{\beta} + MG_{21}\beta + M\tilde{D}_2 = G_{11}\beta + \beta F^T + D_1, \tag{124}$$

which can be solved after computing M. The boundary conditions for Eqs. (123) and (124) are obtained by comparing Eq. (104) with Eq. (120). Thus

$$M(t_0) = -\mu_0\mu_0^T \tag{125}$$

and

$$\beta(t_0) = V_0C^T(t_0) \tag{126}$$

are the initial conditions. Equations (123) and (124) can be integrated forward from these conditions to obtain M and β. Then Eq. (101) is integrated backwards from Eq. (105), using the substitution indicated by Eq. (120). Thus, a solution for $\Lambda(t)$ is obtained.

The two methods of solution that have been presented are for the special case where Eq. (106) is true. One may wonder what circumstances would give rise to this. From Eqs. (64) and (66) it is not hard to see that Eq. (106) will be satisfied provided that U_f and S_f are scalars times the identity matrix, and F is of the form

$$F(t) = F_0(t) + f(t)I, \tag{127}$$

where

$$F_0(t) = -F_0^T(t) \tag{128}$$

and f(t) is a scalar. For example, a matrix of the form

$$F = \begin{bmatrix} -\rho & \omega \\ -\omega & -\rho \end{bmatrix} \tag{129}$$

would be acceptable.

C. *CONVENIENT STRUCTURAL ASSUMPTIONS*

It is obvious that one may not always be able to satisfy Eq. (106). The following generalization may therefore be useful. The F matrix is assumed to be of the form

$$F = \begin{bmatrix} f_1 & & & & & & \\ & \ddots & & & & 0 & \\ & & f_q & & & & \\ & & & [F_1] & & & \\ & 0 & & & [F_2] & & \\ & & & & & \ddots & \\ & & & & & & [F_r] \end{bmatrix}, \tag{130}$$

where the f_i terms are scalar and the blocks labeled F_i are each assumed to be 2×2 matrices of a form similar to Eq. (129). The real and complex eigenvalues of F are separated in this way. There are r complex conjugate pairs and q real eigenvalues, and $2r + q = \ell$. Accordingly, P_{xf} and Λ are partitioned by columns as

$$\left[P^1_{xf} \middle| \cdots \middle| P^q_{xf} \left[\bar{P}^1_{xf} \right] \left[\bar{P}^2_{xf} \right] \cdots \left[\bar{P}^r_{xf} \right] \right], \tag{131}$$

where the terms P^i_{xf} are the first q columns of P_{xf}, and the blocks $\bar{P}^i_{xf}$ represent the last r pairs of columns. The matrix Λ is partitioned in the same way. If U_f and S_f are also of the form indicated by Eq. (130), i.e.,

$$U_f = \begin{bmatrix} u_1 & & & & & \\ & \ddots & & & 0 & \\ & & u_q & & & \\ & & & [U_1] & & \\ & 0 & & & [U_2] & \\ & & & & & \ddots \; [U_r] \end{bmatrix} \tag{132}$$

and similarly for S_f, Λ_{ff} will be of this form also. It is required that U_i and S_i be scalars times I_2. One may write a set of scalar equations

$$u_i = \Lambda_{ff}^i \alpha_i, \quad i = 1, 2, \ldots, q \tag{133}$$

and 2×2 matrix equations

$$U_i \left[\overline{\Lambda}_{ff}^i\right]^{-1} = \overline{\alpha}_i \, I_2, \quad i = 1, 2, \ldots, r, \tag{134}$$

where Λ_{ff} has been partitioned as

$$\Lambda_{ff} = \begin{bmatrix} \Lambda_{ff}^1 & & & & & \\ & \ddots & & & 0 & \\ & & \Lambda_{ff}^q & & & \\ & & & [\overline{\Lambda}_{ff}^1] & & \\ & 0 & & & [\overline{\Lambda}_{ff}^2] & \\ & & & & & \ddots \; [\overline{\Lambda}_{ff}^r] \end{bmatrix} \tag{135}$$

and where each $\overline{\alpha}_i$ is a scalar. Then one can write equations for the columns P_{xf}^i and Λ_{xf}^i, which are of the form

$$\begin{bmatrix} \dot{P}_{xf}^i \\ \hline \dot{\Lambda}^i \end{bmatrix} = \left[\begin{array}{c|c} G_{11}^i & G_{12} \\ \hline G_{21} & G_{22}^i \end{array}\right] \begin{bmatrix} P_{xf}^i \\ \hline \Lambda^i \end{bmatrix} + \begin{bmatrix} D_1^i \\ \hline \tilde{D}_2^i \end{bmatrix}; \; i = 1, 2, \ldots, q, \tag{136}$$

where D_1^i is the ith column of D_1, $\tilde{D}_2^i$ is the ith column of $\tilde{D}_2$, and

$$G_{11}^i = \bar{A} + f_i I, \tag{137}$$

$$G_{22}^i = -\bar{A}^T + (\alpha_i + f_i)I. \tag{138}$$

In the same way, one can take the blocks of column pairs $\bar{P}_{xf}^i$ and $\bar{\Lambda}^i$ and write equations

$$\begin{bmatrix} \dot{\bar{P}}_{xf}^i \\ \hline \dot{\bar{\Lambda}}^i \end{bmatrix} = \left[\begin{array}{c|c} G_{11} & G_{12} \\ \hline G_{21} & G_{22}^i \end{array}\right] \begin{bmatrix} \bar{P}_{xf}^i \\ \hline \bar{\Lambda}^i \end{bmatrix} + \begin{bmatrix} \bar{P}_{xf}^i \\ \hline \bar{\Lambda}^i \end{bmatrix} F_i^T + \begin{bmatrix} \bar{D}_1^i \\ \hline \bar{D}_2^i \end{bmatrix},$$

$$i = 1, 2, \ldots, r. \tag{139}$$

The column pairs $\bar{P}_{xf}^i$ are located as indicated in Eq. (131), while $\bar{\Lambda}^i$, $\bar{D}_1^i$, and $\bar{D}_2^i$ indicate column pairs taken from the same location in matrices Λ, D_1, and $\tilde{D}_2$, respectively. The matrix G_{22}^i is defined as

$$G_{22}^i = -\bar{A}^T + \bar{\alpha}_i I. \tag{140}$$

Equations (136) and (139) both represent sets of equations that can be solved easily using the state transition matrix approach or the Riccati transformation as illustrated in this section. If one chose the latter method, the design of the optimal gain for a reduced-order filter with q real eigenvalues and r complex conjugate pairs would involve solving q + r decoupled Riccati equations, each having $n(n + 1)/2$ variables. This is the advantage of choosing the F matrix as indicated by Eq. (130) instead of the most general form that would require solving one matrix Riccati equation having $(2r + q)n[(2r + q)n + 1]/2$ variables.

Obviously, the form of Eq. (130) may not be the one obtained by looking at the Kalman filter for a reduced-order model. The same eigenvalues may be maintained, however, and one is really suggesting a change in the state space in which the filter is designed to achieve the simplicity achieved by Eq. (130). Some remarks about such a change in filter state variables are in order. Suppose it is given that the filter is designed with a different set of state variables $\hat{y}^*$ linearly related to the original choice $\hat{y}$ so that

$$\hat{y}^* = \Theta\hat{y}. \tag{141}$$

Then if one desires that $\hat{y}$ be a good estimate of $y = Cx$, then $\hat{y}^*$ should be a good estimate of some other vector y^*, which is not the output,

$$y^* = \Theta y = \Theta Cx \triangleq C^* x, \tag{142}$$

and this fact must be observed in the design of the filter in the new state space.

In order to use the development presented here, one must also assume that the measurement driving the filter in the new state space is modified to be of the form

$$m^* = \Theta m = \Theta Cx + \Theta v = C^* x + v^*. \tag{143}$$

Based on the material presented here, one may conclude that the design of an optimal ROF is a reasonable task for a rather wide range of performance measures.

D. *A SPECIAL CASE OF INTEREST*

One special case of interest occurs when U_f is zero as in a terminal estimation problem. In this case it is easy to see that a reduced-order filter can provide an estimate that is as accurate as that of a full-order Kalman filter at the terminal time. In fact the reduced-order filter provides exactly the same estimate as the Kalman filter. Although the design could be achieved using the optimization approach presented herein, it will be seen that an ordinary application of linear systems theory provides a simple approach which does not require that one solve a two-point boundary-value problem.

Obviously it is possible to estimate the output of a system using a full order Kalman filter. The Kalman filter to estimate the full state when Eqs. (1), (4), and (8) are the describing equations is of the form

$$\dot{\hat{x}} = \left(A - K^*_{KF}C\right)\hat{x} + K^*_{KF}m, \tag{144}$$

and the optimal Kalman estimate of y is then given by

$$\hat{y}_{KF}(t) = C(t)\hat{x}(t). \tag{145}$$

At the terminal time, the optimum estimate is

$$\hat{y}_{KF}(t_f) = C(t_f)\phi_{KF}(t_f, t_0)\mu_0 + C(t_f)\int_{t_0}^{t_f} \phi_{KF}(t_f, \tau)K^*_{KF}(\tau)m(\tau)\,d\tau, \tag{146}$$

where

$$\dot{\phi}_{KF}(t, \tau) = \left[A(t) - K^*_{KF}(t)C(t)\right]\phi_{KF}(t, \tau) \tag{147}$$

and

$$\phi_{KF}(t, t) = I. \tag{148}$$

The reduced-order filter of Eq. (12) produces an estimate at t_f

$$\hat{y}(t_f) = \Phi_F(t_f, t_0)\hat{y}(t_0) + \int_{t_0}^{t_f} \Phi_F(t_f, \tau)K(\tau)m(\tau)\,d\tau, \tag{149}$$

where

$$\dot{\Phi}_F(t, \tau) = F(t)\Phi_F(t, \tau) \tag{150}$$

and

$$\Phi_F(t, t) = I. \tag{151}$$

It is evident that with the proper choice of initial condition and gain, the reduced-order filter can produce the same estimate as the Kalman filter at the terminal time, but with far less computation. Specifically, the gain should be selected as

$$K(t) = \Phi_F(t, t_f)C(t_f)\phi_{KF}(t_f, t)K^*_{KF}(t), \tag{152}$$

and the initial condition should be selected so that

$$\hat{y}(t_0) = \Phi_F(t_0, t_f)C(t_f)\phi_{KF}(t_f, t_0)\mu_0. \tag{153}$$

Hence it would be foolish to build a Kalman filter to estimate the output at some given time if computational complexity is an issue.

In closing this section on computation, it is worth noting that there is always a greater amount of design computation for the reduced-order filter than for the Kalman filter. The values needed for the gain are precomputed and stored, and the computational savings is thus obtained for the on-line processing of the data. Since the gain values are precomputed, it is clear that the type of approach suggested here has no application for nonlinear extended Kalman filter problems.

V. EXAMPLE

In this section, the specific example described by Eqs. (15) and (16) is considered in detail. The initial statistics are assumed to be given as

$$\begin{aligned} E\{y(0)\} &= E\{\xi(0)\} = 0, \\ \operatorname{var}\{y(0)\} &= \operatorname{var}\{\xi(0)\} = 1, \\ \operatorname{cov}\{y(0),\ \xi(0)\} &= 0, \end{aligned} \tag{154}$$

while the plant noise parameter is

$$E\{w(t)w(\tau)\} = \delta(t - \tau). \tag{155}$$

The first problem considered is to design a model of the form indicated by Eq. (20) to minimize the performance measure

$$J_m = E\left\{\int_0^1 e_m^2(t)\,dt + Se_m^2(1)\right\}, \tag{156}$$

where $e_m = y - y_m$. The state vector is defined as

$$\begin{bmatrix} x_1 \\ x_2 \end{bmatrix} = \begin{bmatrix} y \\ \xi \end{bmatrix}. \tag{157}$$

Accordingly, Eqs. (36) and (37) are solved for the Lagrange multipliers giving the required expressions

$$\Lambda_{mm}(t) = \frac{1}{2} + \left(S - \frac{1}{2}\right)\exp^{[2(t-t_f)]} \tag{158}$$

and

$$\Lambda_{xm_2}(t) = \frac{1}{2(\alpha + 1)}\left[1 - \exp^{\{(\alpha+1)(t-t_f)\}}\right]$$

$$+ \frac{\left(S - \frac{1}{2}\right)}{(1 - \alpha)}\left[\exp^{\{(\alpha+1)(t-t_f)\}} - \exp^{\{2(t-t_f)\}}\right]. \tag{159}$$

The optimal gain for the model is then

$$K_m(t) = \Lambda_{xm_2}(t)/\Lambda_{mm}(t). \tag{160}$$

Table I. Model Performance for S = 1.

α	$J_1{}^a$	$J_2{}^b$
0.5	0.91970	1.4178
1.5	0.74846	1.0313
2.5	0.68910	0.87741
3.5	0.66120	0.80026
4.5	0.64459	0.75441

[a] J_1, *performance with gain optimized.*

[b] J_2, *performance with singular perturbation model.*

The performance measure J_m is evaluated for several values of α and S in Tables I and II. Also presented is the performance measure when the model Eq. (17) is used. The improvement that is possible when Eq. (20) is used is of course less impressive for larger values of α as one would expect. The difference in performance is pronounced when there is a large value of S so that terminal behavior is much more important than behavior over the interval. From this experiment, the indication is

Table II. Model Performance for α = 1.5.

S	$J_1{}^a$	$J_2{}^b$
0.01	0.50994	0.64424
0.1	0.53308	0.67943
1.0	0.74846	1.0313
10.0	2.7190	4.54989

[a] J_1, *performance with gain optimized.*

[b] J_2, *performance with singular perturbation model.*

that one should consider optimizing the model if either the noise bandwidth is not consistent with a white noise modeling assumption, or if modeling error is more important at certain times than at others.

If one uses a reduced-order model to design a reduced-order filter, it is often, but not always, true that filtering performance will be enhanced because the gain of the reduced-order model was optimized. Several first-order filters and the optimal second-order filter were designed for the example with dynamics indicated by Eqs. (15) and (16) and measurement indicated by Eq. (18). The measurement noise parameter for the example is

$$E\{v(t)v(\tau)\} = \delta(t - \tau), \tag{161}$$

and the performance measure is

$$J = E\left\{\int_0^1 e_f^2(t)\,dt + Se_f^2(1)\right\}, \tag{162}$$

where $e_f = y - \hat{y}$. The linear two-point boundary-value problem described by Eqs. (69) and (70) with boundary conditions indicated by Eqs. (67) and (76) was solved for this example. The results were substituted in Eq. (68) to give the optimal filter gain. This was done for values of F corresponding to a Kalman filter designed for the singular perturbation model, and for values of F corresponding to a Kalman filter designed for the optimal reduced-order model. The performance measure was calculated using Eq. (60). The results of this evaluation are summarized in Table III for several values of α with $S = 1$, and in Table IV for several values of S with $\alpha = 1.5$. In each case the first-order filter with the best performance is that based on both optimal reduced-order modeling and optimal reduced-order filtering labeled by J_3. It appears that gain

Table III. Filter Performance for S = 1.

α	J_1[a]	J_2[b]	J_3[c]	J_4[d]	J_5[e]
0.5	0.76531	0.74960	0.74490	0.74605	0.74445
1.5	0.71481	0.71706	0.70944	0.71015	0.70922
2.5	0.75425	0.75702	0.75149	0.75195	0.75310
3.5	0.79729	0.79882	0.79526	0.79556	0.79509
4.5	0.83169	0.83242	0.83008	0.83029	0.82944

[a] J_1, first-order Kalman filter, model optimized.
[b] J_2, first-order Kalman, singular perturbation model.
[c] J_3, optimized first-order filter, model also optimized.
[d] J_4, optimized first-order filter, singular perturbation model.
[e] J_5, optimal second-order Kalman filter.

Table IV. Filter Performance for α = 1.5.

S	J_1[a]	J_2[b]	J_3[c]	J_4[d]	J_5[e]
0.01	0.44987	0.45096	0.44796	0.44823	0.44789
0.1	0.47419	0.47515	0.47174	0.47208	0.47165
1.0	0.71481	0.71706	0.70944	0.71015	0.70922
10.0	3.0994	3.1362	3.08653	3.08807	3.08490

[a] J_1, first-order Kalman filter, model optimized.
[b] J_2, first-order Kalman, singular perturbation model.
[c] J_3, optimized first-order filter, model also optimized.
[d] J_4, optimized first-order filter, singular perturbation model.
[e] J_5, optimal second-order Kalman filter.

optimization for the reduced-order filter is more significant than reduced-order model optimization. This is understandable since the former is a direct approach, while the latter is indirect. The one data point that is surprising is the case where $S = 1$ and $\alpha = 0.5$. In this instance optimization of the reduced-order model did not lead to a better result when the associated first-order Kalman filter was used. In all other cases, prior optimization of the model was helpful. Optimization of the filter gain is helpful in all cases. For example, when $\alpha = 1.5$ and $S = 1$, the performance loss relative to the second-order Kalman filter is 0.031% when the reduced-order filter gain is optimized, while there is a 1.105% loss when the first-order Kalman filter designed for the reduced-order model is used.

VI. DISCRETE PROBLEMS

A. *REDUCED-ORDER MODELING*

Since ultimately most solutions are to be implemented on the digital computer, it is worthwhile to know how one proceeds with reduced-order modeling and filtering under the circumstances where the discretization is done at the beginning of the problem.* The material presented in this section is also useful for circumstances in which the natural format of the problem is discrete so that no conversion from a continuous to discrete problem is necessary. The discrete dynamical model is assumed to be of the form

$$x(j + 1) = A(j)x(j) + B(j)w(j), \tag{163}$$

**Note that the matrices A, B, C, Q, etc., for the discrete problem are not the same as those matrices for the continuous case.*

where $w(j)$ is zero-mean discrete white noise having covariance

$$E\{w(j)W^T(k)\} = Q(j)\delta_{jk}. \tag{164}$$

The model output is

$$y(j) = C(j)x(j), \tag{165}$$

and the initial statistics are

$$\begin{aligned} E\{x(0)\} &= \mu_0, \\ \text{var } x(0) &= V_0. \end{aligned} \tag{166}$$

The reduced-order model is assumed to have the structure

$$y_m(j + 1) = F_m(j)y_m(j) + K_m(j)w(j). \tag{167}$$

The design objective in the discrete problem is the minimization of the performance measure

$$J_m = E\left\{\sum_{j=0}^{N-1} e_m^T(j)U_m(j)e_m(j) + e_m^T(N)S_m e(N)\right\}, \tag{168}$$

where

$$e_m(j) \triangleq y(j) - y_m(j), \tag{169}$$

and $U_m \geq 0$, $S_m > 0$. The objective is to be accomplished by optimally selecting K_m given that a policy for F_m has been chosen *a priori*. One methodology for choosing F_m involves a single stage optimization [11] and of course there are any number of other possibilities. If one is only going to optimize with respect to K_m, the mathematics turns out to be linear, and therefore a solution is obtained with a reasonable effort. In the reduced-order discrete problem, one has the advantage of knowing precisely how much on-line computational savings will be achieved, whereas in the continuous time version of the problem, the computational savings is a function of the integration method that is used.

As a first step in solving the problem posed here, it is necessary to write the performance measure in terms of the seond moment of the error.

$$J_m = \mathrm{tr}\left\{\sum_{i=0}^{N-1} U_m(j)P_{mm}(j) + S_m P_{mm}(N)\right\}, \tag{170}$$

where

$$P_{mm}(j) \underline{\Delta} E\left\{e_m(j)e_m^T(j)\right\}. \tag{171}$$

The error propagates according to the difference equation

$$e_m(j+1) = B_m(j)x(j) + F_m(j)e_m(j) + \beta_m(j)w(j), \tag{172}$$

where

$$B_m(j) \underline{\Delta} C(j+1)A(j) - F_m(j)C(j) \tag{173}$$

and

$$\beta_m(j) \underline{\Delta} C(j+1)B(j) - K_m(j). \tag{174}$$

From Eqs. (163) and (172) it is easy to write the associated moment propagation equations. These are

$$\begin{aligned} P_{mm}(j+1) = {} & B_m(j)P_{xx}(j)B_m^T(j) + B_m(j)P_{xm}(j)F_m^T(j) \\ & + F_m(j)P_{mx}(j)B_m^T(j) + F_m(j)P_{mm}(j)F_m^T(j) \\ & + \beta_m(j)Q(j)\beta_m^T(j), \end{aligned} \tag{175}$$

$$\begin{aligned} P_{xm}(j+1) = {} & A(j)P_{xx}(j)B_m^T(j) + A(j)P_{xm}(j)F_m^T(j) \\ & + B(j)Q(j)\beta_m^T(j), \end{aligned} \tag{176}$$

and

$$P_{xx}(j+1) = A(j)P_{xx}(j)A^T(j) + B(j)Q(j)B^T(j). \tag{177}$$

In the above equations, the definitions

$$P_{xm}(j) \underline{\Delta} E\left\{x(j)e_m^T(j)\right\} \tag{178}$$

and

$$P_{xx}(j) \underline{\Delta} E\{x(j)x^T(j)\}$$

have been used. The initial conditions for Eqs. (175) and (176) are assumed to be given and interestingly they do not affect the solution.

The Hamiltonian for this problem is

$$H_j = \mathrm{tr}\Big\{U_m(j)P_{mm}(j) + P_{mm}(j+1)\Lambda_{mm}^T(j+1) + P_{xm}(j+1)\Lambda_{xm}^T(j+1) + P_{mx}(j+1)\Lambda_{mx}^T(j+1)\Big\}, \tag{180}$$

where Λ_{mm} and Λ_{xm} are matrices of Lagrange multipliers associated with P_{mm} and P_{xm}, respectively, and Λ_{mx} is the transpose of Λ_{xm}. After substituting from Eqs. (175) and (176), the costate equations can be found from Eq. (180). These are found to be of the form

$$\Lambda_{mm}(j) = \frac{\partial H_j}{\partial P_{mm}(j)} = U_m(j) + F_m^T(j)\Lambda_{mm}(j+1)F_m(j) \tag{181}$$

and

$$\Lambda_{xm}(j) = \frac{\partial H_j}{\partial P_{xm}(j)} = B_m^T(j)\Lambda_{mm}(j+1)F_m(j) + A^T(j)\Lambda_{xm}(j+1)F_m(j). \tag{182}$$

The boundary conditions for Eqs. (181) and (182) are

$$\Lambda_{mm}(N) = S_m$$

and

$$\Lambda_{xm}(N) = 0. \tag{184}$$

If one sets the gradient of H_j with respect to $K(j)$ equal to zero, the resulting expression is

$$\Lambda_{mx}(j+1)B(j)Q(j) + \Lambda_{mm}(j+1)[C(j+1)B(j) - K(j)]Q(j) = 0. \tag{185}$$

The gain may be expressed as

$$K(j) = \left[C(j+1) + \Lambda_{mm}^{-1}(j+1)\Lambda_{mx}(j+1)\right]B(j), \tag{186}$$

when $Q(j)$ and $\Lambda_{mm}(j+1)$ are nonsingular.

Hence the optimum gain is computed from Eq. (186) after having solved Eqs. (181) and (182) backwards from the boundary conditions indicated by Eqs. (183) and (184). It is interesting to note that there is no two-point boundary-value problem to be solved and that the situation is quite similar to the continuous time problem. One difference worth noting is that the matrix F_m appears quadratically in the Hamiltonian in the discrete version of the problem so that optimization with respect to F_m appears to be feasible also. It can be shown, however, that this problem is much more difficult.

B. *REDUCED-ORDER FILTERING*

The discrete reduced-order filtering problem is based on the model indicated by Eqs. (163) and (165) with measurements of the form

$$m(j) = y(j) + v(j), \tag{187}$$

where $v(j)$ is zero-mean white noise with covariance matrix

$$E\{v(j)v^T(k)\} = R(j)\delta_{jk}. \tag{188}$$

The problem is to design a filter of the linear form

$$\hat{y}(j+1) = F(j)\hat{y}(j) + K(j)m(j+1), \tag{189}$$

where $\hat{y}(j)$ is to be a good estimate of $y(j)$. The matrix $K(j)$ is to be selected to minimize the performance measure

$$J = E\left\{\sum_{j=0}^{N-1} e_f^T(j)U_f(j)e_f(j) + e_f^T(N)S_f e(N)\right\}, \tag{190}$$

where

$$e_f(j) \triangleq y(j) - \hat{y}(j), \tag{191}$$

and $U_f \geq 0$, $S_f > 0$. It is assumed that $F(j)$ has been selected *a priori*, as when one includes this selection as a part of the optimization problem, a very difficult nonlinear two-point boundary-value problem results.

The algebraic and notational complexity of the discrete problem tends to be greater than for the continuous time version of the problem. Here the purpose is to indicate the procedure of solution and the fact that only linear mathematics is required to solve the problem. Greater algebraic detail may be found in [24].

The performance measure Eq. (190) can be written in terms of the second moment of the error.

$$J = \operatorname{tr}\left\{\sum_{j=0}^{N-1} U_f P_{ff}(j) + S_f P_{ff}(N)\right\}, \tag{192}$$

where

$$P_{ff}(j) \triangleq E\left\{e_f(j) e_f^T(j)\right\}. \tag{193}$$

The error propagates according to the difference equation

$$e_f(j+1) = B_f(j)x(j) + \beta_f(j)w(j) + F(j)e_f(j) - K(j)v(j+1), \tag{194}$$

where

$$B_f(j) \triangleq [C(j+1) - K(j)C(j+1)]A(j) - F(j)C(j) \tag{195}$$

and

$$\beta_f(j) \triangleq [I - K(j)]C(j+1)B(j). \tag{196}$$

The second moment equation for the error may therefore be written as

$$\begin{aligned} P_{ff}(j+1) &= B_f(j)P_{xx}(j)B_f^T(j) + B_f(j)P_{xf}(j)F^T(j) \\ &\quad + K(j)R(j+1)K^T(j) + F(j)P_{fx}(j)B_f^T(j) \\ &\quad + F(j)P_{ff}(j)F^T(j) + \beta_f(j)Q(j)\beta_f^T(j), \end{aligned} \tag{197}$$

where

$$P_{xf}(j) \triangleq E\left\{x(j)e_f^T(j)\right\} = P_{fx}^T(j). \tag{198}$$

The matrix P_{xf} satisfies the equation

$$\begin{aligned} P_{xf}(j+1) &= A(j)P_{xx}(j)B_f^T(j) + A(j)P_{xf}(j)F^T(j) \\ &\quad + B(j)Q(j)\beta_f^T(j). \end{aligned} \tag{199}$$

The problem is then to minimize the performance measure Eq. (192) subject to the constraints imposed by the moment equations. The solution procedure is just as in the discrete reduced-order modeling problem except that the structure of the result is quite different. In this case one does wind up with a true two-point boundary-value problem, although it is a linear one and therefore computationally more feasible than a nonlinear result would be.

The costate equations for the filtering problem are

$$\Lambda_{ff}(j) = U_f(j) + F^T(j)\Lambda_{ff}(j+1)F(j) \tag{200}$$

and

$$\begin{aligned} \Lambda_{xf}(j) &= B_f^T(j)\Lambda_{ff}(j+1)F(j) \\ &\quad + A^T(j)\Lambda_{xf}(j+1)F(j), \end{aligned} \tag{201}$$

where Λ_{ff} is the Lagrange multiplier matrix associated with P_{ff} and Λ_{xf} is associated with P_{xf}. The terminal conditions for these equations are

$$\Lambda_{ff}(N) = S_f, \tag{202}$$

$$\Lambda_{xf}(N) = 0. \tag{203}$$

It is clear that Λ_{ff} can be precomputed and regarded as a known quantity since F has been assumed known. The equation for the gain is determined by minimizing the Hamiltonian with respect to K. This leads to the necessary condition for optimality

$$\Lambda_{ff}K(j)\left[CP_{xx}C^T + R\right] = \Lambda_{fx}P_{xx}C^T + \Lambda_{ff}\Big\{F(j)[P_{fx}(j) - C(J)P_{xx}(j)]A^T(j) + CP_{xx}\Big\}C^T. \tag{204}$$

For the terms in Eq. (204), the time index is at j + 1 unless otherwise indicated. If the required inverses exist, one may solve for $K^T(j)$ in Eq. (204) obtaining a result of the form

$$K^T(j) = \Xi_1(j)P_{xf}(j)\Xi_2(j) + \Xi_3(j)\Lambda_{xf}(j+1)\Xi_4(j) + \Xi_5(j), \tag{205}$$

where the matrices labeled Ξ_i are all known or precomputable, and easily obtained from Eq. (204). Upon substituting from Eq. (205) in Eqs. (199) and (201), one obtains the linear equations

$$\begin{aligned} P_{xf}(j+1) &= \Omega_1(j)P_{xf}(j)\Omega_2(j) + \Omega_3(j)\Lambda_{xf}(j+1)\Omega_4(j) + \Omega_5(j), \\ \Lambda_{xf}(j) &= \Omega_6(j)P_{xf}(j)\Omega_7(j) + \Omega_8(j)\Lambda_{xf}(j+1)\Omega_9(j) + \Omega_{10}(j), \end{aligned} \tag{206}$$

where the matrices Ω_i are easily obtained by inspection of Eqs. (199) and (200). The terminal condition is indicated by

Eq. (203) and the initial condition is

$$P_{xf}(0) = V_0C^T(0). \qquad (207)$$

Equation (207 is true if $\hat{y}(0)$ is selected as

$$\hat{y}(0) = C(0)\mu_0, \qquad (208)$$

which results in an unbiased initial estimate but is in general suboptimal. Optimization of $\hat{y}(0)$ may also be included in the problem if desired. The solution to the above linear two-point boundary-value problem may be obtained in general by using the column-stack operation and the Kronecker algebra. The design computation performed off-line will be significant. A solution procedure for a special case where the design calculations simplify and the problem can be solved in its matrix format is presented in [24].

VII. SUMMARY AND CONCLUSIONS

A. *SUMMARY*

In this article, emphasis has been directed toward those aspects of reduced-order filtering and modeling problems that can be treated within the framework of linear mathematics. Although reduced-order modeling is a rather broad topic, it was investigated here because it can provide some guidance in the design of reduced-order filtering algorithms. Specifically, in the design of reduced-order filters one can use the system matrix F associated with the Kalman filter designed for the reduced-order model. The selection of the remaining parameters for the filter, the gain K, and the initial condition is then based on direct optimization. The resulting two-point boundary-value problem is linear. The optimal gain offers an improvement in performance relative to the performance of the Kalman gain designed for the reduced-order model.

Solving the reduced-order modeling problem allows the designer to construct a model that is more accurate during some portions of the time interval than at other times. This may be desirable for some applications. Similarly, one can construct reduced-order filters whose performance is emphasized at certain times. The reduced-order modeling and filtering problems are similar, however the modeling optimization problem results in only a single-point linear boundary-value problem, while the filtering problem leads to a linear two-point boundary-value problem. This difference is important enough to motivate consideration of the problems separately. It should be clear that it is not a prerequisite that one solve a reduced-order modeling problem before solving the filtering problem. This was only suggested as one reasonable approach and because the modeling problem is also an interesting exercise.

The methods used in this article can be applied to some slightly different cases than have been treated here. In particular, bilinear problems with white state dependent noise terms are also amenable to solution [25,26]. There are still many unresolved issues associated with the linear problems however, so perhaps future research should focus primarily on the remaining questions associated with linear models.

B. AN AREA FOR FUTURE RESEARCH

Reduced-order problems are some of the most interesting and challenging problems that remain to be solved within the general area of linear systems theory. In a sense, engineers are often dealing with this issue, but not explicitly. The approximation of infinite order, distributed systems by finite order lumped models is one example that might be thought of as the ultimate reduced-order problem. It appears that the area is somewhat unyielding and mathematically difficult with regard to some of the most important issues. As a case in point, the selection of the matrix F(t) for the reduced-order filtering problems is a difficult and unresolved matter in general. The correct choice is clear for the full-order problem, but not for the reduced-order problem. It may be seen that the F matrix appears linearly in the Hamiltonian so that an optimal solution might contain impulses. This undesirable conclusion indicates that the performance measure considered herein may not be well posed with respect to the selection of an optimal F. Further research is needed in this area.

There are three issues that must be considered if one is to be involved with reduced-order problems. One must give consideration to the on-line computational savings, the difficulty of the *a priori* design procedure, and the performance loss relative to the optimal solution. In reality there will always be a certain amount of art, and a certain amount of systematic procedure in obtaining a reasonable solution. It has been the goal of this paper to demonstrate that there are certain areas, in particular gain improvement, where it is possible to proceed in a systematic way.

REFERENCES

1. N. WIENER, "Extrapolation, Interpolation, and Smoothing of Stationary Time Series," MIT Press, Cambridge, Massachusetts, 1949.

2. G. C. NEWTON, L. A. GOULD, and J. F. KAISER, "Analytical Design of Linear Feedback Controls," Wiley, New York, 1957.

3. B. R. EISENBERG and A. P. SAGE, "Closed Loop Optimization of Fixed Configuration Systems," *Int. J. Control 3*, 183-194 (1965).

4. C. S. SIMS and J. L. MELSA, "Sensitivity Reduction in Specific Optimal Control by the Use of a Dynamical Controller," *Int. J. Control 8*, 491-501 (1968).

5. C. S. SIMS and J. L. MELSA, "A Survey of Specific Optimal Results in Control and Estimation," *Int. J. Control 14*, 299-308 (1971).

6. C. S. SIMS and J. L. MELSA, "Specific Optimal Estimation," *IEEE Trans. Auto. Control AC-14*, 183-186 (1969).

7. D. JOHANSEN, "Optimal Control of Linear Stochastic Systems with Complexity Constraints," *in* "Control and Dynamic Systems" (C. T. Leondes, ed.), Vol. 4, pp. 181-278, Academic Press, New York, 1966.

8. C. S. SIMS and J. L. MELSA, "A Fixed Configuration Approach to the Stochastic Linear Regulator Problem," *Proc. 1970 Joint Auto. Control Conf.* (June 1970).

9. T. R. DAMIANI, "A General Partial State Estimator," *Proc. Midwest Symp. Circuit Theory* (May 1971).

10. C. E. HUTCHINSON and J. A. D'APPOLITO, "Minimum Variance Reduced State Filters," *Proc. IEEE Conf. Decision Control* (December 1972).

11. C. S. SIMS, "An Algorithm for Estimating a Portion of a State Vector," *IEEE Trans. Auto. Control AC-19*, 391-393 (1974).

12. C. T. LEONDES and L. M. NOVAK, "Optimal Minimal-Order Observers for Discrete-Time Systems--A Unified Theory," *Automatica 8*, 379-387 (1972).

13. C. E. HUTCHINSON, J. A. D'APPOLITO, and K. J. ROY, "Application of Minimum Variance Reduced-State Estimators," *IEEE Trans. Aeros. Electr. Sys. AES-11*, 785-794 (1975).

14. J. I. GALDOS and D. E. GUSTAFSON, "Information and Distortion in Reduced-Order Filter Design," *IEEE Trans. Inform. Theory IT-23*, 183-194 (1977).

15. R. B. ASHER, K. D. HERRING, and J. C. RYLES, "Bias, Variance, and Estimation Error in Reduced-Order Filters," *Automatica* (November 1976).

16. B. J. UTTAM and W. F. O'HALLORAN, JR., "On the Computation of Optimal Stochastic Observer Gains," *IEEE Trans. Auto. Control AC-20*, 145-146 (1975).

17. C. S. SIMS and R. B. ASHER, "Optimal and Suboptimal Results in Full and Reduced-Order Linear Filtering," *IEEE Trans. Auto. Control AC-23*, 469-472 (1978).

18. A. H. HADDAD, "Linear Filtering of Singularly Perturbed Systems," *IEEE Trans. Auto. Control AC-21*, 515-519 (1976).

19. G. OBINATA and H. INOOKA, "A Method of Reducing the Order of Multivariable Stochastic Systems," *IEEE Trans. Auto. Control AC-22*, 676-677 (1977).

20. Y. BARAM and Y. BE'ERI, "Simplification of High Order and Time Varying Linear Systems," *Proc. Joint Auto. Control Conf.* (August 1980).

21. C. S. SIMS, "Reduced-Order Modeling," *Proc. Asilomar Conf. Circuits, Systems, and Computers* (November 1979).

22. M. ATHANS, "The Matrix Minimum Principle," *Inform. Control 11*, 592-606 (1968).

23. R. BELLMAN, "Introduction to Matrix Analysis," McGraww-Hill, New York, 1960.

24. C. S. SIMS and L. G. STOTTS, "Linear Discrete Reduced-Order Filtering," *Proc. IEEE Conf. Decision Control* (January 1979).

25. R. B. ASHER and C. S. SIMS, "Reduced-Order Filtering with State Dependent Noise," *Proc. Joint Auto. Control Conf.* (October 1978).

26. C. S. SIMS, "Discrete Reduced-Order Filtering with State Dependent Noise," *Proc. Joint Auto. Control Conf.* (August 1980).

Modeling Techniques for Distributed Parameter Systems

GEORGE R. SPALDING
Department of Engineering
Wright State University
Dayton, Ohio

I. INTRODUCTION

Computer control of complex dynamic systems is currently widespread and is destined to increase greatly in the next few years. Computer control techniques utilize a stored model of the system, generally a model that can be improved, i.e., updated, by measurement of system response. Thus, as system parameters change or other effects occur, the model's integrity is maintained. The control process, therefore, generally includes transducers at each response point, as well as actuators at each input point.

ISBN 0-12-012718-0

When the systems are spatially distributed, the potential number of input and output points is large, in theory being infinite. In the interest of economy, however, it is necessary to keep these points as few as possible. Thus the models of these systems are discrete, i.e., a system classically represented by a single partial differential equation is modeled by a finite set of ordinary differential equations.

There are numerous methods for obtaining the sets of differential equations that represent a particular distributed parameter system. Each has its strengths and shortcomings, depending upon the application at hand.

This chapter presents a particular technique for modeling and identifying distributed parameter systems, which is generally applicable and which differs from previous work in three significant respects:

(1) The modeling and identification process need not be based on a detailed knowledge of the system dynamics. The basis of the model construction is an orthogonal vector space that is independent of the system but can be tailored to the system by utilizing prior knowledge.

(2) The art of modeling is transferred from writing a partial differential equation, or a set of ordinary differential equations, to choosing a Green's function for the system (see Appendix A). A set of differential equations comes out of the modeling process somewhat automatically. However, the coefficients of the equations are directly related to the system Green's function, which must be estimated.

(3) The basis vectors of the space are derived from Jacobi polynomials. They offer accuracy and computational ease not generally obtained from other approximating methods.

The Jacobi family of vectors and the Green's function have been used to implement identification processes [1] and as a vehicle for modeling distributed parameter systems [2]. These two processes are combined to develop effective models when the extent of *a priori* knowledge is limited.

In the next section, the Jacobi polynomials are defined, followed by the development of the Jacobi vectors. Then a discrete uncoupled model of a distributed system is derived, starting with the partial differential equation model. This illustrates the procedure for a simple system. Following this, the identification problem, i.e., the problem of specifying a suitable model using measured data, is presented. The chapter concludes with a discussion of some practical considerations and a more general illustration.

II. JACOBI POLYNOMIALS

The Jacobi polynomials are a family of complete function sets characterized by a weighted orthogonality over the region $-1 \leq x \leq 1$. This region can be shifted by a change of independent variable. The region of orthogonality here is $0 \leq x \leq 1$. The relationship is

$$\int_0^1 P_n(x) P_m(x) \omega(x) dx = 0, \qquad m \neq n, \qquad (1)$$
$$= \delta_n, \qquad m = n,$$

where

$$\omega(x) = (1 - x)^{r} x^{s}, \qquad (2)$$

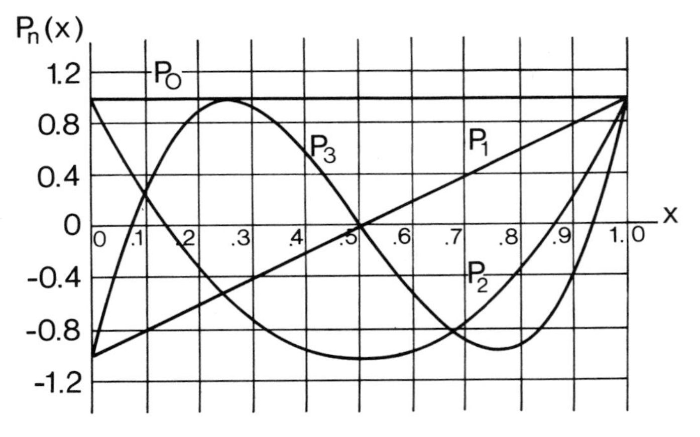

Fig. 1. First four Jacobi polynomials, $r = s = -1/2$.

and

$$\delta_n = \frac{2^{r+s+1}}{r + s + 2n + 1} \frac{\Gamma(r + n + 1)\Gamma(s + n + 1)}{n!\Gamma(r + s + n + 1)}. \tag{3}$$

These polynomials obey a recurrence relation

$$a_{1n}P_{n+1}(x) = [a_{2n} + (2x - 1)a_{3n}]P_n(x) - a_{4n}P_{n-1}(x), \tag{4}$$

where

$$\begin{aligned} a_{1n} &= 2(n + 1)(n + r + s + 1)(2n + r + s), \\ a_{2n} &= (2n + r + s + 1)(r^2 - s^2), \\ a_{3n} &= (2n + r + s)(2n + r + s + 1)(2n + r + s + 2), \\ a_{4n} &= 2(n + r)(n + s)(2n + r + s + 2). \end{aligned} \tag{5}$$

By defining $P_{-1}(x) \equiv 0$ and $P_0(x) \equiv 1$, all other members of the set can be computed directly from the recurrence relation Eq. (4) after r and s have been chosen. For example, if $r = s = 0$, the Legendre polynomials are obtained from Eqs. (4) and (5). If $r = s = -1/2$, the Chebyshev polynomials are obtained.

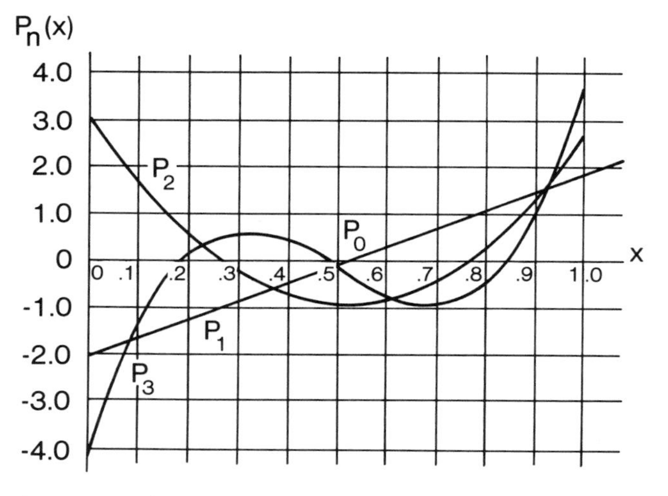

Fig. 2. First four Jacobi polynomials, $r = s = 1$.

It should be noted that the shape and the axis crossings of the polynomial set can be controlled to a significant extent by properly choosing r and s. Figures 1, 2, and 3 show the first four Jacobi polynomials for several values of r and s.

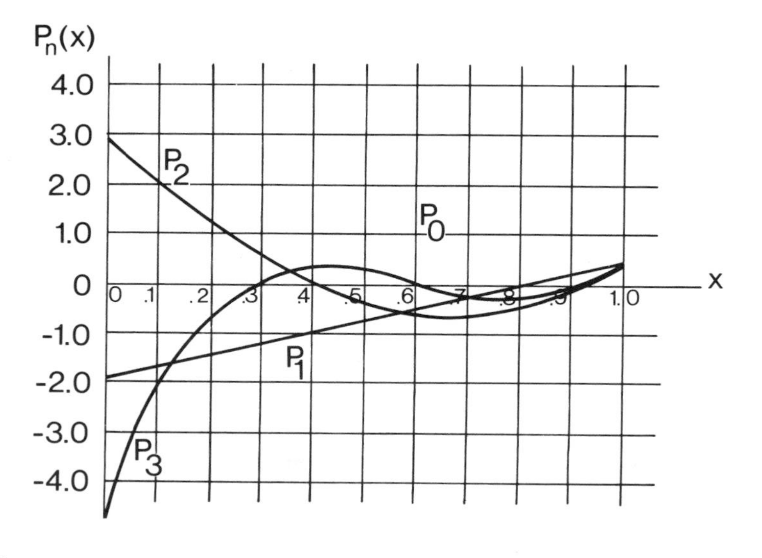

Fig. 3. First four Jacobi polynomials, $r = -1/2$, $s = 1$.

III. JACOBI VECTORS

The Jacobi vectors are a discrete, and thus finite, form of the Jacobi polynomials. These vectors were developed by C. Lanczos [3].

The recurrence relation Eq. (4) can be put in the form

$$a_{n-1}P_{n-1}(x) + (b_n - x)P_n(x) + a_nP_{n+1}(x) = 0. \tag{6}$$

By choosing x as one of the N roots of

$$P_N(x) = 0 \tag{7}$$

and defining $P_{-1}(x) \equiv 0$, Eq. (6) is

$$\begin{bmatrix} (b_0 - x_i) & a_0 & 0 & \cdots & 0 \\ a_0 & (b_1 - x_i) & a_1 & & \vdots \\ 0 & a_1 & (b_2 - x_i) & & 0 \\ \vdots & & & & a_{N-2} \\ 0 & 0 & \cdots & a_{N-2} & (b_{N-1} - x_i) \end{bmatrix} \times \begin{bmatrix} P_0(x_i) \\ P_1(x_i) \\ \vdots \\ P_{N-1}(x_i) \end{bmatrix} = 0. \tag{8}$$

The x_i are the eigenvalues of the square matrix and the $P_n(x_i)$, $(n = 0, 1, \ldots, N - 1)$, are the corresponding eigenvectors. The symmetry of the square matrix indicates that the eigenvectors are orthogonal.

Defining the normalized eigenvectors as

$$\sum_{n=0}^{N-1} \phi_n^2(x_i) = 1 \tag{9}$$

yields the orthonormal modal matrix

$$M = \begin{bmatrix} \phi_0(x_1) & \phi_0(x_2) & \cdots & \phi_0(x_N) \\ \phi_1(x_1) & \phi_1(x_2) & & \\ \vdots & & & \vdots \\ \phi_{N-1}(x_1) & & \cdots & \phi_{N-1}(x_N) \end{bmatrix}. \tag{10}$$

Using $MM^T = M^TM = I$ gives the discrete or summation orthogonality relationships

$$\begin{aligned} \sum_{i=1}^{N} \phi_n(x_i)\phi_m(x_i) &= 1, \quad n = m, \\ &= 0, \quad n \neq m. \end{aligned} \tag{11}$$

IV. MODELING

The concepts involved in Jacobi vector modeling are demonstrated by an example using the wave equation. Let the actual system model be

$$\frac{\partial^2 y}{\partial t^2} - \frac{\partial^2 y}{\partial x^2} = f(x, t), \quad 0 \leq x \leq 1, \tag{12}$$

with boundary conditions $y(0, t) = y(1, t) = 0$. To avoid measuring spatial derivatives of the dependent variable, Eq. (12) is multiplied by a known smooth function $g(x)$ and then integrated from 0 to 1; i.e.,

$$\int_0^1 \left[g(x)\frac{\partial^2 y}{\partial t^2} - g(x)\frac{\partial^2 y}{\partial x^2} \right] dx = \int_0^1 g(x) f(x, t)\, dx. \tag{13}$$

Note that this step projects the system model on $g(x)$.

If the second term on the left-hand side is integrated by parts two times,

$$\left[g(x)\frac{\partial y}{\partial x} - y\frac{dg}{dx} \right]_0^1 + \int_0^1 y\frac{d^2 g}{dx^2}\, dx$$

is obtained. If $g(x)$ is chosen to satisfy the boundary conditions of Eq. (12), the boundary terms above are zero, and Eq. (13) becomes

$$\int_0^1 \left[g(x)\frac{\partial^2 y}{\partial t^2} - \frac{d^2 g}{dx^2} y(x, t) \right] dx = \int_0^1 g(x) f(x, t)\, dx. \tag{14}$$

As a discrete system model is sought, the integration in Eq. (14) is replaced by a Gauss-type quadrature, i.e.,

$$\int_0^1 \omega(x) f(x)\, dx \simeq \sum_{i=1}^{N} \nu_i f(x_i), \tag{15}$$

where $\omega(x)$ is the weighting function for the Jacobi polynomials, the ν_i are discrete weights, and the x_i are the roots of the Nth order polynomial. Applying Eq. (15), Eq. (14) becomes

$$\sum_{i=1}^{N} \nu_i \frac{g(x_i)\ddot{y}_i(t) - g''(x_i) y_i(t) - h(x_i) f_i(t)}{\omega(x_i)} \simeq 0. \tag{16}$$

This is a single equation that must be converted to N simultaneous equations. To do this the coefficients of the time functions in Eq. (16) are expanded in a series of normalized Jacobi vectors, i.e.,

$$\frac{\nu_i g(x_i)}{\omega(x_i)} = \sum_{n=0}^{N-1} \alpha_n \phi_n(x_i),$$

$$\frac{\nu_i g''(x_i)}{\omega(x_i)} = \sum_{n=0}^{N-1} \beta_n \phi_n(x_i), \qquad (17)$$

$$\frac{\nu_i h(x_i)}{\omega(x_i)} = \sum_{n=0}^{N-1} \gamma_n \phi_n(x_i).$$

Substituting Eq. (17) into Eq. (16) yields

$$\sum_{n=0}^{N-1} \sum_{i=1}^{N} \phi_n(x_i)[\alpha_n \ddot{y}_i - \beta_n y_i - \gamma_n f_i(t)] \simeq 0. \qquad (18)$$

The corresponding matrix equation is

$$\alpha^T M\ddot{y} - \beta^T My \simeq \gamma^T Mf(t), \qquad (19)$$

where M is defined by Eq. (10), and

$$\alpha = \begin{bmatrix} \alpha_0 \\ \alpha_1 \\ \vdots \\ \alpha_{N-1} \end{bmatrix}, \quad \text{and} \quad y = \begin{bmatrix} y_1 \\ y_2 \\ \vdots \\ y_N \end{bmatrix}.$$

The first term of Eq. (19) can be expanded:

$$\alpha_0[\phi_0(x_1) \cdots \phi_0(x_N)]\begin{bmatrix}\ddot{y}_1\\ \ddot{y}_2\\ \vdots\\ \ddot{y}_N\end{bmatrix} + \alpha_1[\phi_1(x_1) \cdots \phi_1(x_N)]\begin{bmatrix}\ddot{y}_1\\ \ddot{y}_2\\ \vdots\\ \ddot{y}_N\end{bmatrix}$$

$$+ \cdots + \alpha_{N-1}[\phi_{N-1}(x_1) \cdots \phi_{N-1}(x_N)]\begin{bmatrix}\ddot{y}_1\\ \ddot{y}_2\\ \vdots\\ \ddot{y}_N\end{bmatrix}.$$

It is seen that the system response vector is projected on the orthogonal rows of M. This provides the basis for decomposition. Equation (19) can be written as

$$\alpha_n\phi_n^T\ddot{y} - \beta_n\phi_n^T y \simeq \gamma_n\phi_n^T f(t), \quad (n = 0, 1, \ldots, N-1). \tag{20}$$

Decoupling is accomplished by letting

$$\phi_n^T y = z_n$$

and

$$\phi_n^T f(t) = F_n(t);$$

then

$$\alpha_n\ddot{z}_n - \beta_n z_n = \gamma_n F_n(t), \quad (n = 0, 1, \ldots, N-1). \tag{21}$$

Equation (21) can be put into state form by letting $q_n = z_n$ and $q_{n+3} = \dot{z}_n$. The state equations are then

$$\dot{q}_n = q_{n+3},$$
$$\dot{q}_{n+3} = \frac{\beta_n}{\alpha_n}q_n + \frac{\gamma_n}{\alpha_n}F_n(t). \tag{22}$$

The corresponding vector equation is

$$\dot{q} = \begin{bmatrix} 0 & I \\ \Lambda & 0 \end{bmatrix} q + \begin{bmatrix} 0 \\ b \end{bmatrix} F. \tag{23}$$

Figure 4 shows a simple system (vibrating string, for example). Figure 5 illustrates how the system response is used to obtain the uncoupled state variables, and Fig. 6 shows the corresponding system model.

V. IDENTIFICATION

Once it has been established that the general form of the model is a set of second-order differential equations with constant coefficients (if the system is time-varying the coefficients will be time-varying), the identification problem consists of determining those coefficients.

If Eq. (17) is multiplied by $\phi_m(x_i)$ and both sides summed over i, the coefficients are obtained, i.e.,

$$\alpha_m = \sum_{i=1}^{N} \frac{\nu_i g(x_i)}{\omega(x_i)} \phi_m(x_i). \tag{24}$$

Thus the coefficients of the model are the projections of the spatial coefficients of the original equation, multiplied by g(x) or its derivatives, on the coordinate Jacobi vectors.

In the previous section, g(x) was required to match the boundary conditions of the original problem and also to be at least twice differentiable. The nature of g(x) is seen by comparing the left-hand sides of Eqs. (13) and (14). One is the spatial adjoint of the other. Thus g(x) and y(x) can be related generally in the following manner. Let L_x be any linear differential operator and let L_x^* be its adjoint operator. Consider the response of the two systems to point source

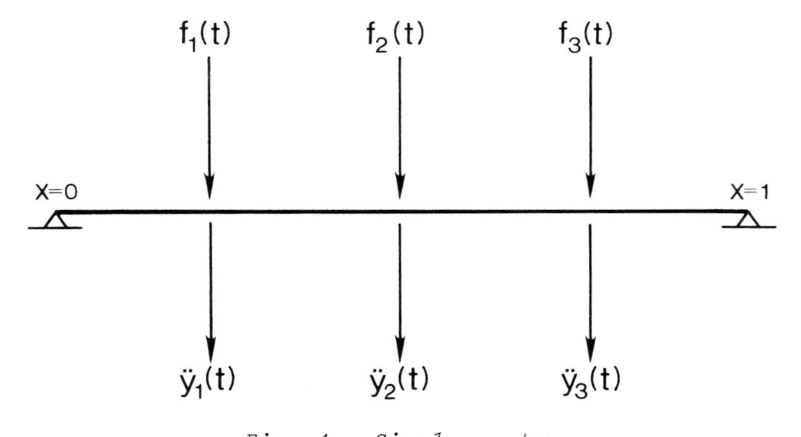

Fig. 4. Simple system.

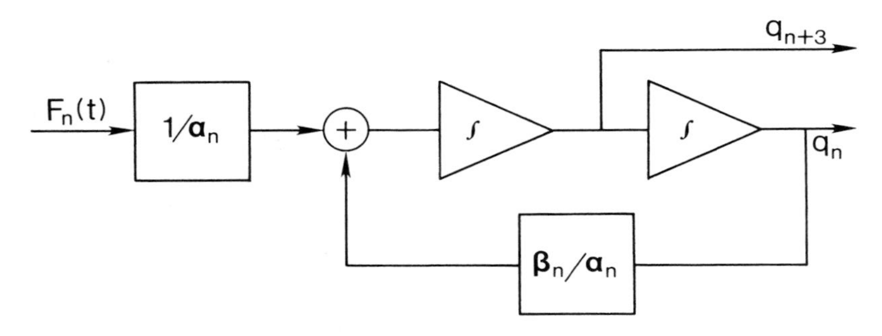

Fig. 5. State variables.

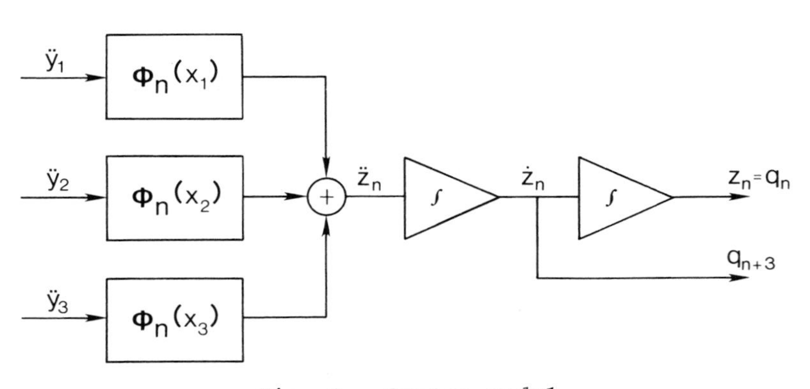

Fig. 6. System model.

inputs, i.e.,

$$L_x y(x, \eta) = \delta(x - \eta) \tag{25}$$

and

$$L_x^* g(x, \xi) = \delta(x - \xi). \tag{26}$$

Multiplying Eq. (25) by $g(x, \xi)$ and Eq. (26) by $y(x, \eta)$, subtracting the second resulting equation from the first, and integrating from 0 to 1 yields

$$\int_0^1 \left[g(x, \xi) L_x y(x, \eta) - y(x, \eta) L_x^* g(x, \xi)\right] dx$$

$$= \int_0^1 g(x, \xi)\delta(x - \eta)dx - \int_0^1 y(x, \eta)\delta(x - \xi)dx. \tag{27}$$

The left-hand side of Eq. (27) is equal to zero by definition of the adjoint operator. Then the right-hand side yields

$$g(\eta, \xi) = y(\xi, \eta). \tag{28}$$

That is, $g(x)$ should be the reciprocal function of $y(x)$; and for self-adjoint systems,

$$g(x) = y(x). \tag{29}$$

Thus the identification is accomplished by measuring $y(x_i)$ and using it to replace $g(x_i)$ in Eq. (24).

The function satisfying Eq. (25) is the Green's function for the spatial system L_x. It is the system response at a point x due to a unit input applied at point η. Thus, it is the influence function from which influence coefficients can be obtained.

VI. APPLICATION TECHNIQUES

As an illustration of this modeling technique, consider the system shown in Fig. 4 and modeled exactly by Eq. (12). A sixth-order discrete model of this system is shown in Fig. 7.

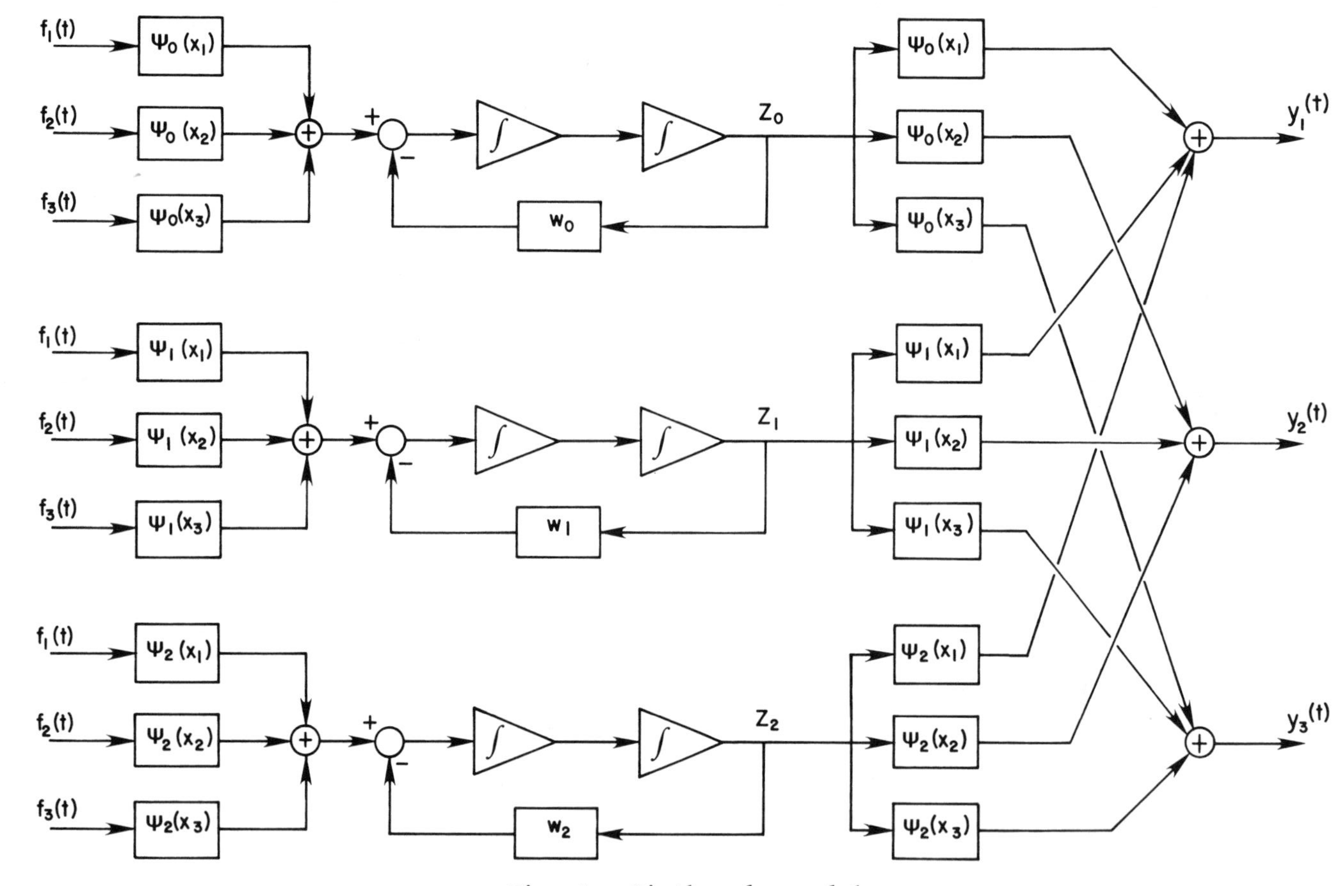

Fig. 7. Sixth-order model.

If this model was based on the natural modes of the system, $\psi_n(x)$ would be sin $n\pi x$, ω_n would be $[(n + 1)\pi]^2$, and the x_i would be the zeros of sin $4\pi x$ between zero and one. On the other hand, the $\psi_n(x_i)$ can equally well (but not equally accurately) be normalized Jacobi vectors. The ω_n would then be computed from Eq. (24), and the x_i would be the zeros of $P_3(x)$.

The ω_n are the eigenvalues of the model, and it is critical that they match those of the actual system. Calculating them from Eq. (24) will yield only approximate values that will decrease in accuracy with increasing mode number. In any realistic model the feedback coefficients must be tuned on the basis of observation of the actual system response.

A convenient expression for the response of an output of the Fig. 7 model is the convolution integral

$$y_1 = \int_0^t [h_0(\tau)h_1(\tau)h_2(\tau)] \times \begin{bmatrix} \psi_{01}^2 & \psi_{01}\psi_{02} & \psi_{01}\psi_{03} \\ \psi_{11}^2 & \psi_{11}\psi_{12} & \psi_{11}\psi_{13} \\ \psi_{21}^2 & \psi_{21}\psi_{22} & \psi_{21}\psi_{23} \end{bmatrix} \begin{bmatrix} f_1(t-\tau) \\ f_2(t-\tau) \\ f_3(t-\tau) \end{bmatrix} d\tau, \quad (30)$$

where $h_n(t)$ is the impulse response of the nth order oscillator and ψ_{ni} indicates $\psi_n(x_i)$. The Jacobi model can be compared with the model based on system modes by comparing the square coefficient matrices in Eq. (30) for the two cases.

To illustrate, if $r = s = 1$, the zeros of $P_3(x)$ are 0.173, 0.500 and 0.827, leading to a Jacobi vector matrix

$$\phi_n(x_i) = \begin{bmatrix} 0.484 & 0.729 & 0.484 \\ -0.707 & 0 & 0.707 \\ 0.516 & -0.683 & 0.516 \end{bmatrix}, \quad (31)$$

and the corresponding coefficient matrix for Eq. (30) is

$$C_1 = \begin{bmatrix} 0.234 & 0.353 & 0.234 \\ 0.500 & 0 & -0.500 \\ 0.266 & -0.353 & 0.266 \end{bmatrix}, \tag{32}$$

where C_1 denotes the coefficient matrix for response point 1. The corresponding modal matrices are

$$\psi_n(x_i) = \begin{bmatrix} 0.517 & 1.00 & 0.517 \\ 0.885 & 0 & -0.885 \\ 0.998 & -1.00 & 0.998 \end{bmatrix} \tag{33}$$

and

$$C_1' = \begin{bmatrix} 0.267 & 0.517 & 0.267 \\ 0.783 & 0 & -0.783 \\ 0.996 & -0.998 & 0.996 \end{bmatrix}. \tag{34}$$

Comparing Eqs. (32) and (34), it is apparent that, while the Jacobi model Eq. (32) has good first and second mode strength compared to the modal model Eq. (34), the third mode is weaker. Figure 8 shows the response of the two models compared with the actual system response at x_1 due to unit impulse inputs at x_1 and x_2.

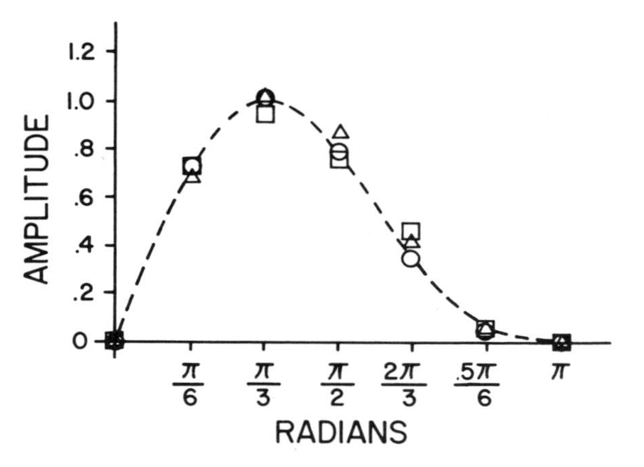

Fig. 8. Response of models and actual system. ○, modal model; □, actual response; △, Jacobi vector model.

VII. ILLUSTRATIVE EXAMPLE

In this section the model of a more complex system is developed so that special cases in the process can be described. The partial differential equation for the illustrative system is

$$\frac{\partial^2}{\partial x^2}\left[K(x)\frac{\partial^2 w}{\partial x^2}\right] + c(x)\frac{\partial^3 w}{\partial x^2 \partial t} + m(x)\frac{\partial^2 w}{\partial t^2} = v(x, t), \tag{35}$$

with boundary conditions

$$\begin{aligned} w(0, t) &= \Delta(t), \quad (\partial^2 w/\partial x^2)(1, t) = 0, \\ (\partial w/\partial x)(0, t) &= 0, \quad (\partial^3 w/\partial x^3)(1, t) = 0. \end{aligned} \tag{36}$$

These equations model the oscillations of a slender cantilever beam with strain-rate damping (see Fig. 9). The system contains self-adjoint and nonself-adjoint terms, variable coefficients and nonhomogeneous boundary conditions. As the treatment to follow is primarily directed toward the spatial variable, this system is quite general. Also note that the work is readily extended to more than one spatial dimension, and that the coefficients $K(x)$, $c(x)$, and $m(x)$ can also be functions of time.

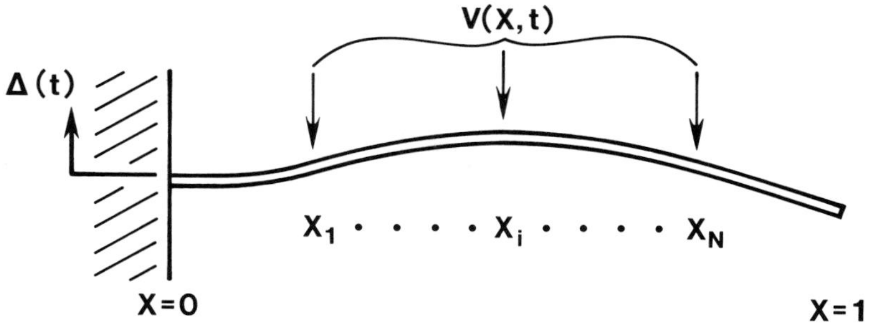

Fig. 9. Illustrative system.

To obtain homogeneous boundary conditions, the change of variables

$$w(x, t) = y(x, t) + \Delta(t) \tag{37}$$

is made, and Eqs. (35) and (36) become

$$\frac{\partial^2}{\partial x^2}\left[K(x)\frac{\partial^2 y}{\partial x^2}\right] + c(x)\frac{\partial^3 y}{\partial x^2 \partial t} + m(x)\frac{\partial^2 y}{\partial t^2} = -m(x)\frac{d^2}{dt^2}\Delta(t) + v(x, t) \tag{38}$$

and

$$\begin{gathered} y(0, t) = 0, \quad (\partial^2 y/\partial x^2)(1, t) = 0, \\ (\partial y/\partial x)(0, t) = 0, \quad (\partial^3 y/\partial x^3)(1, t) = 0. \end{gathered} \tag{39}$$

Employing the procedures developed in the section on modeling, Eq. (38) is multiplied by $g(x)$ and integrated from $x = 0$ to $x = 1$. Integration by parts is then carried out to remove differentiation with respect to x from $y(x, t)$. However, this system differs from the previous one in that there is no choice of $g(x)$ that will make the boundary terms, resulting from integration by parts, go to zero. If $g(x)$ is chosen to match the boundary conditions, i.e., Eq. (39), the result is

$$\begin{aligned} &\int_0^1 \left[g(x)m(x)\frac{\partial^2 y}{\partial t^2} - \frac{d^2}{dx^2}[g(x)c(x)]\frac{\partial y}{\partial t} + \frac{d^2}{dx^2}\left[K(x)\frac{d^2 g}{dx^2}\right]y\right]dx \\ &\quad = \int_0^1 \left[g(x)v(x, t) - g(x)m(x)\frac{d^2\Delta}{dt^2} \right. \\ &\qquad \left. + \delta(x - 1)\frac{d}{dx}[g(x)c(x)]\frac{\partial y}{\partial t}\right]dx. \end{aligned} \tag{40}$$

The right-hand side is composed of terms that are known, or terms that must be measured: $(d^2\Delta/dt^2)(t)$ is the acceleration of the base; $(\partial y/\partial t)(1, t)$ is the velocity of the tip.

Applying Eq. (15) to Eq. (40) yields

$$\sum_{i=1}^{N} \frac{\nu_i}{\omega(x_i)} \left[g(x_i) m(x_i) \ddot{y}_i - \frac{d^2}{dx^2}[g(x)c(x)]_{x=x_i} \dot{y}_i + \frac{d^2}{dx^2}\left[K(x)\frac{d^2 g}{dx^2}\right]_{x=x_i} y_i \right]$$

$$= \sum_{i=1}^{N} \frac{\nu_i}{\omega(x_i)} [g(x_i) v_i(t) - g(x_i) m(x_i) \ddot{\Delta}(t)] + \frac{d}{dx}[g(x)c(x)]_{x=1} \dot{y}(1, t). \tag{41}$$

The coefficients of the terms on the left-hand side can be expanded in a series of Jacobi vectors, i.e.,

$$\frac{\nu_i g(x_i) m(x_i)}{\omega(x_i)} = \sum_{n=0}^{N-1} \alpha_n \phi_n(x_i),$$

$$\frac{\nu_i}{\omega(x_i)} \frac{d^2}{dx^2}[g(x)c(x)]_{x=x_i} = \sum_{n=0}^{N-1} \eta_n \phi_n(x_i), \tag{42}$$

$$\frac{\nu_i}{\omega(x_i)} \frac{d^2}{dx^2}\left[K(x)\frac{d^2 g}{dx^2}\right]_{x=x_i} = \sum_{n=0}^{N-1} \beta_n \phi_n(x_i).$$

The first term on the right-hand side can also be expanded in the same fashion, i.e.,

$$\frac{\nu_i g(x_i)}{\omega(x_i)} = \sum_{n=0}^{N-1} \gamma_n \phi_n(x_i), \tag{43}$$

and the second term has the same expansion as the first relation in Eq. (42).

The last term on the right-hand side of Eq. (41) is a forcing term proportional to tip velocity. These forcing functions arise when nonself-adjoint terms are integrated by parts to remove differentiation. Just as the acceleration of the base $\ddot{\Delta}(t)$ must be measured, so should the tip velocity $\dot{y}(1, t)$. However, as this term is not applied at one of the points x_i, it must be replaced by equivalent forcing functions at the points $x_1, x_2, \ldots, x_N$. This is accomplished by requiring the response vector $[y_1(t), \ldots, y_N(t)]^T$ to be the same for an input vector $[f_1(t), \ldots, f_N(t)]^T$ as for $(d/dx)[g(x)c(x)]_{x=1}$ $\dot{y}(1, t)$.

Due to forcing at the tip, the response vector is

$$\begin{bmatrix} y_1(t) \\ \vdots \\ y_N(t) \end{bmatrix} = \begin{bmatrix} g(x_1, 1) \\ \vdots \\ g(x_N, 1) \end{bmatrix} \frac{d}{dx}[g(x)c(x)]_{x=1}\dot{y}(1, t). \tag{44}$$

The same response must be obtained from the input vector, i.e.,

$$\begin{bmatrix} y_1(t) \\ \vdots \\ y_N(t) \end{bmatrix} = \begin{bmatrix} g(x_1, x_1) & \cdots & g(x_1, x_N) \\ \vdots & & \vdots \\ g(x_N, x_1) & \cdots & g(x_N, x_N) \end{bmatrix} \begin{bmatrix} f_1(t) \\ \vdots \\ f_N(t) \end{bmatrix}. \tag{45}$$

Thus the vector $[f(t)]$ is obtained from

$$[f(t)] = [G]^{-1}[g]\frac{d}{dx}[g(x)c(x)]_{x=1}\dot{y}(1, t). \tag{46}$$

This vector can be represented by a Jacobi vector expansion as

$$f_i(t) = \sum_{n=0}^{N-1} \zeta_n \phi_n(x_i)\dot{y}(1, t). \tag{47}$$

Substituting Eqs. (42), (43), and (47) into Eq. (41) yields

$$\sum_{i=1}^{N} \sum_{n=0}^{N-1} [\gamma_n \phi_n(x_i)\ddot{y}_i - \eta_n \phi_n(x_i)\dot{y}_i + \beta_n \phi_n(x_i) y_i]$$

$$\simeq \sum_{i=1}^{N} \sum_{n=0}^{N-1} [\gamma_n \phi_n(x_i) v_i(t) - \alpha_n \phi_n(x_i)\ddot{\Delta}(t)$$

$$+ \zeta_n \phi_n(x_i)\dot{y}(1, t)]. \tag{48}$$

The corresponding matrix equation is

$$\alpha^T M\ddot{y} - \eta^T M\dot{y} + \beta^T My$$

$$\simeq \gamma^T Mv(t) - \alpha^T M\ddot{\Delta}(t) + \zeta^T M\dot{y}(1, t). \tag{49}$$

This can be decomposed, in the manner used to obtain Eq. (20), to yield the vector equations

$$\alpha_n \phi_n^T \ddot{y} - \eta_n \phi_n^T \dot{y} + \beta_n \phi_n^T y \simeq \gamma_n \phi_n^T v(t) - \alpha_n \phi_n^T \ddot{\Delta}(t)$$

$$+ \zeta_n \phi_n^T \dot{y}(1, t), \quad (n = 0, 1, 2, \ldots, N - 1). \tag{50}$$

Decoupling is again accomplished by letting

$$\phi_n^T y = z_n$$

and

$$\phi_n^T v(t) = V_n(t)$$

$$\phi_n^T \ddot{\Delta}(t) = \ddot{\Delta}_n(t), \tag{51}$$

$$\phi_n^T \dot{y}(1, t) = Y_n(t)$$

to yield the scalar equations

$$\alpha_n \ddot{z}_n - \eta_n \dot{z}_n + \beta_n z_n \simeq \gamma_n V(t) - \alpha_n \ddot{\Delta}_n(t) + \zeta_n Y_n(t),$$

$$(n = 0, 1, \ldots, N - 1). \tag{52}$$

This model is now in the same form as the one shown in Fig. 7, except that there is velocity feedback in the oscillators and the forcing terms are general.

VIII. CONCLUSIONS

An approach to modeling distributed parameter systems has been presented and briefly compared to a model based on the natural modes of the system. The system used in the comparison is ideally suited to modal modeling, and the purpose was only to show that the Jacobi model could give comparable results.

The benefits of the Jacobi model are as follows:

(1) transducer locations can be chosen without particular knowledge of the system dynamics;

(2) an orthonormal coordinate system, with all of its incumbent advantages, is obtained; and

(3) comparable accuracy, in establishing $y(x)$, is obtained with about half the number of transducer locations that would otherwise be required.

The first two of these advantages have been discussed. The third should be emphasized. Transducers and activators are expensive and if their number can be reduced, without loss of accuracy, considerable cost savings will result. The use of sample points located at the zeros of Jacobi polynomials will yield comparable accuracy with about half as many data points as other integration techniques. There is an intimate relationship between the Jacobi polynomials and Gaussian quadrature. This subject is not within the scope of this chapter (for excellent engineering treatments see [4] and [5]); however the general idea behind Gaussian quadrature is discussed in Appendix B.

Figure 8 shows excellent agreement between the Jacobi model and the actual system response, particularly in light of the difference between their coefficient matrices, i.e., Eqs. (32) and (34). Equal excitation was applied only at points x_1 and x_2 to excite both odd and even modes. This resulted in almost complete cancellation of the third mode, which was where the Jacobi model differed most significantly from the actual system.

REFERENCES

1. G. R. SPALDING, "Distributed System Identification: A Green's Function Approach," *ASME J. Dyn. Syst., Measure. Control 98*, Series G, No. 2, 146 (1976).

2. G. R. SPALDING, "A State-Space Model for Distributed Systems," Proceedings of the 1978 AIAA/AAS Symposium on Astrodynamics, Palo Alto, California.

3. C. LANCZOS, "Applied Analysis," Prentice-Hall, New Jersey, 1957.

4. F. B. HILDEBRAND, "Introduction to Numerical Analysis," McGraw-Hill, New York, 1974.

5. L. FOX and I. B. PARKER, "Chebyshev Polynomials in Numerical Analysis," Oxford University Press, London and New York, 1968.

APPENDIX A. THE GREEN'S FUNCTION

The Green's function is a representation of system response to a unit point source of arbitrary location within the system domain. Thus it encompasses all of the system characteristics, i.e., it yields the response at any point due to an input at any point. As an example, the Green's function for the wave equation is the solution of the equation

$$\frac{\partial^2 y}{\partial t^2} - \frac{\partial^2 y}{\partial x^2} = \delta(x - \xi)\delta(t), \qquad 0 \leq x, \xi \leq 1. \tag{A.1}$$

Taking the Laplace transform in time and the sine transform in the spatial variable x yields a solution

$$y(x, \xi; t) = \sum_{n=1}^{\infty} \frac{2}{n\pi} \sin(n\pi x) \sin(n\pi\xi) \sin(n\pi t), \qquad (A.2)$$

which is the Green's function for the wave equation. The function g(x) used in this chapter has the form

$$g(x) = \sum_{n=1}^{N} \frac{2}{n\pi} \sin(n\pi x) \sin(n\pi x_0). \qquad (A.3)$$

APPENDIX B. GAUSSIAN QUADRATURE

The following is a brief outline of Gaussian quadrature (see Hildebrand [4]). Let

$$\pi(x) = (x - x_1), \ldots, (x - x_n) \qquad (B.1)$$

and

$$\pi^*(x_i) = (x_i - x_1), \ldots, (x_i - x_{i-1}),$$

$$(x_i - x_{i+1}), \ldots, (x_i - x_n). \qquad (B.2)$$

Then

$$\ell_i(x) = \frac{\pi(x)}{(x - x_i)\pi^*(x_i)}, \qquad (B.3)$$

which is a polynomial of order n - 1, and

$$\ell_i(x_j) = \delta_{ij} = 1, \quad i = j,$$

$$= 0, \quad i \neq j. \qquad (B.4)$$

Thus a function f(x) that takes on the values $f(x_1), \ldots, f(x_n)$ can be approximated by

$$y(x) = \sum_{k=1}^{n} \ell_k(x) f(x_k). \qquad (B.5)$$

This approximation is exact at the n sample points and is exact for all points if $f(x)$ is of polynomial order $n - 1$ or less.

If not only $f(x)$ but also $f'(x)$ (the derivative of f with respect to x) is known at the points $x_1, \ldots, x_n$, then 2n parameters are available, enough to specify exactly a polynomial of degree $2n - 1$. Thus an improved representation of $f(x)$ is

$$y(x) = \sum_{k=1}^{n} h_k(x) f(x_k) + \sum_{k=1}^{n} \hat{h}_k(x) f'(x_k), \tag{B.6}$$

where $h_k(x)$ and $\hat{h}_k(x)$ should have characteristics similar to $\ell_i(x)$ but be of order $2n - 1$. For

$$y(x_i) = f(x_i),$$

$$h_k(x_j) = \delta_{ij}, \quad \text{and } \hat{h}_k(x_j) = 0. \tag{B.7}$$

For

$$y'(x_i) = f'(x_i),$$

$$h'_k(x_j) = 0, \quad \text{and } \hat{h}'_k(x_j) = \delta_{ij}. \tag{B.8}$$

As $\ell_i(x)$ is of order $n - 1$, $h_k(x)$ and $\hat{h}_k(x)$ can be formed by

$$h_k(x) = r_k(x)[\ell_k(x)]^2, \tag{B.9}$$

$$\hat{h}_k(x) = s_k(x)[\ell_k(x)]^2, \tag{B.10}$$

where $r_k(x)$ and $s_k(x)$ are linear functions and can be determined from the conditions specified in Eqs. (B.7) and (B.8).

Viewing Eq. (B.6) as an approximation of $f(x)$, multiplying the equation by $\omega(x)$ and integrating yields

$$\int_0^1 \omega(x) f(x)\,dx \simeq \sum_{k=1}^{n} H_k f(x_k) + \sum_{k=1}^{n} \hat{H}_k f'(x_k), \tag{B.11}$$

where

$$H_k = \int_0^1 h_k(x)\,\omega(x)\,dx, \tag{B.12}$$

$$\hat{H}_k = \int_0^1 \hat{h}_k(x)\,\omega(x)\,dx. \tag{B.13}$$

Now, the $2n - 1$ accuracy level of (B.11) can be retained with only n sample values of $f(x)$ if the sample points can be chosen so that the $\hat{H}_k \equiv 0$. And they can be. Applying Eqs. (B.7) and (B.8) to Eqs. (B.9) and (B.10) yields

$$s_k(x) = x - x_k. \tag{B.14}$$

Substituting this and Eq. (B.10) into Eq. (B.13) yields

$$\hat{H}_k = \int_0^1 \omega(x)(x - x_k)[\ell_k(x)]^2\,dx. \tag{B.15}$$

Using the definition of $\ell_k(x)$, Eq. (B.15) can be rewritten as

$$\hat{H}_k = \frac{1}{\pi^*(x_k)} \int_0^1 \omega(x)\,\pi(x)\,\ell_k(x)\,dx. \tag{B.16}$$

Thus the $\hat{H}_k$ will be identically zero if $\pi(x)$ and $\ell_k(x)$ are orthogonal with respect to $\omega(x)$ over the interval 0 to 1. This requirement leads to the zeros of the Jacobi polynomials.

Application of Singular Perturbations to Optimal Control

KAPRIEL V. KRIKORIAN
Aerospace Groups
Hughes Aircraft Company
Culver City, California
and
C. T. LEONDES
School of Engineering and Applied Science
University of California
Los Angeles, California

I. INTRODUCTION

In many engineering applications the occurrence of high-dimensional systems generates a tremendous amount of computations. This may not be tolerable from a practical standpoint. The method of singular perturbations is an analysis tool that can reduce the order of the system, thus simplifying the computational complexities. It may be accomplished by techniques such as separating the system slow modes (of large time constants) from the fast modes (of small time constants) and applying boundary-layer corrections in separate time scales or

ISBN 0-12-012718-0

approximating the given system model by two lower order models, one of which is the sensitivity of the reduced-system model.

Problems with slow and fast states arise in most engineering fields [1]. In mechanical systems the small masses and stiff springs (small time constants) may be scaled relative to large masses and weak springs (large time constants). They give rise to fast and slow subsystems. In electrical systems the inductors and capacitors can behave similarly. Generally, in coupled electrical/mechanical systems, the electrical system represents the fast subsystem. Also, when modeling various other problems, such as flight dynamics, chemical reactions, heat transfer, and economic phenomena, fast and slow mode interactions are often encountered.

Singular perturbation techniques developed in mathematics have formed the necessary foundation for works developed in this area in various engineering fields. There is a great deal of literature in mathematics covering different issues on singular perturbations. Among these are the papers by Chang [2] and Vasileva [3] and the texts by Cole [4] and O'Malley [5].

The paper by Chang [2] provides a nonsingular transformation that expresses a singularly perturbed system into a purely diagonalized form. His technique has been used in the solution of singularly perturbed optimal control and linear filtering problems.

Vasileva's paper [3] examines a differentiability of the variables with respect to a small parameter μ that changes the system order at $\mu = 0$.

Cole [4] uses asymptotic expansions in terms of a small parameter μ ($\mu \geq 0$) to approximate the solution of various boundary-value problems. Inner and outer expansions (matched asymptotic expansions) are successfully applied to obtain solutions. Adding the inner and outer expansions and subtracting the common part of the two expansion (in the matching process), composite uniformly valid expansions are constructed. Because of difficulties near the boundaries, different asymptotic expansions valid in different regions are obtained. Determination of the unknown constants is done by matching these expansions in an overlap domain. Two-variable expansion method to solve initial value problems on semi-infinite intervals are also discussed. Asymptotic expansions of general form using two time variables (two time scales, long and short) are generated.

O'Malley [5] also treats singular perturbation problems by constructing asymptotic expansions. However, in regions of nonuniform convergence, instead of constructing the inner and outer expansions, the outer expansions are improved by applying "boundary-layer" corrections. Thus as $\mu \to 0$ the solution is the sum of the outer expansion in the given time scale and the boundary-layer correction (a second asymptotic series) in the stretched time scale.

The singular perturbation approach has been used in various areas of optimal control and linear filtering. Some of the main areas and contributors include: Kokotovic and Sannuti [6,7,8], who have reduced model order in optimal control using an optimally sensitive design. The linear quadratic regulator problem was studied by Kokotovic and Yackel [9], O'Malley [5], etc.; Calise [10] and others have investigated trajectory optimization problems; controllability and time optimal control have been studied by Kokotovic and Haddad [11,12]; linear filtering and smoothing was studied by Haddad and Kokotovic [13,14], Rauch [15], and others.

Further extensions of previous studies and new areas of applicability of the singular perturbation method need exploration. This chapter further expands the use of the singular perturbation method to the optimal control of deterministic systems. It first develops a general solution for the linear singularly perturbed minimum control effort problem with constraints on the control variable (Section II). It then develops a solution for singularly perturbed linear systems using dynamic programming (Section III). Examples are given to illustrate these developments.

Besides expanding on the work done so far and utilizing singular perturbation methods in engineering systems, it is the intent of this work to present relatively simple obtainable solutions that can be easily applied to reduce the computational complexities of engineering systems.

II. THE SINGULARLY PERTURBED MINIMUM CONTROL EFFORT PROBLEM

Among the most typical optimal control problems are the minimum time and the minimum control effort (fuel) problems. A near-optimum solution to the singularly perturbed linear time-invariant minimum time problem is obtained in [11] using the two time-scale method. In this chapter we extend the approach and apply singular perturbations to the minimum fuel problem for linear time-invariant systems with constrained control [18]. First the controllability properties and the existence of singular intervals are explored. Then, based on the separability of the time scales, controls for the slow and fast modes are obtained.

A. *THE SYSTEM MODEL*

Given the linear time-invariant system with large and small time constraints,

$$\dot{x}(t) = A_{11}x(t) + A_{12}z(t) + B_1u(t), \tag{1}$$

$$\mu\dot{z}(t) = A_{21}x(t) + A_{22}z(t) + B_2u(t), \tag{2}$$

where x (n-dimensional) and z (m-dimensional) compose the state vector, u (r-dimensional) is the control vector, and μ is a small parameter ($\mu \geq 0$), which by its presence changes the order of the system.

In this chapter we assume:

(a) The inverse of matrix A_{22} exists,

(b) The matrix A_{22} is asymptotically stable.

For $\mu = 0$, the reduced system is obtained:

$$\overline{z} = -A_{22}^{-1}(A_{21}\overline{x} + B_2\overline{u}), \tag{3}$$

which when substituted into (1) yields

$$\dot{\overline{x}} = A_0\overline{x} + B_0\overline{u}, \tag{4}$$

where

$$A_0 = A_{11} - A_{12}A_{22}^{-1}A_{21} \tag{5}$$

$$B_0 = B_1 - A_{12}A_{22}^{-1}B_2. \tag{6}$$

Separation of the fast and slow subsystems may be performed as in [2] and [11]. Introducing variables η (the fast state) and ξ (the slow state), this transformation may be represented as

$$\begin{bmatrix} \xi \\ \eta \end{bmatrix} = \begin{bmatrix} I_n - \mu HL & -\mu H \\ L & I_m \end{bmatrix} \begin{bmatrix} x \\ z \end{bmatrix}, \tag{7}$$

where

$$L = A_{22}^{-1}A_{21} + \mu G,$$

$$G = A_{22}^{-2}A_{21}A_0 + 0(\mu),$$

$$H = A_{12}A_{22}^{-1} + \mu\left[A_0A_{12}A_{22}^{-2} - A_{12}A_{22}^{-2}A_{21}A_{12}A_{22}^{-1} + 0(\mu)\right].$$

$0(\mu)$ represents a function of μ with norm less than $c\mu$, where c and μ are constants; $c > 0$, $\mu \in [0, \mu^*] > 0$. I_n and I_m are, respectively, $n \times n$ and $m \times m$ identity matrices. The inverse of the above transformation is

$$\begin{bmatrix} x \\ z \end{bmatrix} = \begin{bmatrix} I_n & \mu H \\ -L & I_m - \mu LH \end{bmatrix} \begin{bmatrix} \xi \\ \eta \end{bmatrix}. \tag{8}$$

The time-scale separated system equations are shown below. They only have in common the control u.

$$\dot{\xi} = a_0\xi + \beta_0 u, \tag{9}$$

$$\mu\dot{\eta} = a_2\eta + \beta_2 u, \tag{10}$$

where

$$a_0 = A_0 - \mu A_{12}G, \qquad \beta_0 = B_0 - \mu(HLB_1 + MB_2),$$

$$a_2 = A_{22} + \mu LA_{12}, \qquad \beta_2 = B_2 + \mu LB_1,$$

$$M = A_0A_{12}A_{22}^{-2} - A_{12}A_{22}^{-2}A_{21}A_{12}A_{22}^{-1} + 0(\mu).$$

B. *CONTROLLABILITY AND OPTIMAL CONTROL*

The general case minimum fuel problem, controllability properties, and the existence of singular intervals are discussed in [17]. In this section we shall apply singular perturbations to the minimum fuel problem of systems composed of slow and fast modes.

The slow subsystem (9) is a function of μ regular perturbation of the reduced system. As $\mu \to 0$, the eigenvalues of (9) tend to the eigenvalues of the reduced subsystem while the eigenvalues of (10) tend to those of A_{22}/μ. This tendency is analogous to that of the original system. Hence the controllability properties of (1) and (2) are the same as those for (9) and (10) [11]. Therefore, for complete controllability of the system (1) and (2),

$$\operatorname{rank}\left[B_0 : A_0 B_0 : --- : A_0^{n-1} B_0\right] = n$$

and

$$\operatorname{rank}\left[B_2 : A_{22} B_2 : --- : A_{22}^{m-1} B_2\right] = m$$

are only required. In addition, if the conditions

$$\begin{aligned} \operatorname{rank}\left[B_0 : A_0 B_0 : --- : A_0^{n-1} B_0\right] &= n, \\ \operatorname{rank}\left[B_2 : A_{22} B_2 : --- : A_{22}^{m-1} B_2\right] &= m, \end{aligned} \tag{11}$$

and

$$\operatorname{rank}\begin{bmatrix} A_0 & 0 \\ 0 & A_{22} \end{bmatrix} = n + m$$

apply, there are no singular intervals for the minimum fuel optimal control of the given system in the interval $[0, \mu^*]$.

The minimum fuel problem to be solved here is to bring the variables x and z of the system (1) and (2) from an initial state x^0 and z^0 to zero within a given time interval $0 \le t \le T$ using minimum fuel subject to the constraint $|u_e| \le 1$, $e = 1, 2, \ldots, r$. The performance measure for the minimum fuel problem is defined as

$$J = \int_0^T \sum_{e=1}^{r} |u_e(t)|\,dt. \tag{12}$$

The Hamiltonians for Eqs. (9) and (10) are, respectively,

$$H_\varepsilon = \sum_e |u_e| + p^T[a_0\xi + \beta_0 u], \tag{13}$$

$$H_\eta = \sum_e |u_e| + q^T[a_2\eta + \beta_2 u]/\mu, \tag{14}$$

where p and q are the costate vectors for the slow and fast states, respectively. Superscript T denotes the matrix transpose.

The necessary conditions for a minimum include

$$\dot{p} = -(\partial H_\xi/\partial\xi) = -a_0^T p \Rightarrow p = k_1 \exp^{(-a_0^T t)}$$
$$= k_3 \exp^{[a_0^T(T-t)]}, \tag{15}$$

$$\mu\dot{q} = -(\partial H_\eta/\partial\eta) = -a_2^T q \Rightarrow q = k_2 \exp^{(-a_2^T t/\mu)}$$
$$= k_4 \exp^{[a_2^T(T-t)/\mu]}, \tag{16}$$

where k_1, k_2, k_3, and k_4 are constants. From Pontryagin's principle,

$$u_e^* + p^T\beta_{0e}u_e^* \le |u_e| + p^T\beta_{oe}u_e$$

and

$$u_e^* + (q^T/\mu)\beta_{2e}u_e^* \le |u_e| + (q^T/\mu)\beta_{2e}u_e, \tag{17}$$

where the components e of u have been assumed independent of one another, β_{0e} and β_{2e} are the eth columns of β_0 and β_2 matrices, and superscript * denotes optimality.

Clearly the eth component of the optimum control for the overall system [(9), (10)] can be expressed as

$$u_e^*(t) = \begin{cases} 1 \text{ for } p^T\beta_{0e} + \dfrac{q^T}{\mu}\beta_{2e} < -1 \\ 0 \text{ for } -1 < p^T\beta_{0e} + \dfrac{q^T}{\mu}\beta_{2e} < 1 \\ -1 \text{ for } p^T\beta_{0e} + \dfrac{q^T}{\mu}\beta_{2e} > 1 \\ \text{undetermined negative number for} \\ \quad p^T\beta_{0e} + \dfrac{q^T}{\mu}\beta_{2e} = 1 \\ \text{undetermined positive number for} \\ \quad p^T\beta_{0e} + \dfrac{q^T}{\mu}\beta_{2e} = -1 \end{cases} \tag{18}$$

The final states are given by

$$\xi(T) = \exp^{(a_0T)}\xi(0) + \int_0^T \exp^{[a_0(T-t)]}\beta_0 u(t)\,dt,$$

$$\eta(T) = \exp^{(a_2T/\mu)}\eta(0) + \int_0^T \exp^{[a_2(T-t)/\mu]}\beta_2 u(t)\,dt,$$

and they should be zero. The constants k_3 and k_4 in Eqs. (15) and (16) may be determined knowing the initial and final values of the state and that for a fixed final time the Hamiltonian must be a constant when evaluated on the optimal trajectory provided the Hamiltonian does not depend explicitly on time;

$$H_\xi = C_1, \quad t \in [0, T],$$

$$H_\eta = C_2, \quad t \in [0, T].$$

When the eigenvalues of A_{22} have negative real parts, the fast costate (q) in Eq. (18), which depends on $\exp^{[a_2(T-t/\mu)]}$ is significant near the final time T. Thus one may allocate a small time interval near the final time T, $\hat{t} \leq t \leq T$, which is long enough to steer the fast subsystem to zero (i.e., $\hat{t}$ may be selected to be of the order of the fast subsystem settling time). To simplify the discussion, a stretched time variable σ for all $\mu > 0$ may be defined as

$$\sigma = (T - t)/\mu \tag{19}$$

and

$$\sigma^* = (T - \hat{t})/\mu.$$

From the above discussion, it follows that as for the minimum time problem [11], the minimum fuel problem is divided into two parts:

One that considers the slow subsystem or the reduced subsystem and computes an optimal slow control u_s^*, or a reduced control $\bar{u}^*$, respectively, within the time interval $0 \le t < \hat{t}$.

One that computes the fast control u_f^* for the fast subsystem within the time $\hat{t} \le t \le T$.

The slow mode control u_s^* is applied in the interval $0 \le t < T - \mu\sigma^*$. This control takes the original system from the state x^0 to the origin using minimum fuel. The eth element is:

$$u_{se}^* = \begin{cases} 1 \text{ for } p^T\beta_{0e} < -1 \\ 0 \text{ for } -1 < p^T\beta_{0e} < 1 \\ -1 \text{ for } p^T\beta_{0e} > 1 \\ \text{and undetermined for } p^T\beta_{0e} = \pm 1 \end{cases} \tag{21}$$

Similarly, the reduced control $\overline{u}^*$ may be applied to the reduced system. In the interval $0 \leq t < T - \mu\sigma^*$, the eth component is:

$$\overline{u}_e^* = \begin{cases} 1 \text{ for } p_0^T B_{0e} < -1 \\ 0 \text{ for } -1 < p_0^T B_{0e} < 1 \\ -1 \text{ for } p_0^T B_{0e} > 1 \\ \text{and undetermined for } p^T B_{0e} = \pm 1 \end{cases} \tag{22}$$

where $p_0 = k_0 \exp^{(-A_0^T t)}$ is the costate corresponding to the reduced system (4), and k_0 is a constant. Clearly the reduced control may be used adequately instead of the slow mode control for very small values of μ.

To bring the fast state to the origin, the fast mode control u_f^* is applied in the interval $T - \mu\sigma^* \leq t \leq T$

$$u_{fe}^* = \begin{cases} 1 \text{ for } p^T(\hat{t})\beta_{0e} + (q^T/\mu)\beta_{2e} < -1 \\ 0 \text{ for } -1 < p^T(\hat{t})\beta_{0e} + (q^T/\mu)\beta_{2e} < +1 \\ -1 \text{ for } p^T(\hat{t})\beta_{0e} + (q^T/\mu)\beta_{2e} > 1 \\ \text{undetermined elsewhere} \end{cases} \tag{23}$$

where $p(\hat{t})$ is the value of p at time $\hat{t}$.

A near optimum control for the original system (1) and (2) may be written as

$$u^*(t) = \begin{cases} u_s^*(t) \text{ or } \overline{u}^*(t) \quad \text{for } 0 \leq t < T - \mu\sigma^* \\ u_f^*(t) \quad \text{for } T - \mu\sigma^* \leq t \leq T \end{cases} \tag{24}$$

The basis for near optimality is discussed below.

The minimum fuel control is of bang-off-bang type when there are no singular intervals. Since the magnitude is constant, the area (the cost) under the control curve depends on when the control is applied. With A_{22} stable, the fast subsystem control becomes the system driver in the $\mu\sigma^*$ interval before the final time (T). In the optimal case one obtains the combined optimum control and the corresponding cost for the slow and fast subsystems. In the near optimal case [Eq. (24)], the overall cost is the sum of the optimum cost $J(u_s^*)$ to transfer the slow subsystem to the origin in the interval $0 \leq t < T - \mu\sigma^*$ and the optimum cost $J(u_f^*)$ to transfer the fast sybsystem to zero in the interval $T - \mu\sigma^* \leq t < T$. Thus the resulting error in the overall cost using the singular perturbation method is $0(\mu)$.

The final states $\xi(T)$ and $\eta(T)$ may be expressed as

$$\xi(T) = \exp^{(\mu a_0\sigma^*)}\left\{\exp^{(a_0\hat{t})}\xi(0) + \int_0^{\hat{t}} \exp^{[a_0(\hat{t}-t)]}\beta_0 u_s(t)\,dt\right\} + \mu\int_0^{\sigma^*}\exp^{[\mu a_0(\sigma^*-\tau)]}\beta_0 u_f(\tau)\,d\tau, \tag{25}$$

where $\hat{t} = T - \mu\sigma^*$ and $\tau = \sigma^* - \sigma$.

$$\eta(T) = \exp^{(a_2\sigma^*)}\eta^* + \int_0^{\sigma^*}\exp^{[a_2(\sigma^*-\tau)]}\beta_2 u_f(\tau)\,d\tau, \tag{26}$$

where $\eta^* \cong a_2^{-1}\beta_2 u_s(\hat{t}^-)$ is the steady state of η right before the time $\hat{t}$. From Eqs. (25) and (26), we note that $u^*(t)$ produces an error of $0(\mu)$ in $x(t)$ and $z(t)$. The last term in Eq. (25) suggests that a correction to the slow control may be obtained by bringing ξ to

$$\xi(T - \mu\sigma^*) = -\mu \int_0^{\sigma^*} \exp^{[\mu a_0(\sigma^* - \tau)]} \beta_0 u_f(\tau)\, d\tau$$

and then determining $u_s^*(t)$ with the appropriate switching times given by the costate vector p.

C. EXAMPLE

Consider the dynamic system of large and small time constants:

$$\dot{x} = -x + z + u, \qquad \mu\dot{z} = -2x - z + u.$$

It is required to find the control that transfers the system from states x^0, z^0 to the origin minimizing the performance measure

$$J = \int_0^T |u(t)|\, dt.$$

Separation of the slow and fast states yields:

$$\dot{\xi} = (-3 - 6\mu)\xi + (2 + 7\mu)u, \qquad \text{(slow subsystem)}$$

$$\mu\dot{\eta} = (-1 + 2\mu)\eta + (1 + 2\mu)u, \qquad \text{(fast subsystem)}$$

where only the linear terms in μ have been kept. A check of controllability and the esixtence of singular intervals shows that the given system is controllable within an interval $\mu \in [0, \mu^*]$ where $\mu^* > 0$, and there are no singular intervals.

From the necessary conditions for a minimum:

$$p_0 = k_0 \exp^{(3t)}, \qquad p = k_1 \exp^{[(3+6\mu)t]},$$

$$q = k_2 \exp^{\left[\frac{(1-2\mu)}{\mu} t\right]},$$

where k_0, k_1, and k_2 are integration constants.

Using Eqs. (21), (22), and (23), a set of reduced, slow, and fast near-optimum controls are obtained:

$$\overline{u}^* = \begin{cases} 1 \text{ for } 2p_0 < -1 \\ 0 \text{ for } -1 < 2p_0 < 1 \\ -1 \text{ for } 2p_0 > 1 \\ \text{undetermined elsewhere} \end{cases}$$

$$u_s^* = \begin{cases} 1 \text{ for } (2 + 7\mu)p < -1 \\ 0 \text{ for } -1 < (2 + 7\mu)p < 1 \\ -1 \text{ for } (2 + 7\mu)p > 1 \\ \text{undetermined elsewhere} \end{cases}$$

$$u_f^* = \begin{cases} 1 \text{ for } (2 + 7\mu)p(\hat{t}) + \frac{(1 + 2\mu)}{\mu} q < -1 \\ 0 \text{ for } -1 < (2 + 7\mu)p(\hat{t}) + \frac{(1 + 2\mu)}{\mu} q < 1 \\ -1 \text{ for } (2 + 7\mu)p(\hat{t}) + \frac{(1 + 2\mu)}{\mu} q > 1 \\ \text{undetermined elsewhere} \end{cases}$$

where the choice of reduced control or slow mode control may be based on the value of μ.

From the initial time to the time $\hat{t} = T - \mu\sigma^*$, the initial state x^0 of x is brought to the origin by the reduced control or the slow mode control.

The states $\overline{x}(T)$ and $\xi(T)$ may be written as

$$\overline{x}(T) = \exp^{(3u\sigma^*)}\left\{\exp^{(-3t)}\ x_0 + \int_0^t \exp^{[-3(\hat{t}-t)]}(2\overline{u})\,dt\right\}$$

$$+ \mu \int_0^{\sigma^*} \exp^{[-3\mu(\sigma^*-\tau)]}(2u_f)\,d\tau,$$

$$\xi(T) = \exp^{[-(3+6\mu)\mu\sigma^*]}\left\{\exp^{[(-3-6\mu)\hat{t}]}\ \xi(0) + \int_0^{\hat{t}} \exp^{[(-3-6\mu)(\hat{t}-t)]}(2 + 7\mu)u_s dt\right\}$$

$$+ \mu \int_0^{\sigma^*} \exp^{[-(3+6\mu)(\sigma^*-\tau)]}(2 + 7\mu)u_f d\tau.$$

Assuming $x^0 > 0$ and considering the first two terms in each of the last two equations, it can be seen that to bring x^0 within $0(\mu)$ of the origin, the controls $\bar{u}^*$ and u_s^* are either -1 or zero switching to -1. These values for the controls are determined in conjunction with the costates p_0 (reduced system) and p (slow subsystem). The admissable controls for different costates are shown in Fig. 1.

Since $u_s^*(\hat{t}^-) = -1$, the state η at $\hat{t}$ equation is

$$\eta^* = \frac{1 + 2\mu}{1 - 2\mu} > 0.$$

The final state of η is

$$\eta(T) = \exp^{[(-1+2\mu)\sigma^*]}\ \eta^*$$

$$+ \int_0^{\sigma^*} \exp^{[(-1+2\mu)(\sigma^*-\tau)]}(1 + 2\mu)u_f d\tau.$$

Since $\eta(T)$ is to be zero, it follows that

$$u_f^* = -1.$$

Hence a near optimum control u^* for the original system (1) and (2) is composed of the slow control (or reduced control) and the fast control.

$$u_s^* = [-1] \quad \text{or } [0, -1] \quad \text{for } 0 < t \leq \hat{t}$$

$$u_f^* = [-1] \quad \text{for } \hat{t} \leq t \leq T.$$

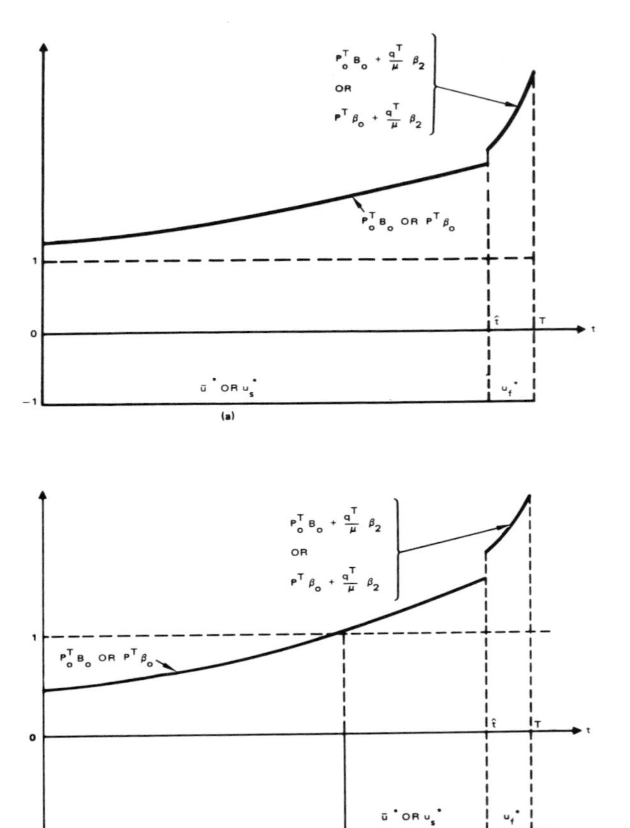

Fig. 1. Relationships between the costates and the acceptable controls as a function of time.

In this section the singular perturbation method was applied to derive a near-optimum control for the linear minimum fuel problem. The near optimality of the solution was established by examining the errors produced in the state and the cost. The technique was illustrated through an example. The results have shown that by using the singular perturbation method, system order reduction can be performed.

In the next section singular perturbations will be applied in dynamic programming.

III. DYNAMIC PROGRAMMING USING SINGULAR PERTURBATIONS

Dynamic programming is extensively used in solving optimal control problems, especially in problems where closed form solutions are hard or impossible to obtain. However, the utilization of dynamic programming in the case of high-dimensional systems may have prohibitive computational and storage requirements. Various techniques (e.g., [19], [20], and [21]) have been developed to reduce the dimensionality problem in dynamic programming. They include state increment dynamic programming, polynomial approximation of the minimum cost function, adaptive data processing, etc.

In this section the singular perturbation method is used in dynamic programming to reduce the order and the computational requirements of linear systems composed of slow and fast modes [18]. After the fast modes are separated, a near-optimum solution is computed at two different iteration rates determined by the slow and fast subsystem dynamics. The result is a reduction in the order of the computational requirement of the given system to that of the slow subsystem.

A. *THE SYSTEM MODEL*

Consider the linear time-invariant system with large and small time constants previously defined as

$$\dot{x} = A_{11}x + A_{12}z + B_1u, \tag{27}$$

$$\mu\dot{z} = A_{21}x + A_{22}z + B_2u, \tag{28}$$

where x is an n vector, z an m vector, u an r vector, A_{11}, A_{12}, B_1, A_{21}, A_{22}, and B_2 are $n \times n$, $n \times m$, $n \times r$, $m \times n$, $m \times m$, and $m \times r$ matrices, respectively, and μ is a small parameter,

$\mu \geq 0$. As in Section II, the reduced system ($\mu = 0$) is

$$\bar{z} = -A_{22}^{-1} A_{21}\bar{x} - A_{22}^{-1} B_2\bar{u}, \tag{29}$$

$$\dot{\bar{x}} = A_0\bar{x} + B_0\bar{u}, \tag{30}$$

where

$$A_0 = A_{11} - A_{12} A_{22}^{-1} A_{21},$$

$$B_0 = B_1 - A_{12} A_{22}^{-1} B_2.$$

Define the fast state {see [1, 11]}

$$\eta = z + A_{22}^{-1} A_{21}x + \mu Gx, \tag{31}$$

where $G = A_{22}^{-2} A_{21}A_0 + 0(\mu)$. $0(\mu)$ is a function of μ which for all $\mu \in [0, \mu^*]$ has the norm less than $c\mu$, where c and μ^* are constants and $c > 0$, $\mu^* > 0$. In (31), $z + A_{22}^{-1} A_{21}x$ is the error in the state z when the reduced system is considered. The quantity G is selected so that it uncouples the fast states from the slow states in (28). The slow and fast modes may then be expressed as

$$\dot{x} = (A_0 - \mu A_{12}G)x + A_{12}\eta + B_1u, \tag{32}$$

$$\mu\dot{\eta} = \left[A_{22} + \mu\left(A_{22}^{-1} A_{21} + \mu G\right)A_{12}\right]\eta$$
$$+ \left[B_2 + \mu\left(A_{22}^{-1} A_{21} + \mu G\right)B_1\right]u. \tag{33}$$

The homogenous part of (32) is an $0(\mu)$ perturbation of the reduced Eq. (30), while the solution of the fast subsystem (33) is an input to the slow subsystem (32). If A_{22} has eigenvalues of negative real parts only, μ is small, and the control stays constant within a considered time interval Δt

sufficiently long so that $\dot{\eta} \to 0$ and $\eta \to \eta_{ss}$ (the steady-state value of η), Eqs. (32) and (33) may be written as

$$\dot{x}(t) = (A_0 - \mu A_{12}G)x(t) + A_{12}\eta_{ss} + B_1 u, \tag{34}$$

$$\eta_{ss} = -a_2^{-1} b_2 u, \tag{35}$$

where

$$a_2 = A_{22} + \mu\left(A_{22}^{-1} A_{21} + \mu G\right)A_{12},$$

$$b_2 = B_2 + \mu\left(A_{22}^{-1} A_{21} + \mu G\right)B_1.$$

Using (35), Eq. (31) may be solved for z:

$$z(t) = -A_{22}^{-1} A_{21}x(t) - \mu Gx(t) - a_2^{-1}b_2 u. \tag{36}$$

B. *DISCRETIZATION AND DYNAMIC PROGRAMMING*

In this section certain assumptions are made:

(a) The matrix A_{22} is asymptotically stable.

(b) The system, (27) and (28), is assumed controllable.

(c) For small iteration times (of the order of the fast subsystem time constant), the slow modes remain effectively unchanged. For large iteration times the fast modes are effectively in steady state and the slow mode dynamics drive the system (e.g., see [22]).

(d) The dynamics of the fast subsystem becomes important near the boundaries where boundary layer corrections may be required (e.g., see [1]).

Consider the performance measure J_{TOT} for the original system in the interval $[0, T + \Delta t_f]$:

$$J_{TOT} = h[x_F, z_F] + \int_0^{T+\Delta t_f} g(x, z, u)dt, \tag{37}$$

where x_F and z_F are the desired final values of the state, g and h are scalar functions (e.g., $h[x_F, z_F]$ may represent a penalty for the error in reaching x_F and z_F), and Δt_f is a small time interval. For the given singularly perturbed system, Eq. (37) will be minimized using dynamic programming.

Suppose the time interval [0, T] is divided in subintervals of Δt duration and the applied control variables are constant within Δt.

Discretization of Eqs. (34), (35), and (36), where time is represented by the index k (k = 0 for t = 0 and k = N for t = T), leads to the following difference equations:

$$x(k+1) = [I + \Delta t(A_0 - \mu A_{12}G)]x(k) - \Delta t\left[A_{12}a_2^{-1}b_2 + B_1\right]u(k), \tag{38}$$

$$\eta_{ss}(k) = -a_2^{-1}b_2u(k), \tag{39}$$

$$z(k) = -\left(A_{22}^{-1} A_{21} + \mu G\right)x(k) - a_2^{-1}b_2u(k). \tag{40}$$

The discretized cost in the interval [0, T] is

$$J_{0,N} = h[x(N), z(N)] + \Delta t \sum_{k=0}^{N-1} g[x(k), z(k), u(k)], \tag{41}$$

where $x(N) = x_F$ and z(N) are obtained using (39) at k = N.

Application of Bellman's optimality principle to Eq. (41) leads to the minimum cost (e.g., [17]) of transferring the state from (N - k) to (N):

$$J^*_{N-K,N}[x(N-k), z(N-k)] = J^*_{N-k+1,N}[x(N-k+1), z(N-k+1)] + \min_{u(N-k)} \{\Delta t\, g[x(N-k), z(N-k), u(N-k)]\}. \tag{42}$$

A computer program which applies singular perturbations to dynamic programming uses Eqs. (38), (39), (40), and (42). Since η is assumed to reach steady state within the iteration interval Δt, the actual quantization levels for the state variable may be set using the quantization levels of x and u in Eq. (40).

The selection of the iteration time Δt requires careful consideration. The iteration time should be greater than the settling time for the fast subsystem such that within the interval Δt it would be justifiable to use the steady state η_{ss} of Eq. (39) to compute the states x [Eq. (38)] and z [Eq. (40)]. Additionally, the iteration interval Δt should be sufficiently small for the slow state Eq. (38) to give results within the desired accuracy. (See the example at the end of this chapter.) The final choice of the iteration interval will depend on the desired accuracy and system computational requirements.

A control $\bar{u}^*$ that steers the state x to a desired final value in [0, T] and minimizes the performance measure is determined by using the above approach. Since the final state of z is also of interest [Eq. (37)], the solution is obtained in two major steps. First the control $\bar{u}^*$ is determined as described, then the fast state is brought to a desired final value in an interval Δt_f, where $\Delta t_f < \Delta t$. During this time interval, state x is assumed unchanged and a fast control u_f^* for the fast subsystem is sought which steers $\eta_{ss}(T)$, Eq. (39), to the desired final value and minimizes the cost (37) in the interval $[T, T + \Delta t_f]$.

Depending on the performance measure and the length of time Δt_f, the steering of the fast state may be performed in one stage (i.e., when Δt_f is approximately the settling time of the fast subsystem), or in more than one stage (when Δt_f is less than the settling time of the fast subsystem) in which case Δt_f is divided in I subintervals of $\Delta t'$ duration. Obviously the latter results in a smaller error in x. To transfer $\eta_{ss}(T)$ to the desired final value $[\eta(T + \Delta t_f)]$, the descritized version of (33) is used:

$$\eta(i + 1) = \eta(i) + (\Delta t'/\mu)a_2\eta(i) + (\Delta t'/\mu)b_2u(i), \tag{43}$$

where i is the time index. The discretized cost in the interval $[T, T + \Delta t_f]$ is

$$J_{0,I} = h[x_F, z_F] + \Delta t' \sum_{i=0}^{I-1} g(x, z, u), \tag{44}$$

where z is obtained from η using (31). Application of Bellman's optimality principle on (44) and relation (43) yield u_f^* and $J_{0,I}^*$.

Note that an interval $[0, \Delta t_0]$, where $\Delta t_0 < \Delta t$, may be imposed to perform an initial value-boundary correction for the fast subsystem. This can be done in the same manner as the final value-boundary correction [Eq. (43) and (44)] in the interval $[T, T + \Delta t_f]$. However, for the performance measure (37) considered here, this does not significantly affect the solution.

Obviously, in the interval $[T, T + \Delta t_f]$ the error in the slow continuous subsystem (32) due to variations in the fast subsystem is $O(\mu)$. For stringent boundary conditions, it is possible to compensate for this error. If x_F is the desired final value of x, x(T) is computed applying Bellman's optimality principle on (44) and the high-dimensional system

Eqs. (27) and (28). Here $\Delta t_f < \Delta t$ does not apply. The vector $x(T)$ thus obtained is used to generate $\overline{u}^*$ in the interval $[0, T]$ as described.

Errors in the state x are also caused by the fact that η (which strictly represents the fast subsystem) is assumed in steady state throughout the iteration interval Δt (transients of η are ignored). However, selecting Δt as described and noting that A_{22}/μ has eigenvalues of large negative real parts, the dynamics of the overall system is approximated to within $0(\mu)$ of the actual dynamics given by (27) and (28). Its effect on the minimum cost (42) is an $0(\mu)$ error per stage.

Thus using Bellman's optimality principle and the two iteration rates a near-optimum control u^* for the systems (27) and (28) is determined in the sense that it produces an $0(\mu)$ error in the state. That is;

$$u^* = \begin{cases} \overline{u}^* & 0 \leq t < T \\ u_f^* & T \leq t < T + \Delta t_f \end{cases} \tag{45}$$

and the cost is

$$J^*_{TOT} = J^*_{0,N} + J^*_{0,I} - h[x^*(N), z^*(N)], \tag{46}$$

where

$$J^*_{0,N} = \min_{\overline{u}(0)\cdots\overline{u}(N-1)} \left\{ h[x(N), z(N)] + \Delta t \sum_{k=0}^{N-1} g[x(k), z(k), u(k)] \right\},$$

$$J^*_{0,I} = \min_{u_f(0)\cdots u_f(I-1)} \left\{ h[x_F, z_F] + \Delta t' \sum_{i=0}^{I-1} g[x(i), z(i), u(i)] \right\}.$$

Depending on the order of the fast subsystem, there can be large computational and memory savings by singular perturbations. For an m-dimensional fast mode and an n-dimensional slow mode, the amount of required computations and memory allocations is reduced by power of $n/(n + n)$. For the case where $n = m = 3$ states and $r = 1$ control with $q = 10$ quantization levels per state and control, the amount of computations and storage per iteration are reduced from an order of $q^{(n+m)}q^r = 10^7$ to $q^n q^r = 10^4$, a factor of 1000 per iteration. The savings become greater as the dimension of the fast subsystem and the number of quantization levels increase.

This work is also applicable to linear time-varying systems when the coefficients of x, z, and u can be assumed constant within the iteration interval Δt. The selection of the iteration interval may then be made a function of the change of system coefficients with time, in addition to being a function of the slow and fast subsystem time constants.

C. *EXAMPLE*

Consider the dynamic system of large and small time constants:

$$\dot{x} = -x + z + u,$$

$$\mu\dot{z} = -4x - 8z + 3u.$$

Assume the state variable is constrained, $-0.5 \leq x \leq 0.5$, and is quantized in increments of 0.1. The control is constrained, $-0.3 \leq u \leq 0.3$, and is also quantized in increments of 0.1. The performance measure to be minimized is

$$J_{TOT} = 2x^2(T + \Delta t_f) + \int_0^{T+\Delta t_f} [x^2(t) + z^2(t) + u^2(t)]dt.$$

It is desired to determine a near-optimum trajectory, control, and cost for the given system to reach the origin from the initial state $x(0) = +0.5$ as a function of the system parameter μ $(0 \leq \mu \leq 0.2)$.

To discretize the given system, Eqs. (38), (39), and (40) are used. The performance measure is discretized using the dynamic recurrence relation (42).

The quantization levels and the constraints on $z(k)$ are determined from the quantization and the constraints on $x(k)$ assuming $\mu = 0$ and $u(k) = 0$. Hence, $-0.25 < z < 0.25$ with the levels of z in increments of 0.05.

The choice of the iteration interval Δt is made by considering the eigenvalues of the continuous system for the case where the fast subsystem gives the slowest response (in this example the highest μ is 0.2). The eigenvalues are -1.5 and -39.5. The settling times for the slow and fast subsystem are approximately 2.0 and 0.07. Based on these, an iteration time of $\Delta t = 0.2$ is selected.

Table I summarizes near-optimum controls and costs that transfer the system from the initial state at $x = 0.5$ to zero for system parameters $\mu = 0$, 0.001, 0.01, 0.05, 0.1, and 0.2. At stage N, $z = -0.05$. Selecting $\Delta t_f = 0.07$ (the settling time of the fast subsystem), $u_f^* = 0$ and the fast subsystem settles at the origin in one stage. The near-optimum trajectories x and z are shown in Fig. 2. Here $\Delta t_f = 0.07 < \Delta t$. Note that $x(T)$ and $x(T + \Delta t_f)$ are at the same quantization level and no error in $x(T)$ is introduced by the extension to $T + \Delta t_f$.

Table I. Near-Optimum Control $\bar{u}^*$ and Cost J^* for the Trajectories in Fig. 2.

	Last		Last two		Last three		Last four	
Stage	$\bar{u}^*$	$J_{4,3}$	$\bar{u}^*$	$J_{4,2}$	$\bar{u}^*$	$J_{4,1}$	$\bar{u}^*$	$J_{4,0}$
$\mu = 0$ (reduced)	-0.1	0.0055	0	0.0155	0	0.0380	-0.1	0.1066
$\mu = 0.001$	-0.1	0.0055	0	0.0155	0	0.0380	0	0.1005
$\mu = 0.01$	-0.1	0.0055	0	0.0155	0	0.0381	0	0.1006
$\mu = 0.05$	-0.1	0.0056	0	0.0156	0.1	0.0382	0	0.1009
$\mu = 0.1$	-0.1	0.0056	0	0.0157	0.1	0.0383	0	0.1013
$\mu = 0.2$	-0.1	0.0057	0	0.0158	0.1	0.0385	0	0.1020

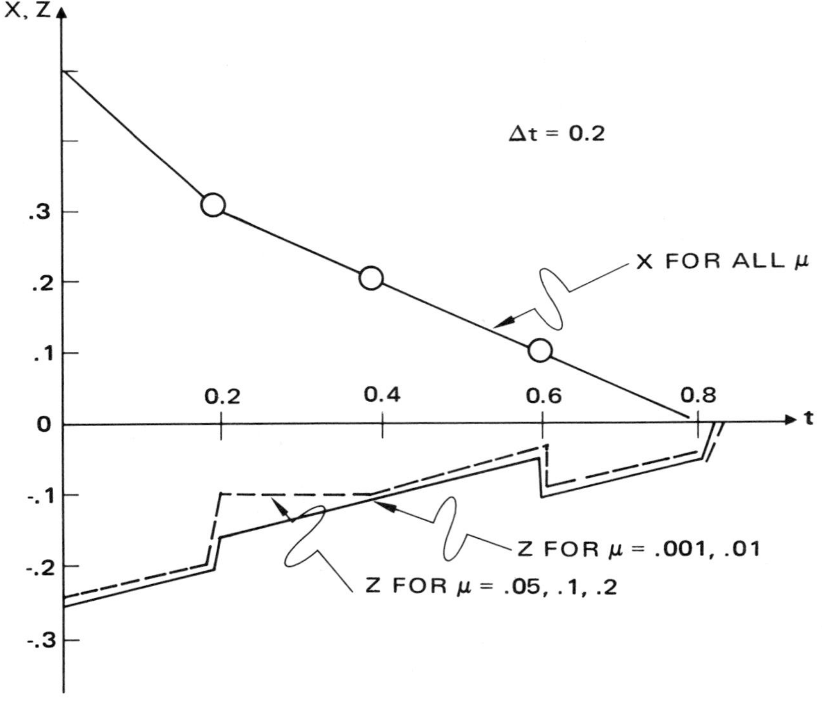

Fig. 2. The quantized near-optimum trajectories for x and z.

If dynamic programming were to be applied directly to the original system, much smaller iteration intervals (of the order of the fast subsystem smallest time constant) would be required. It would also require finer quantization of the states in order to ensure adequate sensitivity to small iteration intervals.

Thus, in this section we have found a near-optimum control by the singular perturbation method for the given system in the time interval $[0, T + \Delta t_f]$ by

(a) Separating the fast states, dividing the interval $0 \leq t < T$ in stages and using dynamic programming to solve for the control $\bar{u}^*$ and the cost $J^*_{0,N}$. Here $x(T)$ is the final value of x.

(b) Determining the fast control u^*_f in $T \leq t < T + \Delta t_f$, which transfers the fast state to the desired final state, computing the cost $J^*_{0,I}$, and adding to the corresponding cost in (a), to obtain J^*_{TOT}.

We have also indicated that $x(T + \Delta t_f)$ can be made to coincide with the final value of x. In this case the fast control (u^*_f) is determined first, then $x(T)$ and the slow mode control $(\bar{u}^*)$ are computed.

IV. SUMMARY

Singular perturbation techniques were used to study system order reduction in the optimal control of deterministic systems. First, a brief introduction describing the applicability of singular perturbations to various engineering systems and the main themes of the chapter were presented. Then the singular perturbation method was applied to the minimum control effort problem. Separation of time scales was used to determine the slow mode (or reduced system) and fast mode controls for two different time scales. From the necessary conditions for a minimum, a near-optimum control was obtained.

Separation of the fast modes was used in the determination of a near-optimum control of linear systems using dynamic programming. Two different sampling rates have been used. When the fast modes were not monitored closely, the larger sampling interval of the continuous system was selected so that the steady-state value for the fast subsystem could be used.

One of the major objectives of the chapter was to extend the use of singular perturbations and to develop practical techniques for system order reduction when dealing with complex problems. Here some basic problems were examined. Several problems remain to be investigated, some of which are listed below.

In the optimal control of deterministic systems with constrained controls, the singular perturbation method needs to be applied to solve problems that also involve constraints on the state vector. In this context effects of weighting the elapsed time need also to be included. Wherever possible, general solutions to optimal control problems need to be formulated.

Estimating one dominant value of the fast state within the iteration interval of dynamic programming can lift the restrictions on the large iteration interval. In addition, dynamic programming techniques using singular perturbation techniques should be extended to nonlinear systems. Methods should be developed for close monitoring of the fast modes when such modes are of interest.

Stochastic optimal control problems by singular perturbations require further investigation. Order reduction using dynamic programming in stochastic control systems needs to be studied.

The chapter has concerned itself with some areas of singular perturbations. In dealing with complex systems, the availability of analysis and design simplification tools, such as those discussed above, may prove to be essential in problem solving.

REFERENCES

1. P. V. KOKOTOVIC, R. E. O'MALLEY, JR., and P. SANNUTI, *Automatica 12*, 123 (1976).
2. K. W. CHANG, *SIAM J. Math. Anal. 3*, 520 (1972).
3. A. D. VASILEVA, *Russian Math Surveys 18*, 13 (1963).
4. J. D. COLE, "Perturbation Methods in Applied Mathematics," Blaisdell, Waltham, 1968.
5. R. E. O'MALLEY, JR., "Introduction to Singular Perturbations," Academic Press, New York, 1974.
6. P. V. KOKOTOVIC and P. SANNUTI, *IEEE Trans. Auto. Control AC-13*, 377 (1968).
7. P. SANNUTI, *Proc. JACC*, Atlanta, Georgia, 489 (1970).
8. P. SANNUTI and P. V. KOKOTOVIC, *Automatica 5*, 773 (1969).
9. P. V. KOKOTOVIC and R. A. YACKEL, *IEEE Trans. Auto. Control AC-17*, 29 (1972).
10. A. F. CALISE, *Proc. JACC*, San Francisco, California, 1248 (1977).
11. P. V. KOKOTOVIC and A. H. HADDAD, *IEEE Trans. Auto. Control AC-20*, 111 (1975).
12. P. V. KOKOTOVIC and A. H. HADDAD, *IEEE Trans. Auto. Control AC-20*, 163 (1975).
13. A. H. HADDAD and P. V. KOKOTOVIC, "Proc. 1974 Symposium on Nonlinear Estimation and Its Applications," San Diego, California, 96 (1974).
14. A. H. HADDAD and P. V. KOKOTOVIC, *IEEE Trans. Auto. Control AC-22*, 815 (1977).

15. H. E. RAUCH, "Proc. 12th Albertson Conference on Circuit and System Theory," University of Illinois, 718 (1973).

16. J. J. ALLEMONG and P. V. KOKOTOVIC, *IEEE Trans. Auto. Control AC-25*, 821 (1980).

17. D. E. KIRK, "Optimal Control Theory. An Introduction," Prentice Hall, Englewood Cliffs, New Jersey, 1970.

18. K. V. KRIKORIAN, "Application of Singular Perturbation to Deterministic and Stochastic Systems With and Without Time Delay," Ph.D. Dissertation, University of California, Los Angeles, 1978.

19. R. E. LARSON, *IEEE Trans. Auto. Control AC-10*, 135 (1965).

20. R. E. LARSON, *IEEE Trans. Auto. Control AC-12*, 767 (1967).

21. G. L. NEMHAUSER, "Introduction to Dynamic Programming," Wiley, New York, 1966.

22. M. SANDELL, P. VARAIYA, M. ATHANS, and M. SAFONOR, *IEEE Trans. Auto. Control AC-23*, 108 (1978).

Design Techniques for Multivariable Flight Control Systems

C. A. HARVEY

R. E. POPE

Aerospace and Defense Group
Honeywell Systems and Research Center
Minneapolis, Minnesota

I. INTRODUCTION

A multivariable flight control system is one in which there are multiple interacting control loops. This interaction is dictated by the dynamic coupling resulting from the aircraft design. Until recently, aircraft were designed to minimize dynamic coupling.

For the design of flight control systems with little dynamic coupling, traditional or so-called classical design and analysis techniques are more than adequate and provide key insights into the fundamental design issues of feedback control systems. The design and analysis techniques discussed in this chapter are directed at systems whose control loop interaction extend the utility of classical techniques to the point

ISBN 0-12-012718-0

where they are not only cumbersome to use as design tools but produce flight control system designs with undesirable performance characteristics. The inadequacies of classical design techniques are by no means accepted facts. There has been continuous debate over the two last decades as to the utility of classical techniques versus the utility of nonclassical or modern techniques.

Table I. Aircraft Control Systems--Early to Current Operational Systems.

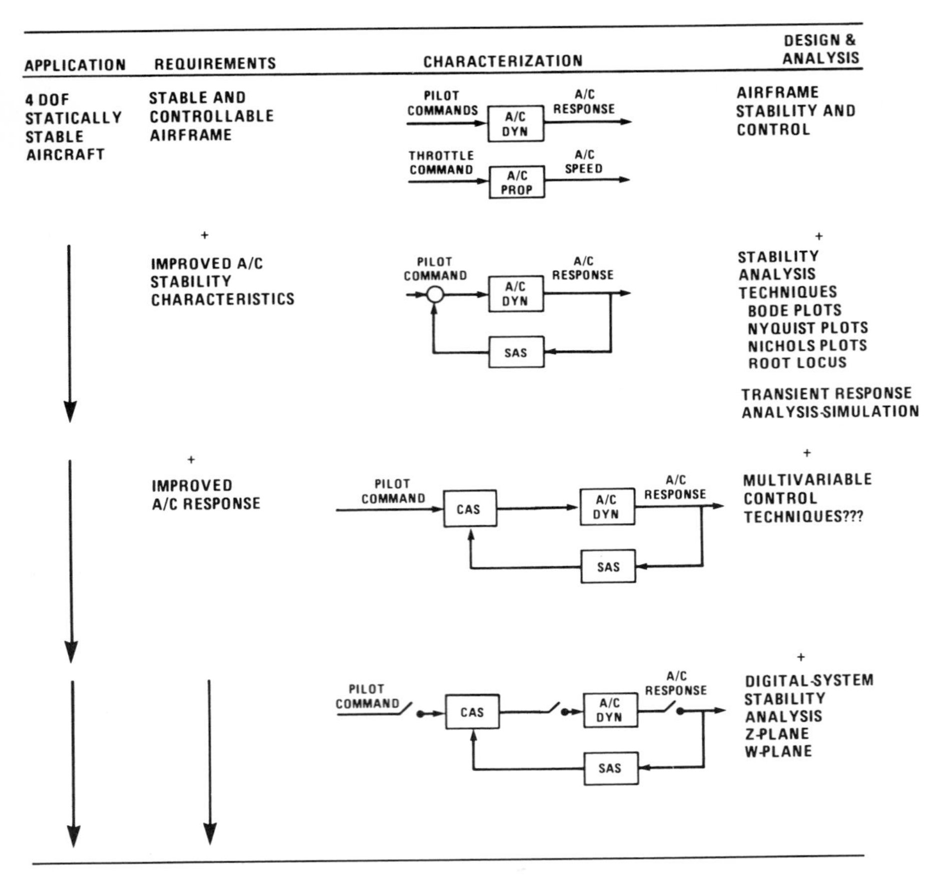

APPLICATION	REQUIREMENTS	CHARACTERIZATION	DESIGN & ANALYSIS
4 DOF STATICALLY STABLE AIRCRAFT	STABLE AND CONTROLLABLE AIRFRAME	PILOT COMMANDS → A/C DYN → A/C RESPONSE; THROTTLE COMMAND → A/C PROP → A/C SPEED	AIRFRAME STABILITY AND CONTROL
	+ IMPROVED A/C STABILITY CHARACTERISTICS	PILOT COMMAND → A/C DYN → A/C RESPONSE; SAS	+ STABILITY ANALYSIS TECHNIQUES: BODE PLOTS, NYQUIST PLOTS, NICHOLS PLOTS, ROOT LOCUS; TRANSIENT RESPONSE ANALYSIS-SIMULATION
	+ IMPROVED A/C RESPONSE	PILOT COMMAND → CAS → A/C DYN → A/C RESPONSE; SAS	+ MULTIVARIABLE CONTROL TECHNIQUES???
		PILOT COMMAND → CAS → A/C DYN → A/C RESPONSE; SAS	+ DIGITAL-SYSTEM STABILITY ANALYSIS Z-PLANE W-PLANE

It is useful to view the utility issue from the perspective of the flight control system application, particularly as flight control systems have evolved over the years. In depicting that history, the essential items to consider are the aircraft application, the performance requirements, and the control approach. These items determine the utility of the design and analysis techniques.

The first row of Table I represents the aircraft application, performance requirements, control approach, and design and analysis techniques for early aircraft up through aircraft built in the 1950s. For these applications the airframe was designed to provide stability and control for the three attitude degrees of freedom (DOF) and the propulsion system was designed for speed control. The control approach was open loop and design and analysis techniques were airframe oriented. Feedback control design techniques representative of classical techniques are shown for systems described by row 2 of Table I. In this case also, the airframe was designed to provide 3 DOF attitude stability and control, and the propulsion system provide speed control. Inadequate airframe designs or the promise of improved performance resulted in feedback systems that were used to augment stability. The most prevalent example of such a stability augmentation system is a yaw damper. The introduction of feedback control required additional design and analysis techniques, particularly those which addressed the stability characteristics of feedback control systems. Because the feedback control design was very simple, involving only one sensor variable and one surface command, classical techniques were very effective and led to the acceptance of stability margins as flight control system design specifications

Additional demands were then placed on flight control systems in the form of command augmentation systems as shown in row 3 of Table I. The airframe application still remained the same with or without a need for stability augmentation. Handling quality tests determined that command augmentation provided better handling qualities as exemplified by the acceptance of rate command systems or C^* systems. Traditional design techniques were still very adequate for design since, despite an increase in the number of sensor and surface pairs, the design could be performed one pair at a time because of the loose dynamic coupling.

The introduction of command augmentation, however, initiated the application of modern multivariable techniques, particularly model following approaches. These techniques promised to facilitate flight control system design thus producing better designs. Despite the promise, they were not widely accepted by practical control system designers.

Fly-by-wire (FBW) systems, particularly digital FBW systems, as shown in row 4 of Table I, brought new issues to flight control design. New techniques were developed and utilized to ensure that digital systems performed as closely as possible to their analog counterparts. In addition, the availability of a digital computer and its associated "unlimited" computational capability on board the aircraft encouraged more application of modern techniques which promised better performance. Again these techniques were not widely accepted by practical control designers.

The systems described by row 4 of Table I represent the state of the art of today's production aircraft. Table II presents characterizations for current experimental and prototype aircraft and projected production systems. The introduction of the control configured vehicle (CCV) concept has had a dramatic effect on flight control systems. In a CCV aircraft, the flight control system is not merely augmenting stability or improving performance, but is providing a flight critical stabilizing and control function. The criticality of the flight control system in a CCV application has intensified the need for efficient and reliable design and analyses techniques. CCV aircraft, in themselves however, do not possess dynamic coupling levels, which make classical design techniques intractable. In addition, classical techniques directly address stability issues and have therefore been much more attractive to a designer for CCV control designs.

Direct force control, made possible by additional surfaces and thrust vectoring of the propulsion system, as characterized by row 2 of Table II introduced a flight control design application that benefits from multivariable control techniques. In this application, it is difficult to eliminate closely coupled dynamics in the airframe design. Interaction with the propulsion system can magnify the coupling. For this application, the large number of control inputs and the close coupling of dynamics can easily overwhelm classical "one loop at a time" techniques.

The problem is projected to worsen with the integration of flight control and other avionics subsystems as shown in row 3 of Table II and Fig. 1. In the sections that follow, we discuss design and analysis techniques that have been

Table II. Aircraft Control Systems--Current Experimental Systems and Projected.

APPLICATION	CURRENT REQUIREMENTS	CHARACTERIZATION	DESIGN & ANALYSIS
CCV	STABLE AND CONTROLLABLE CLOSED LOOP AIRCRAFT SYSTEM	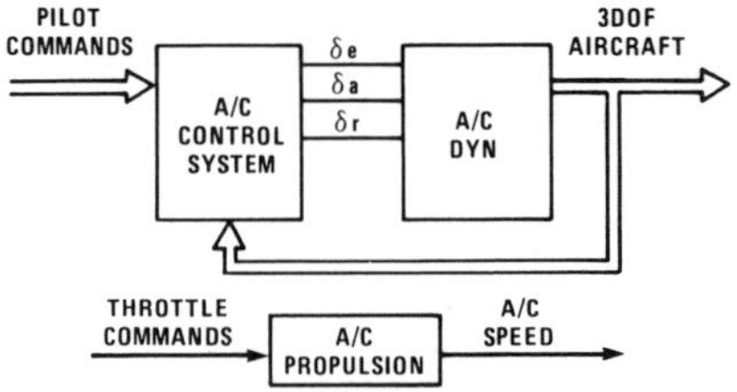 PILOT COMMANDS → A/C CONTROL SYSTEM → δe, δa, δr → A/C DYN → 3DOF AIRCRAFT (feedback to A/C CONTROL SYSTEM); THROTTLE COMMANDS → A/C PROPULSION → A/C SPEED	CLASSICAL STABILITY ANALYSIS TECHNIQUES, MULTIVARIABLE CONTROL TECHNIQUES??, TRANSIENT RESPONSE ANALYSIS-SIMULATION
CCV + 6DOF	+ IMPROVED MANEUVERABILITY CAPABILITY	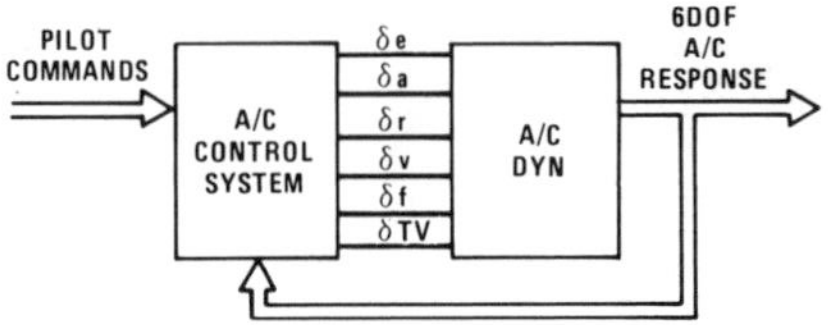 PILOT COMMANDS → A/C CONTROL SYSTEM → δe, δa, δr, δv, δf, δTV → A/C DYN → 6DOF A/C RESPONSE (feedback to A/C CONTROL SYSTEM)	MULTIVARIABLE CONTROL TECHNIQUES?
↓	INTEGRATE WITH GUIDANCE FIRE CONTROL PROPULSION CONTROL WEAPON DELIVERY STRUCTURAL CONTROL FOR IMPROVED PERFORMANCE AND RELIABILITY AT LOWER POWER, SIZE, WEIGHT AND COST	SEE FIGURE 1	MULTIVARIABLE CONTROL TECHNIQUES MULTIVARIABLE OPTIMIZATION TECHNIQUES

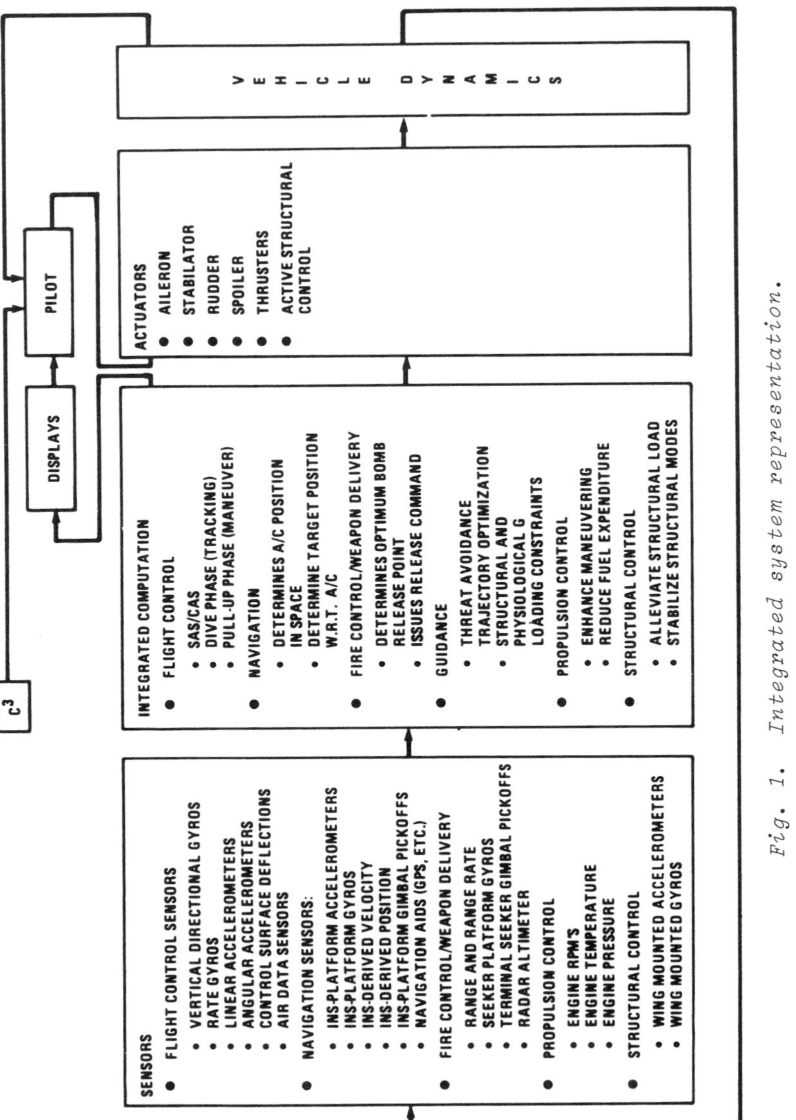

Fig. 1. Integrated system representation.

developed to address these multivariable control problems. In the development of these techniques, which is still ongoing, the goal has been an efficient flight control design and analysis capability which addresses the following:

(i) multi-input closely coupled dynamic systems;

(ii) conventional design specifications, particularly stability margin and high frequency attenuation;

(iii) the impact of unmodeled or uncertain dynamics on system performance;

(iv) digital realizations.

These items are addressed in varying degrees of detail in the techniques that are described and illustrated by design examples.

II. DISCUSSION OF TECHNIQUES

Various approaches exist for the design of multivariable flight control systems. This section presents a brief review of certain of these approaches followed by summary descriptions of the techniques with illustrative examples.

A. *REVIEW OF APPROACHES*

One approach is to use a classical single-input, single-output (SISO) technique for the design of one control loop at a time. In this approach the design of each individual loop is carried out on the basis of that loop's input-output pair and its effect on the input-output pairs for the other loops. This approach can be useful for certain problems, but its capability is severely limited for highly coupled multivariable systems. Furthermore, analysis of multivariable feedback systems with SISO techniques can give misleading results.

An approach to extending SISO techniques to multivariable systems makes use of approximate decoupling and the Inverse Nyquist Array (INA) methodology [1,2]. This approach can be useful for systems that are sufficiently decoupled to be naturally diagonally dominant. But, forcing the loop dynamics to be diagonally dominant appears to impose undue restrictions on the design. Furthermore, the diagonalization or near-diagonalization process can yield highly misleading conclusions concerning robustness with respect to design model uncertainty, by which we mean all the uncertainty between the design model and the actual operating system.

Another approach to extending SISO frequency domain techniques to multivariable systems uses the Characteristic Loci methodology [3,4]. In this approach characteristic gain functions and characteristic frequency functions are defined in terms of appropriate matrix-valued rational functions of a complex variable to generalize the Nyquist-Bode and root locus methods used for SISO systems. This approach is appealing because of the mathematical insights that it can provide. But, it also can yield misleading conclusions concerning robustness with respect to design model uncertainty. Another drawback to this technique and the INA technique is that they are limited to square systems, that is, systems that have the same number of outputs as inputs. This could represent a severe limitation of the design process.

Alternatives to the frequency domain methods include the modal control approach and the linear-optimal approach. Both of these approaches commonly deal with state space formulations which permit treatment of nonsquare systems with ease.

But, the deficiencies associated with these approaches are generally related to key issues that are most naturally expressed in frequency domain terms.

The modal control approach consists of choosing feedback gains so that the closed-loop system has desired eigenvalues and eigenvectors [5,6]. This approach can be useful when design requirements can be easily expressed in terms of desired closed-loop modal characteristics. But, in many cases the system design requirements cannot be so simply expressed. This is especially true for design requirements associated with tolerance to design model uncertainty.

In the linear-optimal control approach, controllers are determined that minimize a performance index which is the integral of a cost function. This cost function is a sum of quadratic terms in the states and controls. The controller uses feedback of all the states of the system or estimates of these states if all the states cannot be measured. This approach appears to be the most widely applicable, but there are certain difficulties involved in its use. The first issue that arises is that of selecting appropriate weighting matrices for the cost function. A second issue that arises is that linear-optimal controllers often have inadequate high-frequency attenuation and excessive bandwidth. Another issue is the care required in the design of an estimator or observer so that the guaranteed stability margins of optimal full state feedback control are nearly preserved when the output of the estimator or observer is used in the feedback path.

A major deficiency in each of the above approaches is the lack of guaranteed robustness with respect to design model uncertainty. A viable approach to the analysis of this robustness is the use of singular values and certain characterizations of the design model uncertainty. Since there is no existing synthesis technique that corresponds directly to this analysis technique, it appears that such an analysis must be incorporated in the design process. That is, preliminary designs should be subjected to such an analysis, and if this analysis shows the design to be inadequate, then further iterations of the design should be made guided by the robustness analysis results.

B. DESCRIPTION OF LINEAR-MULTIVARIABLE DESIGN TECHNIQUES AND ILLUSTRAVE EXAMPLES

The robustness of several of the illustrative examples will be analyzed using singular value analysis. This method will be briefly summarized before proceeding to the description of the design techniques.

As indicated above, a critical property of feedback systems is their ability to maintain performance in the face of uncertainties. In particular, it is important that a closed-loop system remain stable despite differences between the model used for design and the actual plant. These differences result from variations in modeled parameters as well as plant elements which are either approximated, aggregated, or ignored in the design model. The robustness requirements of a linear feedback design are often specified in terms of desired gain and phase margins and bandwidth limitations associated with loops broken at the input to the plant actuators [7,8]. These specifications reflect in part the classical notion of designing controllers that are adequate for a set of plants constituting a frequency-domain envelope of transfer functions [9].

The bandwidth limitation provides insurance against the uncertainty that grows with frequency due to unmodeled or aggregated high-frequency dynamics.

The Nyquist or Inverse Nyquist diagram (polar plots of the loop transfer function) provides a means of assessing stability and robustness at a glance for SISO systems. Multivariable generalizations of the scalar Nyquist, Inverse Nyquist, and Bode analysis methods can be developed from a basic result on robustness properties of linear systems expressed in terms of singular values.

Detailed discussions of the concepts of singular values and singular vectors are given in [10] and [11]. For simplicity, these concepts will be briefly described here for square matrices, although the concepts apply to arbitrary matrices.

1. Singular Values

The singular values σ_i of a complex $n \times n$ matrix A are the nonnegative square roots of the eigenvalues of A^*A, where A^* is the conjugate transpose of A. The (right) eigenvectors v_i of A^*A and r_i of AA^* are the right and left singular vectors, respectively, of A. These may be chosen such that

$$\sigma_i r_i = Av_i, \quad i = 1, \ldots n \tag{1}$$

$$\sigma_1 \leq \sigma_2 \leq \ldots \leq \sigma_n$$

and the $\{r_i\}$ and $\{v_i\}$ form orthonormal sets of vectors. The singular value decomposition of A is

$$A = R \, \Sigma \, V^*, \tag{2}$$

where

$$R = [r_1, r_2, \ldots, r_n], \quad V = [v_1, v_2, \ldots, v_n], \tag{3}$$

and Σ is the diagonal matrix with diagonal elements σ_1, $\sigma_2, \ldots, \sigma_n$. The minimum and maximum singular values have special significance and will be denoted here by $\underline{\sigma}(A)$ and $\overline{\sigma}(A)$, respectively. These singular values derive their special significance from the relations

$$\underline{\sigma}(A) = \min_{\|x\|=1} \|Ax\|, \quad \overline{\sigma}(A) = \max_{\|x\|=1} \|Ax\|, \tag{4}$$

and $\overline{\sigma}(A)$ is the spectral norm of A. The singular values give an accurate measure of how close A is to being singular. The ratio $\underline{\sigma}/\overline{\sigma}$ is known as the condition number with respect to inversion. The eigenvalues of A generally fail to provide such an accurate measure. The magnitudes of the eigenvalues of A are bounded below by $\underline{\sigma}(A)$ and above by $\overline{\sigma}(A)$. But, the magnitude of the smallest eigenvalue can be much larger than $\underline{\sigma}(A)$, and the magnitude of the largest eigenvalue can be much smaller than $\overline{\sigma}(A)$. This key difference between singular values and eigenvalues is the reason that singular value analysis provides an adequate measure of robustness and that eigenvalue analysis is inadequate.

Now consider the linear feedback system shown in Fig. 2 where G(s) is the nominal loop transfer matrix and L(s) is a perturbation matrix that is nominally zero and represents the deviation between G(s) and the real plant. A reasonable

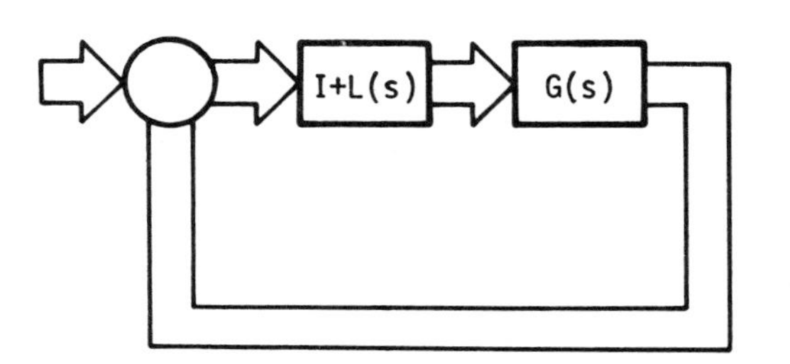

Fig. 2. Linear feedback system with multiplicative perturbation.

measure of the robustness of this feedback system is the magnitude of the perturbation that may be tolerated without causing instability. Taking the magnitude of $L(s)$ to be its spectral norm, the following basic result was obtained by Doyle [12].

If a system, such as that shown in Fig. 2, satisfies the following conditions:

(i) $G(s)$ and $L(s)$ are $n \times n$ rational square matrices,

(ii) $\det G(s) \neq 0$,

(iii) $L(s)$ is stable,

(iv) the nominal closed-loop system $H = G(I + G)^{-1}$ is stable,

then the system is stable for all perturbations that satisfy

$$\underline{\sigma}(I + G^{-1}(s)) > \overline{\sigma}(L(s)) \tag{5}$$

for all s in the classical Nyquist D-contour consisting of the segment of the imaginary axis from $-jR$ to $+jR$ and the semicircle of radius R in the right half-plane with R chosen sufficiently large.

A similar result holds for a system with the perturbation shown in Fig. 3. In this case the quantity $I + G(s)^{-1}$ in (5) is replaced by the quantity $I + G(s)$.

The singular values also have useful graphical interpretations. Consider the dyadic expansion

$$H^{-1} = I + G^{-1} = \sum_{i=1}^{n} \sigma_i r_i v_i^*, \qquad \sigma_1 \leq \sigma_2 \leq \cdots \leq \sigma_n, \tag{6}$$

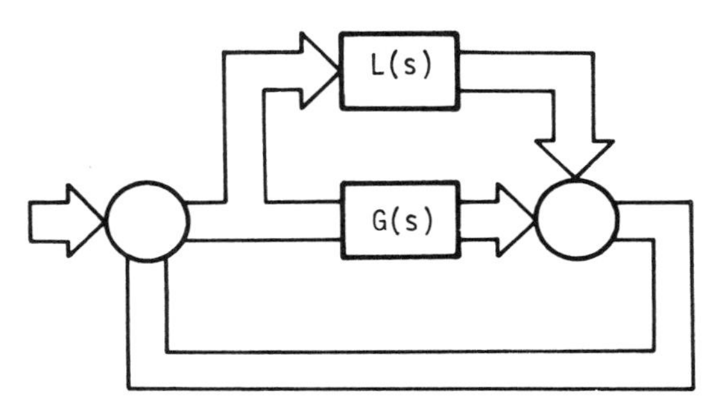

Fig. 3. Linear feedback system with additive perturbation.

where the σ_i, r_i, and v_i are the singular values, and left and right singular vectors, respectively, of $I + G^{-1}$. This is an alternative form of the singular value decomposition in Eq.(2).

It has been shown [13] that the eigenvalues and eigenvectors of a rational matrix are continuous (though generally not rational) functions of frequency. Since singular values and vectors are just special cases, $\sigma_i(j\omega)$, $r_i(j\omega)$, and $v_i(j\omega)$ are also continuous functions of ω.

Since

$$H = (I + G^{-1})^{-1} = \Sigma(1/\sigma_i)v_i r_i^*, \tag{7}$$

the values $1/\sigma_1(j\omega)$ and $1/\sigma_n(j\omega)$ give the maximum and minimum possible magnitude responses to an input sinusoid at frequency ω. In this sense, a plot of these singular values versus frequency may be thought of as a multivariable generalization of the Bode gain plot. Plots of this type will be referred to as σ-plots.

Another useful graphical interpretation analogous to the scalar Inverse Nyquist diagram may be constructed by noting that

$$\begin{aligned} G^{-1} &= \Sigma\sigma_i r_i v_i^* - I = \Sigma\sigma_i r_i v_i^* - \Sigma v_i v_i^* \\ &= \Sigma(\sigma_i r_i - v_i)v_i^* = \Sigma\beta_i g_i v_i^*, \end{aligned} \tag{8}$$

where $\beta_i g_i = \sigma_i r_i - v_i$ with β_i real and $||g_i|| = 1$ for all i.

Let z_i be defined implicitly as a function of σ_i and β_i by the quadratic equation

$$z_i^2 + \left(1 + \beta_i^2 - \sigma_i^2\right)z_i + \beta_i^2 = 0. \tag{9}$$

By plotting the $z_i(j\omega)$ $(i = 1,\ldots, m)$ for frequencies of interest, a plot analogous to the scalar Inverse Nyquist plot is generated. While phase does not have the conventional meaning on these plots, the more important notion of distance from the critical point preserves its importance. These plots will be referred to as z-plots.

Concepts such as M-circles are also obvious in this context. The minimum value of M is given by

$$M_m = \max_{\omega} [1/\sigma_1(j\omega)]. \tag{10}$$

Similar results may be obtained for additive perturbations by working with I + G rather than $I + G^{-1}$. In this case a diagram is generated which is analogous to the scalar Nyquist diagram.

Singular values offer no encirclement condition to test for right half-plane poles. But this is not a major deficiency because there are other simple techniques to assess the stability of the nominal system.

2. *Deficiencies in Existing Approaches*

The single-loop-at-a-time approach involves first of all the selection of loops. That is, for a system with a given number of inputs, it is assumed that there are at least that many outputs of interest and the first question that arises is what outputs should be paired with the inputs. For some systems this choice is obvious from the system characteristics. In highly coupled systems the choice may be difficult. Once the pairs have been selected, it is necessary to choose the sequence of loop closures. This choice also can be clearly dictated for some systems but not for others. The design then proceeds by closing one loop at a time, generally conducted in an iterative fashion. The robustness of the final closed-loop system is then examined by breaking one loop at a time and determining the stability margins.

The following example illustrates that analyzing robustness by breaking one loop at a time can be very misleading. The example was chosen to demonstrate this point only and does not represent any particular physical system. In fact, the controller considered is not representative of a good single-loop-at-a-time design. The loop transfer matrix for the example is

$$G(s) = \frac{1}{s^2 + 100}\begin{bmatrix} s - 100 & 10(s + 1) \\ -10(s + 1) & s - 100 \end{bmatrix}. \tag{11}$$

The open-loop poles are at $\pm 10j$, and with identity feedback the closed-loop poles are both at -1. Single loop breaking analysis for either loop indicates that the phase margin is 90° and the gain margin is $\pm\infty$ db. For comparison, the corresponding z-plot is shown in Fig. 4. It is not a plot of a rational

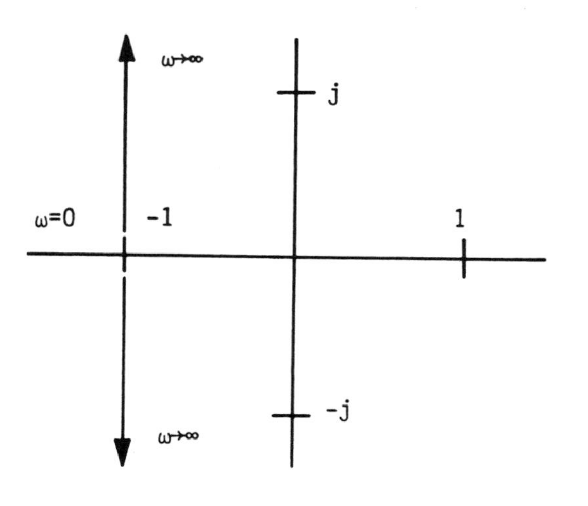

Fig. 4. z-Plot for the first example.

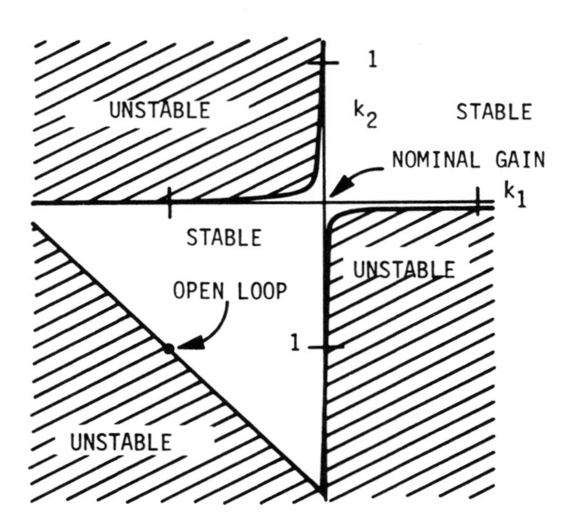

Fig. 5. Stability domain for the first example.

function, so it may appear somewhat unusual. The important feature is the proximity of the plot to the critical point -1, which indicates a lack of robustness.

The discrepancy between these two robustness indications can be easily understood by considering a diagonal perturbation

$$L = \begin{bmatrix} k_1 & 0 \\ 0 & k_2 \end{bmatrix}, \tag{12}$$

where k_1 and k_2 are constants.

Then regions of stability and instability may be plotted in the (k_1, k_2) plane as has been done in Fig. 5. The open-loop point corresponds to $k_1 = k_2 = 0$. Breaking each loop individually examines stability along the k_1, k_2 axes where robustness is good, but misses the close unstable regions caused by simultaneous changes in k_1 and k_2. Thus, single loop analysis is not a reliable way of testing robustness.

The approach taken with the Inverse Nyquist Array (INA) methodology attempts to extend SISO techniques to multivariable systems. In this approach the key feature is the use of diagonal dominance of the INA. Computer-aided displays of the Nyquist plots of the elements of the inverse of the loop transfer matrix can be used to assess the closeness to decoupling. These plots can be examined, and the information obtained can be useful in the selection of appropriate input-output pairs. If diagonal dominance cannot be achieved by this selection process, the methodology suggests techniques for introducing compensation to achieve the desired dominance. Once the diagonal dominance of the INA is achieved, design of diagonal feedback is accomplished with SISO methods. References [1] and [2] are recommended for detailed descriptions of this approach.

The Characteristic Loci methodology uses multivariable generalizations of the open-loop gain as a function of frequency and the closed-loop characteristic frequency as a function of gain. In this technique an inner-loop is designed on the basis of characteristic frequency as a function of gain to serve as a starting point for an outer-loop design that provides sufficient feedback gain to ensure satisfactory performance. This approach also uses computer-aided displays. In this case the displays of interest are the loci of characteristic gain and characteristic frequency. Detailed descriptions of this approach can be found in [3] and [4].

The following simple example was constructed to illustrate deficiencies associated with the Inverse Nyquist Array and Characteristic Loci approaches.

Consider the system with loop transfer matrix

$$G(s) = \frac{1}{(s + 1)(s + 2)} \begin{bmatrix} -47s + 2 & 56s \\ -42s & 50s + 2 \end{bmatrix}. \tag{13}$$

Assuming identity feedback, the closed-loop poles are at -2 and -4. This system may be diagonalized by introducing constant compensation. Let

$$U = \begin{bmatrix} 7 & 8 \\ 6 & 7 \end{bmatrix} \tag{14}$$

and

$$V = U^{-1} = \begin{bmatrix} 7 & -8 \\ -6 & 7 \end{bmatrix}. \tag{15}$$

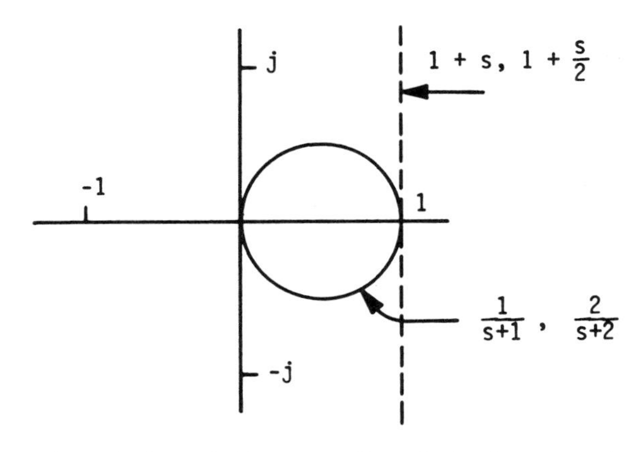

Fig. 6. Nyquist and Inverse Nyquist diagram for the second example.

Then letting

$$\hat{G} = VGU = \begin{bmatrix} \frac{1}{s+1} & 0 \\ 0 & \frac{2}{s+2} \end{bmatrix}, \tag{16}$$

the system may be rearranged so that

$$\begin{aligned} H &= G(I + G)^{-1} = U\hat{G}V(I + U\hat{G}V)^{-1} \\ &= U\hat{G}(I + \hat{G})^{-1}V = U\hat{G}(I + \hat{G})^{-1}V. \end{aligned} \tag{17}$$

This yields a diagonal system that may be analyzed by scalar methods. In particular, under the assumption of identity feedback $\hat{G}$ represents the new loop transfer matrix. Because U and V represent a similarity transformation, the diagonal elements of G are also the eigenvalues of $\hat{G}$ so that the decoupling or dominance approach and eigenvalue or characteristic loci approach would generate the same Nyquist or Inverse Nyquist plot shown in Fig. 6. Only a single locus is shown since the contours of $1/(s + 1)$ and $2/(s + 2)$ are identical. The tempting conclusion that might be reached from these plots

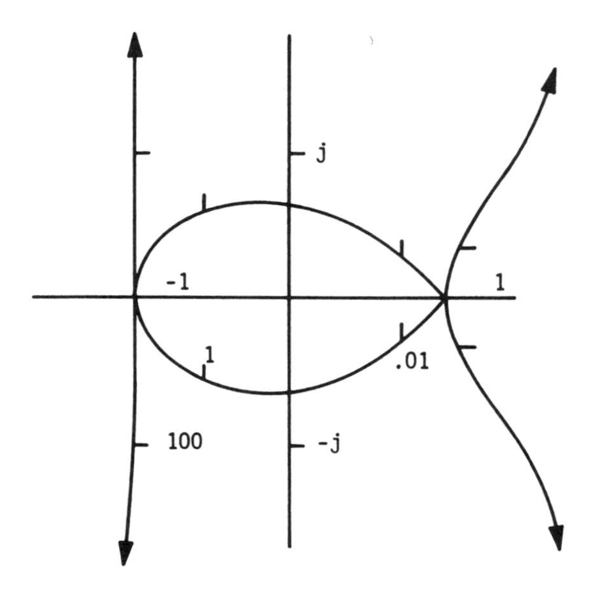

Fig. 7. z-Plot for the second example.

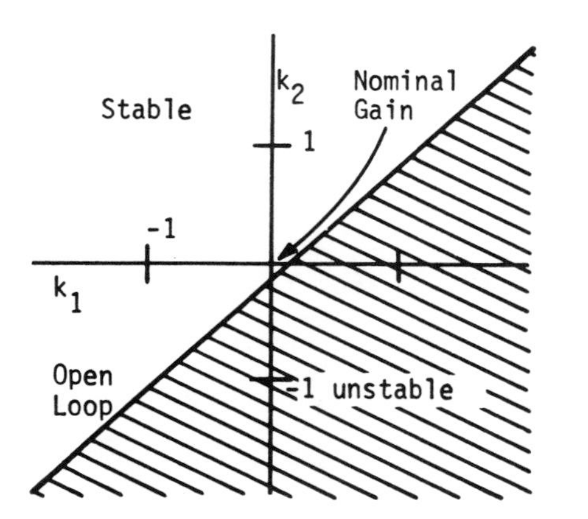

Fig. 8. Stability domain for the second example.

is that the feedback system is very robust with apparent margins of $\pm\infty$ db in gain and 90°+ in phase. The closed-loop pole locations would seem to support this.

This conclusion, however, would be wrong. The z-plot for $I + G^{-1}$ is shown in Fig. 7 and there is clearly a serious lack of robustness. The (k_1, k_2) - plane stability plot for this example is shown in Fig. 8. Neither the diagonal dominance nor eigenvalue approaches indicate the close proximity of an unstable region. This failure can be attributed to two causes.

First, the eigenvalues of a matrix do not, in general, give a reliable measure of its distance (in a parametric sense) from singularity, and so computing the eigenvalues of $G(s)$ [or $I + G(s)$] does not give an indication of robustness. Using eigenvalues rather than singular values will always detect unstable regions that lie along the $k_1 = k_2$ diagonal, but may miss regions such as the one in Fig. 8.

Second, when compensation and/or feedback is used to achieve dominance, the "new plant" includes this compensation and feedback. Because of this, no reliable conclusions may be drawn from this "new plant" concerning the robustness of the final design with respect to variations in the actual plant. It is important to evaluate robustness where there is uncertainty.

3. *Modal Control*

The modal control approach provides a method of finding feedback gains that yield certain desired closed-loop eigenvalues and eigenvectors. The modal concept is common to many flight control applications. For example, the lateral-directional axis of a fixed wing aircraft has three dominant modes (roll, dutch roll, and spiral) and the longitudinal axis has two dominant modes (short period and phugoid). Desired modal properties are often specified in terms of eigenvalues (frequencies, damping ratios, time constants) and associated coupling or decoupling of responses. Handling quality criteria for the lateral axis are expressed in such terms. The dutch roll mode should have an eigenvalue with desired frequency and damping ratio and an eigenvector that ideally contains nonzero components for sideslip and yaw rate only. The roll mode should have an eigenvalue corresponding to a desired time constant and an eigenvector which ideally has roll rate as its only nonzero component. The spiral mode should have a small real eigenvalue and an eigenvector that ideally has only bank angle as its nonzero component. This approach can be used with state feedback or output feedback.

The details of this approach may be summarized as follows. Consider the linear multivariable feedback system

$$\dot{x} = Ax + Bu, \tag{18}$$

$$y = Cx, \tag{19}$$

$$u = -Ky, \tag{20}$$

where the state vector x is n-dimensional, the control vector u is m-dimensional, the output vector y is p-dimensional, and A, B, C, and K are matrices of appropriate dimensions, and it is assumed that

$$\text{rank}(B) = m, \tag{21}$$

$$\text{rank}(C) = p \leq n. \tag{22}$$

An eigenvalue-eigenvector pair for the closed-loop system is denoted as (λ_i, v_i), where

$$A_c v_i = \lambda_i v_i \tag{23}$$

and A_c is the closed-loop system matrix

$$A_c = A - BKC. \tag{24}$$

Only certain pairs are achievable. An achievable pair satisfies

$$Av_i + Bw_i = \lambda_i v_i, \tag{25}$$

where w_i is a vector with dimension = min(m, p).

In the case of state feedback, C may be chosen to be the identity, and (23) and (25) yield

$$w_i = Kv_i. \tag{26}$$

Letting Λ be the diagonal matrix with elements λ_i on the diagonal, V be the matrix with its ith column being the eigenvector v_i, and W be the $(m \times n)$ matrix with its ith column being w_i, Eq. (25) yields

$$V\Lambda - AV = BW, \tag{27}$$

and Eq. (26) yields

$$W = KV, \tag{28}$$

which can be solved for K, i.e.,

$$K = WV^{-1}. \tag{29}$$

Complex eigenvector pairs can be rotated to provide a real matrix K. In the case of output feedback, Eq. (28) is replaced by

$$W = KCV, \tag{30}$$

and if $p < n$, the matrix CV has rank less than n so that (30) cannot be solved for K in general. One method for resolving this difficulty is to select only p eigenvalue-eigenvector pairs with $\hat{\Lambda}$ denoting the $(p \times p)$ diagonal matrix of eigenvalues and $\hat{V}$ denoting the $(n \times p)$ matrix of corresponding eigenvectors, and $\hat{W}$ denote the corresponding $(w \times p)$ matrix. Then K is given by

$$K = \hat{W}(C\hat{V})^{-1}, \tag{31}$$

subject to

$$\hat{V}\hat{\Lambda} - A\hat{V} = B\hat{W}. \tag{32}$$

In this case there is no constraint on the remaining n - p eigenvalue-eigenvector pairs, and they could be undesirable.

The design procedure initially involves choosing portions of v_i in order to eliminate certain state responses from a mode while emphasizing others and letting other responses (control or compensation) react arbitrarily.

For rank(B) = m, m-free parameters can be specified, one of which is the eigenvalue. Equation (25) can be rewritten

$$A_{ci}q_i = [(A - \lambda_i I), B]q_i = 0,$$

$$q = \begin{bmatrix} v_i \\ w_i \end{bmatrix}. \tag{33}$$

q_i is therefore a null space mapping of A_{ci}. A convenient tool for finding the relationship between v_i and w_i is contained in the singular value decomposition of A_{ci};

$$A_{ci} = X_i \Sigma_i Z_i^*, \tag{34}$$

where X_i is an n-by-n matrix containing columns of orthogonal left singular vectors of A_{ci} and Σ_i is an n-by-n + m matrix containing n-singular values σ's of A_{ci}

$$\Sigma_i = \left.\left[\begin{array}{cccc|ccc} \sigma_1 & 0 & \cdots & 0 & 0 & \cdots & 0 \\ 0 & \sigma_2 & & & & & \\ \vdots & & \ddots & & & & \\ 0 & & & \sigma_n & 0 & \cdots & 0 \end{array}\right]\right\}n \qquad \underbrace{}_{n}\ \underbrace{}_{m}$$

$$= [\overline{\Sigma}_i, [0]]; \quad \overline{\Sigma}_i \text{ is } n \times n \text{ diagonal.}$$

Z_i is an n + m-by-n + m matrix containing n + m orthogonal right singular vectors of A_{ci}. By rearranging (34),

$$A_{ci}Z_i = X_i\Sigma_i, \tag{35}$$

and noting that the last m-columns of the $X_i\Sigma_i$ product are null, we find the appropriate null space for A_{ci} by using the last m-columns of Z_i;

$$A_{ci}\overline{Z}_i = 0, \tag{36}$$

where $\overline{Z}_i$ is defined as

$$Z_i = [\underbrace{Z_i}_{n} \quad \underbrace{\overline{Z}_i}_{m}]\}n + m. \tag{37}$$

The matrix $\overline{Z}_i$ is a set of m-orthonormal basis vectors spanning the null space of A_{ci}. Referring to (33), we have

$$q_i = \overline{Z}_i\alpha_i, \tag{38}$$

where α_i is an m-vector of linear coefficients not all of which can be zero.

Now we would like to select desired v_i and solve (33) for the corresponding w_i. This cannot be done in general if $m < n$, but an approximate solution can be obtained by minimizing a performance index

$$J = (v_{di} - v_i)^* Q(v_{di} - v_i) \tag{39}$$

subject to (33), where v_{di} is the desired eigenvector. By this process one can select arbitrarily the desired v_{di} and the resulting v_i that are attainable are the closest to those desired in the sense of (39). This can be accomplished by replacing v_i in (38) with $v_i = EZ_i\alpha_i$, where E is an n-by-n + m matrix and

$$E \triangleq [\underbrace{I}_{n} \quad \underbrace{0}_{m}]\}n,$$

and minimizing (39) with respect to α_i,

$$\alpha_i = \arg\min J = \left(\overline{Z}_i^* E^T Q E \overline{Z}_i\right)^{-1} \overline{Z}_i^* E^T Q V_{di}; \tag{40}$$

the appropriate W_i is found using (38)

$$W_i = \overline{E}\overline{Z}_i\alpha_i = \overline{E}\overline{Z}_i\left(\overline{Z}_i^* E^T Q E \overline{Z}_i\right)^{-1} \overline{Z}_i^* E^T Q V_{di}, \tag{41}$$

where $\overline{E}$ is an m-by-n + m matrix

$$E \triangleq [\underbrace{0}_{n} \quad \underbrace{I}_{m}]\}m.$$

4. *F-4 Design Example*

As an illustrative example this method was applied to the design of an inner-loop control law for the F-4 lateral axis. The data for this example are taken from [14]. The state space representation of this example is in the form of Eq. (18) with

$$x = \begin{bmatrix} p_s \\ r_s \\ \beta \\ \phi \\ \delta_r \\ \delta_a \end{bmatrix} \begin{matrix} \text{stability axis roll rate} \\ \text{stability axis yaw rate} \\ \text{angle of sideslip} \\ \text{bank angle} \\ \text{rudder deflection} \\ \text{aileron deflection} \end{matrix} \qquad u = \begin{bmatrix} \delta_{rc} \\ \delta_{ac} \end{bmatrix} \begin{matrix} \text{rudder command} \\ \text{aileron command} \end{matrix}$$

Matrices A and B are

$$A = \left[\begin{array}{cccc|cc} -0.746 & 0.387 & -12.9 & 0. & 0.952 & 6.05 \\ 0.024 & -0.174 & 4.31 & 0. & -1.76 & -0.416 \\ 0.006 & -0.9994 & -0.0578 & 0.0369 & 0.0092 & -0.0012 \\ 1. & 0. & 0. & 0. & 0. & 0. \\ \hline 0. & 0. & 0. & 0. & -20. & 0. \\ 0. & 0. & 0. & 0. & 0. & -10. \end{array}\right]$$

$$B = \left[\begin{array}{cc} 0. & 0. \\ 0. & 0. \\ 0. & 0. \\ 0. & 0. \\ \hline 20. & 0. \\ 0. & 10. \end{array}\right]$$

Open-loop poles

λ roll subsidence = -0.079

λ dutch roll = $-0.098 \pm j2.079$

λ spiral = -0.0063

λ rudder actuator = -20.0

λ aileron actuator = -10.0

From the point of view of fighter handling qualities, all four of the lateral axis closed-loop roots have desired values that can be taken from MIL-F8785B, as is done, for example, in [14]. The desired roots are

(a) roll subsidence mode = -4.0

(b) dutch roll mode = $-0.63 \pm j2.42$

(c) spiral mode = -0.05

Desired eigenvectors were selected to pair with the desired eigenvalues, and the method described above was used to compute nearest-attainable eigenvectors. The results are

(a) Roll subsidence mode ($e^{-4t}v_1$)

Desired v_{d1} = [1. 0 0 a a a]

Attainable v_1 = [1. -0.007 0 -0.25 0.13 -0.56]

(b) Dutch roll mode, real part $[e^{-0.63t}(\cos 2.42t)v_2]$

Desired v_{d2} = [0 a 1. 0 a a]

Attainable v_2 = [0 15.6 1. 0 7.86 -0.103]

(c) Dutch roll mode, imaginary part $[e^{-0.63t}(\sin 2.42t)v_3]$

Desired v_{d3} = [0 1. a 0 a a]

Attainable v_3 = [0 1. 6.16 0 -9.49 14.6]

(d) Spiral mode $(e^{-0.05t}v_4)$

Desired v_{d4} = [a a 0 1. a a]

Attainable v_4 = [-0.05 0.037 0 1. -0.0014 -0.0079]

A few comments are in order to explain these choices. Consider, for example, the roll subsidence mode. The desired eigenvector is taken to be v_{d1} = (1 0 0 a a a), which means that the mode should show up dominantly on roll rate, but not on yaw rate or sideslip (we want no sideslip buildup during turn entries). These are good basic handling quality considerations. The a's in the vector indicate that we do not care how much of the mode shows up on these components. Certainly, since $\phi = \int p_s dt$, some mode content has to be expected on element a_4 and, similarly, if the surfaces are actually controlling the mode, some mode content should also appear in a_5 and a_6. The linear projection which best achieves these objectives is shown as v_1 above. Note that we can satisfy our desires almost perfectly.

Similar arguments also apply to the dutch roll mode. In this case we want no oscillatory dutch roll content on roll rate and bank angle. This is a key handling quality requirement for all well-behaved lateral control laws.

In the case of the spiral, we want the mode to be predominantly bank angle (corresponding to steady turns) with no substantial sideslip component. The latter is a basic turn coordination requirement.

A sample design was carried out with the modal approach assuming that the output y consisted of the first four states. This output posed little difficulty, since our desire is embodied in four eigenvalue-eigenvector pairs. There is some concern with the remaining two pairs. It is desired to have these poles not be too far left from their open-loop values. This is not a problem in this case, since the trace of the open-loop system matrix is the same as the trace of the closed-loop system matrix. So the sum of the closed-loop eigenvalues is the same as the sum of the open-loop eigenvalues. Thus, for this example the sum of the remaining closed-loop poles is greater than their open-loop sum, and, in fact, each of these closed-loop poles is to the right of their

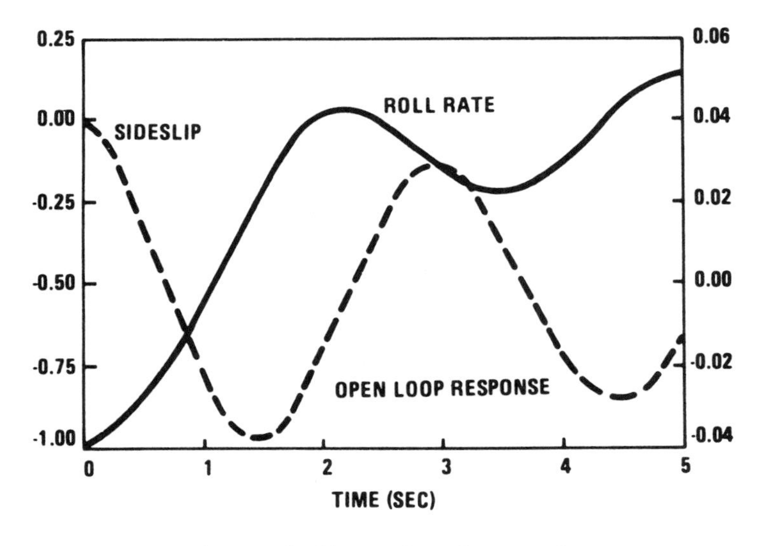

Fig. 9. F-4 lateral-directional open-loop response.

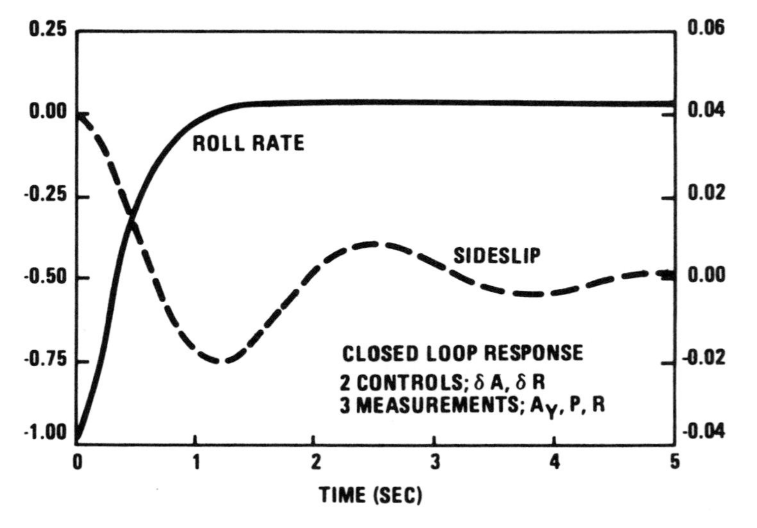

Fig. 10. F-4 lateral-directional closed-loop response.

open-loop values. The closed-loop values of λ_5 and λ_6 are -19.03 and -6.64 in comparison to open-loop values of -20 and -10.

An initial condition response for the open-loop system is shown in Fig. 9. A similar response is shown in Fig. 10 for the closed-loop system. Dramatic improvement in reduced cross coupling between the roll and dutch roll modes is evident. Also evident is the improved dutch roll damping and the roll response time.

It is a simple matter to use this method to determine the effects of different sensor combinations and compensators. This makes it a useful tool for tradeoff studies. Unfortunately there is no robustness guaranteed so that this aspect of the design must be analyzed independently.

5. *Linear-Optimal Control*

The linear-optimal control approach has been extensively discussed in the automatic control literature. The synthesis procedure in this approach usually starts with a mathematical model of the form

$$\dot{x} = Ax + Bu + \xi, \tag{42}$$

$$y = Cx + \theta, \tag{43}$$

where ξ and θ represent system and sensor noise that are assumed to be Gaussian, white, mutually independent, stationary, and zero-mean with covariances,

$$\mathrm{cov}[\xi(t);\ \xi(\tau)] = \Xi\delta(t - \tau),\quad \Xi \geq 0, \tag{44}$$

$$\mathrm{cov}[\theta(t);\ \theta(\tau)] = \Theta\delta(t - \tau),\quad \Theta \geq 0. \tag{45}$$

The control law is obtained by minimizing a performance index of the form

$$J = E\int_0^\infty (x^TQx + u^TRu)\,dt. \tag{46}$$

The resulting control law is

$$u = -Kx = -R^{-1}B^TP\hat{x}, \tag{47}$$

where P satisfies a Riccati equation and $\hat{x}$ is an estimate of x given by

$$\dot{\hat{x}} = A\hat{x} + Bu + F(y - C\hat{x}), \tag{48}$$

with $F = SC^T\Theta^{-1}$ and S satisfies another Riccati equation. In a deterministic version of this approach where it is assumed that there is no noise and C is the identity, the control law is given by (47) with $\hat{x}$ replaced by x. For this latter version, it is known that the controllers possess guaranteed stability margins of at least -6 db and $+\infty$ db in gain and at least 60° in phase at each input if R is chosen to be the identity. It is also known that these controllers have first-order attenuation at high frequencies. This property can be a problem for many flight control applications. Controllers for the nondeterministic version generally have higher order attenuation at high frequencies but they do not possess any guaranteed stability margins.

One difficulty with this approach is that of relating the performance index to design specifications. A method for selecting the weighting matrices Q and R on the basis of desired modal characteristics is described in [15]. This method is based on the asymptotic modal properties of optimal controllers as the weight on the control tends to zero. The designer can choose desired modes the same way as in the modal control approach and construct corresponding weighting matrices. The F-4 lateral-directional example was also treated by this method, and the resulting controller's closed-loop transient response was essentially the same as that shown in Fig. 9. This controller does possess the guaranteed stability margins.

6. *CH-47 Design Example*

In another illustrative example this method was used in conjunction with singular value analysis of robustness. The example treats the longitudinal degrees of freedom of the CH-47 helicopter. In forward flight, this vehicle exhibits coupled pitch attitude and vertical motion dynamics that must be controlled by coordinated action of two inputs. This vehicle is a tandem rotor machine whose physical characteristics and mathematical models are given in [16]. Control over vertical motions is achieved by simultaneous changes of blade angle-of-attack on both rotors (collective), while pitch and forward motions are controlled by changing blade angle differentially between the two rotors (differential-collective). These blade angle changes are transformed through rotor dynamics and aerodynamics into hub forces, which then move the machine.

Our objectives will be to design a command augmentation control law that achieves tight, noninteracting control of the vertical velocity and pitch attitude responses. A small perturbation linearized aircraft model should prove adequate for this purpose and is available from [16]. The state vector consists of the vehicle's basic rigid body variables x where $x = (V, \dot{z}, q, \theta)$ (forward velocity, vertical velocity, pitch rate, pitch angle). Two integrators are appended to achieve integral control of the primary responses, and controls are the collective and differential collective inputs described above, $u = (c, dc)$. Hence, the design model is

$$\begin{aligned} \dot{x} &= Ax + Bu, \quad A, B \text{ in } [16] \\ \dot{x}_5 &= -\dot{z} + \dot{z}_{cmd}, \\ \dot{x}_6 &= -\theta + \theta_{cmd}. \end{aligned} \tag{49}$$

The major approximations associated with this model are due to neglected dynamics of the rotors, to neglected nonlinearities in the blade angle actuation hardware, and to variations of the A and B matrices with operating point (flight condition variations). We shall treat modeling errors due to these approximations as sources of the perturbation $L(s)$ in Fig. 2 and shall attempt to make controllers robust with respect to them.

Elementary dynamic and aerodynamic analyses of rotating airfoils, hinged at the rotor hub, indicate that lift forces will not be transmitted to the hub instantaneously with collective changes in blade angle-of-attack but will appear only when the cone angle of the rotor has appropriately changed. The dynamics of the latter have been shown to be damped second-order oscillations with natural frequency equal to rotor speed and damping determined by somewhat uncertain aerodynamic effects [17]. Hence, rotor dynamics can be crudely represented by second-order transfer functions

$$g_R(s, \zeta) = \omega_R^2 \Big/ \left(s^2 + 2\zeta\omega_R s + \omega_R^2\right), \tag{50}$$

with $\omega_R = 25$ rad/sec and ζ conservatively confined to the range $0.1 \leq 1.0$. Because collective and differential-collective inputs both involve coning motions of the rotors, one such transfer function will appear in each control channel. Since these dynamics are neglected in Eq. (49), it then follows that any perturbed transfer function matrix computed from Fig. 2 will have the form

$$\hat{G} = G(I + L) = G \operatorname{diag}(g_R), \tag{51}$$

and hence,

$$L = \operatorname{diag}(g_R - 1), \tag{52}$$

$$\bar{\sigma}[L] = \max_{\zeta} \left| \frac{s^2 + 2\zeta\omega_R}{\left(s^2 + 2\zeta\omega_R s + \omega_R\right)} \right|. \tag{53}$$

The function was evaluated for a range of $s = j\omega$ values (with brute force maximization over ζ) and is shown by the solid lines in Fig. 11.

Figure 11 also shows an alternate bound for $\bar{\sigma}[L]$ derived from a general method developed by Safonov [18]. This bound is slightly more conservative than that given by (15) because it allows a larger class of perturbations corresponding to nonconstant ζ's in the specified range.

In addition to the dynamics of rotors, each control channel of the CH-47 also exhibits various nonlinearities that are neglected in the nominal design model. Of these, the rate limit nonlinearity imposes the greatest dynamic constraint on performance, and we consider bounds only for this one effect.

An approximate model for rate limits on the CH-47 is given by

$$\dot{u} = R_{lim} \text{SAT}[94(u_c - u)/R_{lim}], \tag{54}$$

where SAT( ) denotes the standard saturation nonlinearity, saturating at +1. Bounds for this model can be developed with Safonov's procedure by treating the SAT element as an uncertain component. These bounds will of course depend on the magnitudes of the arguments of the SAT function. The dashed lines in Fig. 11 depict the bounds for several values of η, where η denotes the magnitude bound assumed for the argument of the SAT function. Note that as η becomes large, the bound approaches unity at all frequencies. This is consistent with physical intuition, since the effective gain across rate-limited nonlinearities will approach zero for large signal levels.

The third major source of model uncertainty is the variation of A and B matrices with flight condition. Such "component" variation could again be translated into an overall bound for L(s) via Safonov's procedure. In this case, however, the result would be unduly conservative because coefficient variations tend to be highly correlated and are not arbitrary dynamical operators. A more direct way to compute the bound is to compute the loop transfer matrix for a number of representative flight conditions and then compute the

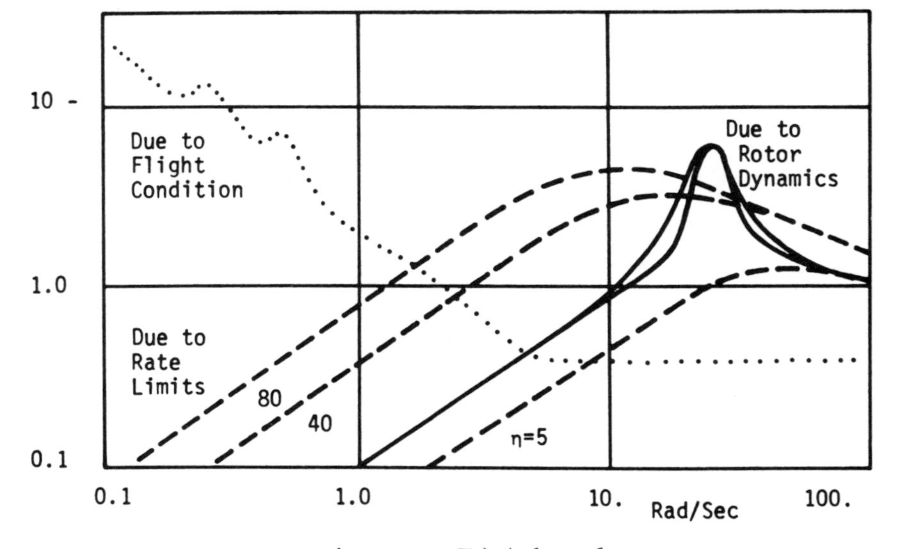

Fig. 11. $\bar{\sigma}(L)$ *bound.*

maximum (singular value) deviation. Results of this process are shown by the dotted line in Fig. 11. We see the initially surprising result that $\bar{\sigma}[L]$ becomes quite large at low frequencies. This happens because the basic helicopter's low-frequency modes are stable at some flight conditions and unstable at others. Theoretically, $\bar{\sigma}[G(j\omega)]$ will approach infinity for frequencies and flight conditions where these modes cross the $j\omega$-axis. This means that the perturbations exhibited by our plant are not necessarily stable and, hence, the stability-robustness result cited earlier fails to apply. We shall see later that stable controllers can still be obtained and that the ability to incorporate unstable L's in a generalized multivariable stability robustness theory appears to be an important research topic. For the moment, however, our designs will be restricted to individual flight conditions for which the dotted L's in Fig. 11 can be disregarded.

The uncertainty bounds shown in Fig. 11 indicate that the robustness criterion given by Eq. (5) imposes a "multivariable bandwidth" limitation on the feedback loop. Magnitudes of $L(j\omega)$ tend to be large beyond certain frequencies that require G^{-1} to be large, and consequently G must be small. This is most readily illustrated with a single-loop example where plots of the function $\underline{\sigma}[1 + g^{-1}]$ reduce to the inverse closed-loop frequency response, i.e.,

$$\underline{\sigma}[1 + g^{-1}] = \left[\frac{1 + g}{g}\right] = \left[\frac{1}{g_{cl}}\right]. \tag{55}$$

The condition that $\underline{\sigma}[1 + g^{-1}]$ be large then translates directly into the high-frequency "roll-off" requirement commonly imposed on classical control loops.

For illustrative purposes, single-loop and multiloop designs will be described.

The vertical velocity and pitch attitude motions of the nominal CH-47 model at hover uncouple naturally into two non-interacting channels: $(\dot{z}, x_5)$ controlled by (c), and (v, q, θ, x_6) controlled by (dc). The hover flight condition thus offers an attractive single-loop design case. Sigma plots for several trial pitch-motion controllers for this case are shown in Fig. 12. These controllers were all designed with the linear-optimal methodology and correspond to the following cost functional:

$$J = \int_0^\infty \left[(57.3\ x_6)^2 + \rho(dc)^2\right]dt, \tag{56}$$

with $\rho = 900.$, 9.0, 0.09, and 1.0, respectively, for the four trials. These weights were selected in accordance with the asymptotic procedure described in [19]. The asymptotic modes corresponding to these weights are a forward-speed mode with

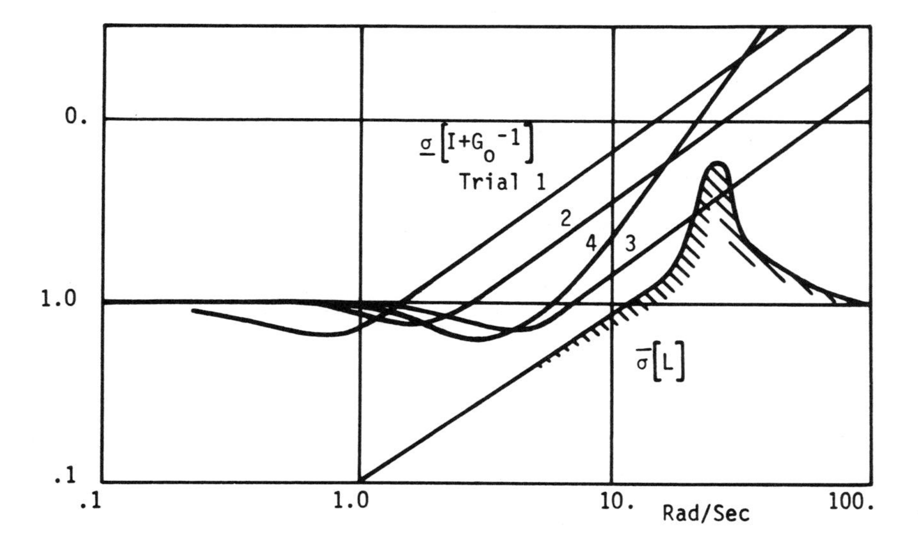

Fig. 12. Trial designs for pitch control at hover.

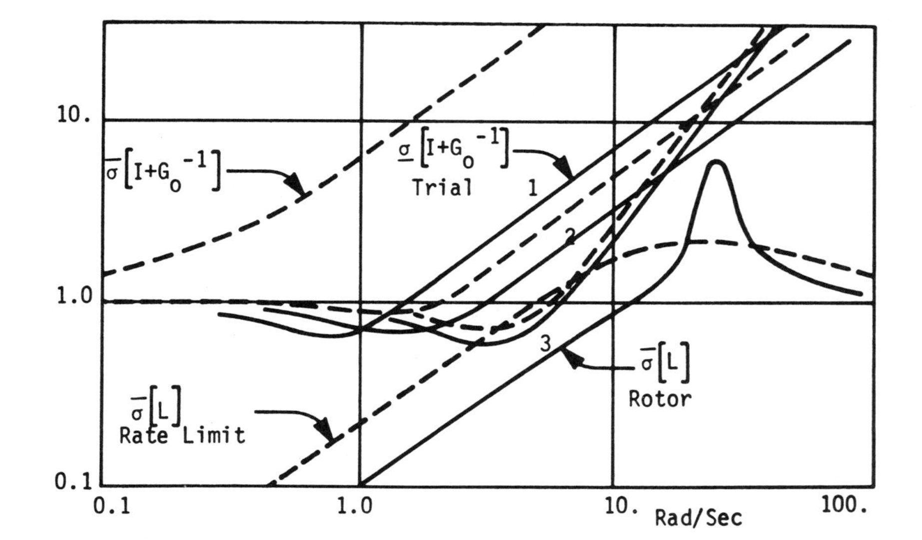

Fig. 13. Trial designs for pitch and vertical velocity control.

with an eigenvalue near the origin and a third-order Butterworth pattern for the remaining modes. As expected, bandwidth of these controllers increases with decreasing ρ and eventually violates the stability-robustness constraint imposed by neglected rotor dynamics (for the moment we ignore rate limit and flight condition variations). That this violation actually produces instabilities was verified by computing closed-loop roots of the trial controllers in the presence of the rotor. Trial 3 is unstable! Our options are therefore to restrict bandwidth approximately to trial 2 or to provide additional roll-off beyond the maximum 20 db/decade attenuation inherent in LQ-design [20]. The latter option is illustrated by trial 4, which uses a ρ-value somewhat smaller than trial 2 but includes a low-pass filter at $\omega = 12$ rad/sec to help avoid the rotor resonance peak. Note that the closed-loop frequency responses[1] are well-shaped for all pure LQ-trials and that trial 4 achieves extra bandwidth at the expense of slightly larger M-peaks.

The beauty of singular values is that the above stability-robustness analyses carry over without change to multivariable systems. This is illustrated in Fig. 13 with some two-channel trial designs at a 40-knot forward-speed flight condition. These controllers are again of the LQ-type, this time using the cost function

$$J = \int_0^\infty \left[(x_5)^2 + (57.3\ x_6)^2 + \rho_1 (c)^2 + \rho_2 (dc)^2 \right] dt, \qquad (57)$$

with $(\rho_1, \rho_2) = (10000, 900)$, $(9.0, 9.0)$, and $(1.0, 1.0)$ for the three trials shown. The distinction between Figs. 12 and

[1] *According to (55), these are given by the σ-plots of Fig. 12 viewed "upside down."*

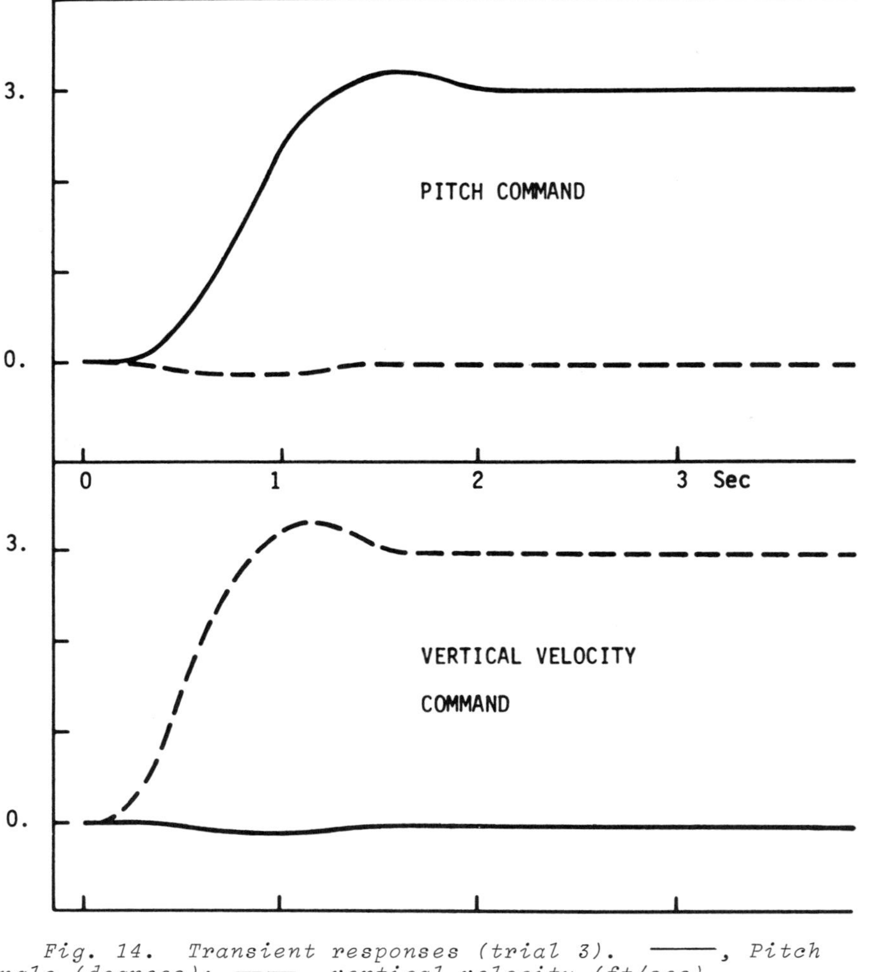

Fig. 14. Transient responses (trial 3). ———, Pitch angle (degrees); — —, vertical velocity (ft/sec).

13 is that Fig. 13 shows two sigma-plots for each trial, corresponding to the two singular values of $(I + G^{-1})$. For stability-robustness, the smaller of these values must fall above the σ-plot of L at all frequencies. The larger value is unspecified. However, in order to maximize bandwidth "in all directions," it is reasonable to adjust the relative weights

(ρ_1, ρ_2) such that the two singular values are approximately equal and then to push them jointly to as high a bandwidth as the $\bar{\sigma}[L]$ plot permits. (For the moment, we again use only neglected rotor dynamics for L.) This design philosphy is incorporated in the three trials of Fig. 13. The first trial has low bandwidth and substantial differences between the two singular values. These differences are reduced and bandwidth is increased in the next trial. The third trial serves to maximize bandwidth by using additional roll-off filters in each control channel.

As seen from these trials, singular value analyses appear to offer a convenient way to maximize multivariable bandwidth subject to stability-robustness limitations. The next design step is to achieve reasonable command responses from the resulting feedback loop. One way to do this is to place a command shaping filter ahead of the loop. For feedback loops with integral control on the primary responses, such sophistication is often unnecessary because commands inserted at the integrators [as shown in Eq. (49)] produce good transients. This is the case here, as evidenced by the responses of trial 3 to step attitude and step velocity commands shown in Fig. 14. Note that the loops are tight, well damped, and noninteracting as desired.

So far we have ignored model uncertainties due to rate limits. This was done because there is no *a priori* way to select the parameter η for Fig. 11, which is determined by the maximum magnitudes of signals in the closed loop. Clearly, for η sufficiently large all our trial designs would violate the resulting $\bar{\sigma}[L]$ bound. That such violations actually correspond to instabilities was verified by repeating the transient responses for trial 3 with progressively larger attitude commands. Unstable behavior occurs for $\theta_{cmd} \geq 18°$, with $\eta \simeq 60$.

In order to improve robustness with respect to rate limits, the following iterative procedure may be used:

(1) Assume a signal level limit $\eta \leq \eta_0$.

(2) Design $I + G^{-1}$ consistent with the resulting $\overline{\sigma}[L]$.

(3) Evaluate the actual maximum signal level η_1 by computing transient responses with worst case commands and/or initial conditions.

(4) If η_1 and η_0 are substantially different, return to step 1 with $\eta_0 = \eta_0 + \varepsilon(\eta_1 - \eta_0)$, where ε is a design parameter. Otherwise STOP.

An illustration of the first iteration of this procedure is given in Fig. 13, where the assumed signal level $\eta_0 = 20$. The dashed $\overline{\sigma}[L]$ curve yields a controller (trial 2) whose actual signal level is $\eta_1 \leq 0.6$. The associated transient responses are slow but stable. To fine tune this design, a second iteration might be taken with $\eta_0 = 5$.

We noted earlier that $\overline{\sigma}[L]$ due to operating point changes becomes quite large at low frequencies because the helicopter's slow modes are not stable at all operating points. At intermediate and high frequency ranges, however, the uncertainty bounds are reasonably small (Fig. 11). This suggests that if the loop transfer matrix G(s) has sufficient low-frequency gain to stabilize the slow modes under all conditions, then the design might well be stable even though the (sufficient) stability-robustness condition fails. This is in fact the case. Both trial designs No. 2 and No. 3 remain stable at eight representative flight conditions ranging from hover to 160 knot forward speed and from +2000 ft/min to -2000 ft/min ascent rates. The intuitive idea that underlies this result (sufficiently high low-frequency gain) may well provide needed insight toward a generalized multivariable robustness theory for unstable perturbations.

This example illustrates that the linear-optimal approach may be combined with singular value analyses to provide a useful multivariable technique. In this illustrative example the use of a state estimator in the loop was not considered. But, in the general linear-optimal approach the inclusion of such an estimator must be considered.

7. *Robust Estimators*

It was shown in [21] that multivariable linear-optimal regulators using full state feedback have impressive robustness properties, including guaranteed gain margins of -6 db and $+\infty$ db and phase margins of 60° in all channels. But, if observers or Kalman filters are used for state estimation in the implementation, there are no guaranteed margins [22]. Fortunately, an adjustment procedure is available [23] for use in the observer or filter design, which makes it possible essentially to recover the guaranteed margins of full state feedback for minimum phase systems. The adjustment procedure involves the introduction of a scalar design parameter q with the property that as q tends to infinity the stability margins tend to the full state margins.

An example that illustrates this adjustment procedure is a linear-optimal regulator designed for flutter suppression with the DAST (Drones for Aerodynamic and Structural Testing) wing.

Table III. Robustness Summary for Dast Example

Controller	q	RMS δ	RMS $\dot{\delta}$	Gain margin				Phase margin				Bandwidth
		Deg	Deg/sec	Db	Hz	Db	Hz	Deg	Hz	Deg	Hz	Hz
State	.0	1.804	164.9	-8.1	9.90	∞	--	-63.7	7.05	+82.3	15.25	23.2
Kalman	.0	2.330	186.2	-5.3	10.28	+6.4	21.87	-42.5	8.26	+32.4	13.64	59.8
Robust	.000001	2.733	196.8	-6.6	9.85	20.0	75.28	-45.4	7.46	+66.2	13.56	19.3
Robust	.00001	3.066	212.0	-7.3	9.87	21.5	103.56	-53.3	7.23	+72.9	14.23	11.1
Robust	.0001	3.534	240.2	-7.7	9.90	24.9	149.65	-57.9	7.16	+76.3	14.64	21.8

A single control surface input and two accelerometer outputs were used in the design. The model consisted of five flexure modes, five aerodynamic lags, a third-order actuator, and a first-order wind gust model. The robustness of five controllers is summarized in Table III and Fig. 15. The first controller is the full state feedback controller. The second uses a Kalman filter for state estimation. The remaining three use filters for state estimation derived using the adjustment procedure with different values of the design parameter q. In addition to the stability margin data, the RMS control surface activity corresponding to an RMS gust input of 1 ft/sec is given as well as the controller bandwidth. The stability margins actually increase monotonically with q with an attendant increase in control surface activity. But the control surace activity corresponding to the largest value of q was within the design specification limits and the stability margins for this value of q are significantly improved over the $q = 0$ case and are nearly as good as those of the full state feedback controller. As shown in the frequency response plots of Fig. 15, the robust estimators smooth out the notching characteristics of the Kalman filter providing better stability margins with only a slight degradation in RMS performance. This example clearly demonstrates the utility of the adjustment procedure.

C. DIGITAL DESIGN

The techniques described above do not explicitly address a digital mechanization of flight control designs. They are, of course, all applicable to digital system design and analysis lacking only in a transformation from continuous to discrete

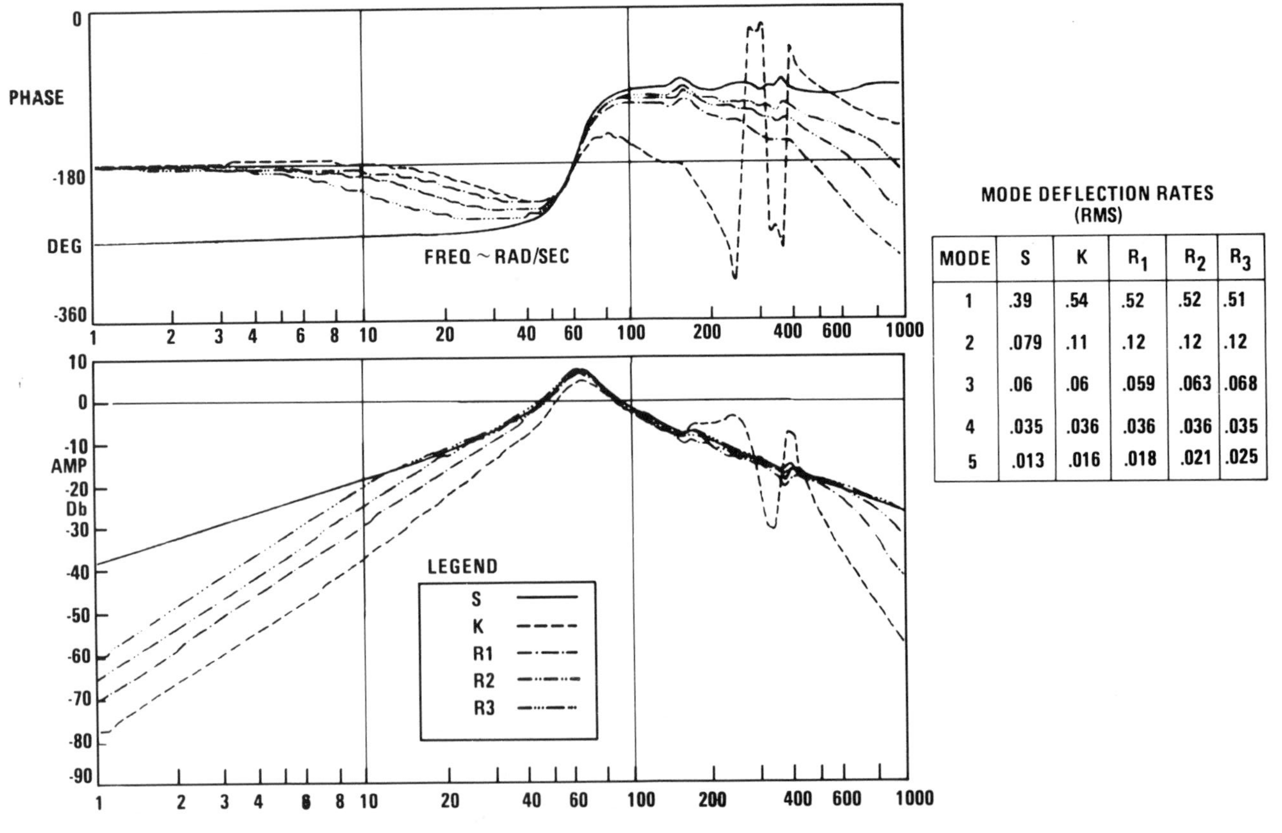

MODE DEFLECTION RATES (RMS)

MODE	S	K	R_1	R_2	R_3
1	.39	.54	.52	.52	.51
2	.079	.11	.12	.12	.12
3	.06	.06	.059	.063	.068
4	.035	.036	.036	.036	.035
5	.013	.016	.018	.021	.025

Fig. 15. Robust estimator characteristics for DAST example.

space. In fact the most common design scenario for a digital flight control law has evolved using continuous system techniques. The steps generally followed are

(1) Design the feedback structures using continuous techniques to achieve performance and sensitivity goals (i.e., stability margins) for the continuous plant-controller

(2) Choose a sample time with a Nyquist frequency well above the control frequencies and discretize the continuous compensators using appropriate algorithms such as Z-transform, Tustin's method, or prewarped Tustin's method.

(3) Choose a first- or second-order continuous prefilter for each sensor to eliminate the impact of aliasing from unmodeled high-frequency dynamics such as structural modes or high-band sensor noise. This is typically chosen conservatively low, i.e., well below the Nyquist frequency.

(4) Select a control command output continuous postfilter (also thought of as an actuator prefilter) with an output hold device to reduce the effects of digital quantization but obtain minimum phase loss at the control frequencies.

Steps 2, 3, and 4 provide a sufficient scenario to implement a digital representation yielding the desired flight control goals of step 1.

The major difficulty with this approach is that these steps all contribute phase lag to the system. This presents little problem if the sample rate is sufficiently higher than the stability crossover frequencies. Lowering the sample rate results in stability difficulties and performance reduction. The net result is a design technique that dictates high sample rates. This represents the state of the art in digital design techniques for production aircraft systems today.

1. Low Sample Rate Design

It has been claimed that designing directly with discrete plant models will produce lower sample rates and minimum compensation and still meet the same design goals as analog conversion techniques. This has led to a number of design techniques and tools which do result in low sample rates [24,25]. The direct digital design technique described in [24] is based on an optimization approach formulated to match a modeled transient response at discrete time points. The feedback control law, gains, and compensation parameters, are designed to force the closed-loop response of a discretized plant (including pre- and postfilters) to best fit a desired response transient at the discretized points. This technique was utilized by Peled to examine prefilter and sample rate selection [26]. As an example of the technique, a direct digital design was performed for a digital feedback around a simple integrator, which attempted to match a desired closed loop exponential response.

As shown in Fig. 16, the optimization technique results in parameter selections for K, α_c, and β_c for a fixed sample time T and prefilter bandwidth a. These parameters are synthesized to fit the discrete points generated at the sample instances of a unit step response of a continuous feedback control design (Fig. 16a). Details are given in [26] with a summary provided in [27].

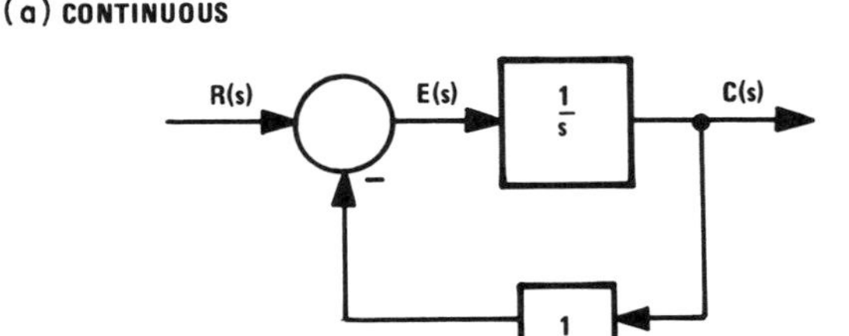

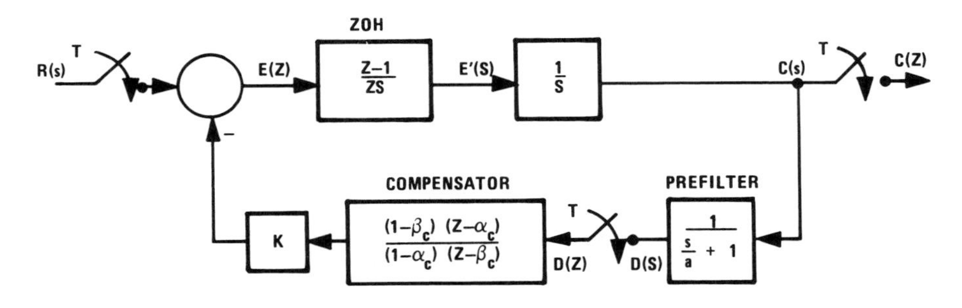

Fig. 16. Peled's first-order example.

The key point for discussion is not the optimization but what it produced. Examining the plant to be operated on

$$\frac{D(z)}{E(z)} = \frac{z-1}{z} \cdot Z\left[\frac{1}{s^2\left(\frac{s}{a}+1\right)}\right]^{1} = \frac{J_1(T)(z-\beta_0)}{(z-1)(z-\alpha_1)},$$

where as shown in [28]

$$J_1(T) = T + \frac{e^{-at}}{a} - \frac{1}{a}, \quad \alpha_1 = e^{-at},$$

$$\beta_0 = \frac{1-\alpha_1-aT\alpha_1}{1-\alpha_1-aT}.$$

The inclusion of an extra discrete zero β_0 along with the prefilter pole α_1 form the basis for the selection of the compensator parameters in Fig. 16b, α_c and β_c.

[1] *Implies Z-transform.*

In almost all cases examined in [26], i.e., choices of T and a, α_c was optimized to cancel β_0 and β_c optimized to cancel α_1. The effect of this cancellation is characterized by lead compensation near the Nyquist frequency and results in total system phase enhancement.

2. F-8 Design Example

The compensation structure suggested by this technique was used to develop a low sample rate design for the NASA Digital Fly-by-wire F-8C CCV aircraft. The F-8 digital system operates at 53.33 Hz due to a remote augmented vehicle (RAV) implementation. Analysis determined that the system could operate at 20 Hz with no significant degradation in aircraft stability and performance characteristics and with no phase enhancement applied through direct digital design techniques. F-8 Flight tests had indicated unsatisfactory performance at 6.7 Hz. Application of phase enhancement through direct digital design permitted a 4-Hz sample rate [28]. A comparison of the continuous and low sample rate control structure block diagrams is shown in Fig. 17. Table IV presents a comparison of pitch axis stability margins for the continuous design and the low sample rate design at four flight conditions. A comparison of frequency response plots for FC#1 is shown in

Table IV. Pitch Axis Stability Margins

Flight conditions	*1*		*5*		*9*		*17*	
	Cont.	*4 Hz*	*Cont.*	*4 Hz*	*Cont.*	*4 Hz*	*Cont.*	*4 Hz*
Gain margin (db)	>40	7.0	>40	7.3	>40	6.4	>40	6.1
Phase margin (deg)	78°	63°	72°	50°	81°	90°	80°	48°

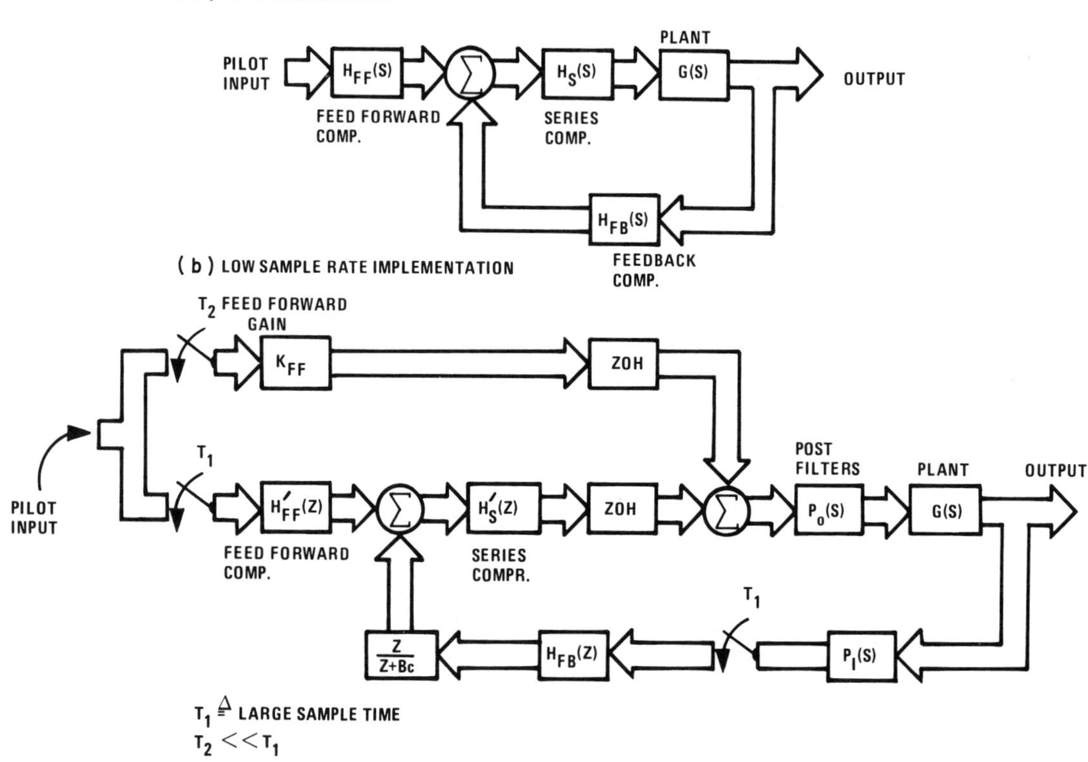

Fig. 17. Low sample rate control implementation. $T_1 \triangleq$ *large sample time;* $T_2 \ll T_1$.

Figs. 18 and 19. Transient response comparisons are given in Figs. 20 and 21.

These results have not been verified in man in the loop simulation of flight test, however they do indicate lower sample rates are achievable based on analytical design criteria. A limiting factor not discussed but which also must be recognized is the effect of output quantization in the actuator command signals. Smoothing techniques will generally provide adequate performance in this area.
recognized is the effect of output quantization in the actuator command signals. Smoothing techniques will generally provide adequate performance in this area.

III. CONCLUSIONS AND RECOMMENDATIONS

Because of the number of control inputs and close dynamic coupling of future aircraft, and requirements that the flight control system provide the stabilizing influence on the vehicle, efficient and reliable control system design and analysis techniques are essential to satisfactory aircraft performance. Several techniques have been discussed in this chapter. Modal control techniques permit the designer to handle a large number of inputs and provide an approach to achieve a desired, within constraints, system response. A technique based on asymptotic regulator properties reduced the weight selection difficulties associated with optimal regulator design in a root locus equivalent design approach. We have shown how to design in stability margins for Kalman filters in the loop with the robust estimator technique. The most powerful technique of all, however, is the use of singular values in analyzing the stability characteristics of multivariable

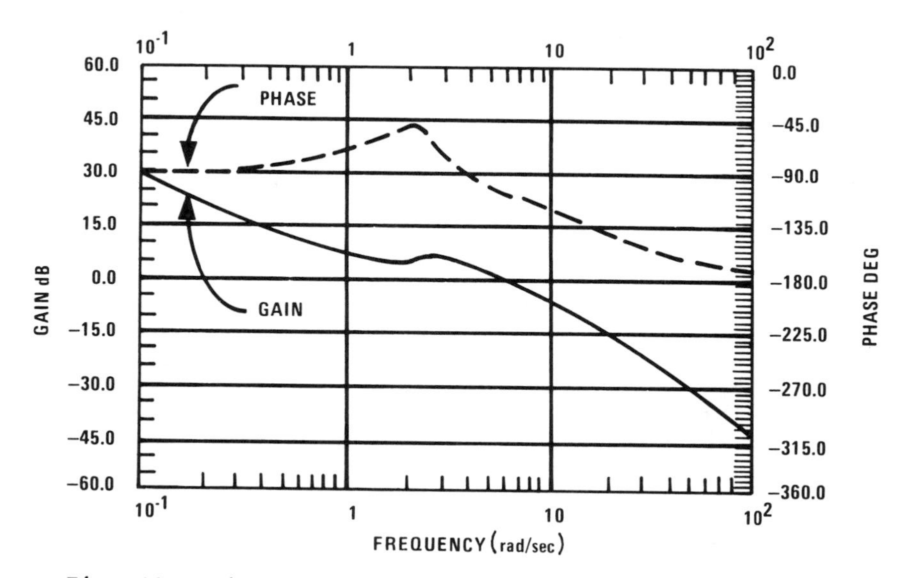

Fig. 18. Pitch axis open-loop response (continuous). Flight condition #1: gain margin = 59 db; phase margin = 78°; crossover frequency = 5.2 rad/sec.

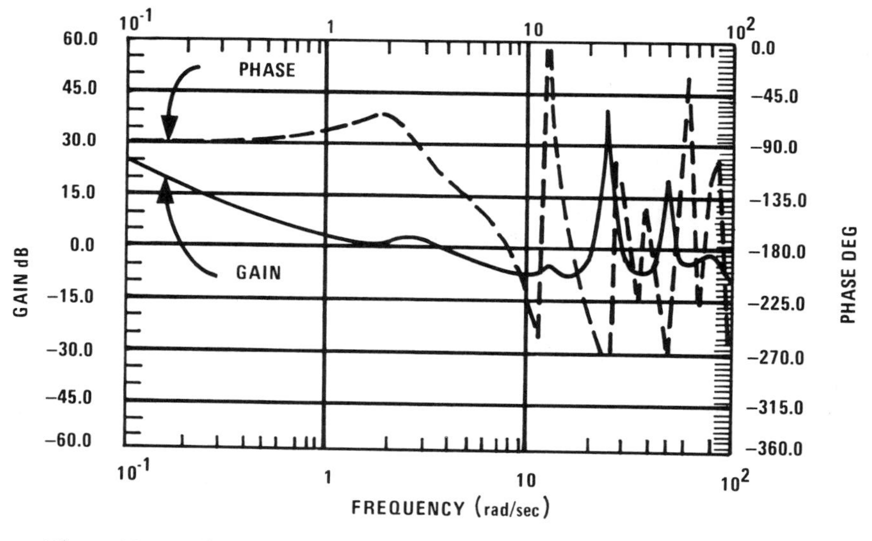

Fig. 19. Pitch axis open-loop frequency response (low sample rate). Low sample rate design: gain margin = 7db; phase margin = 63°; crossover frequency = 3.8 rad/sec.

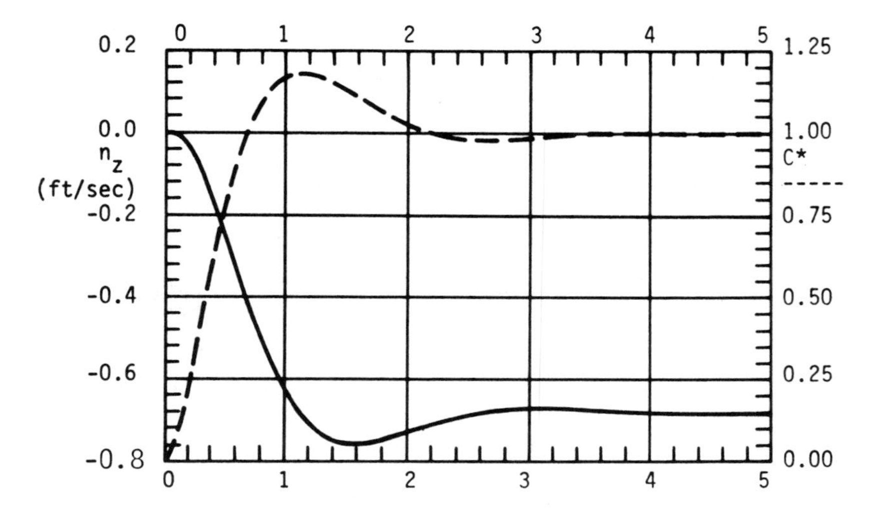

Fig. 20. Pitch axis transient response (continuous).

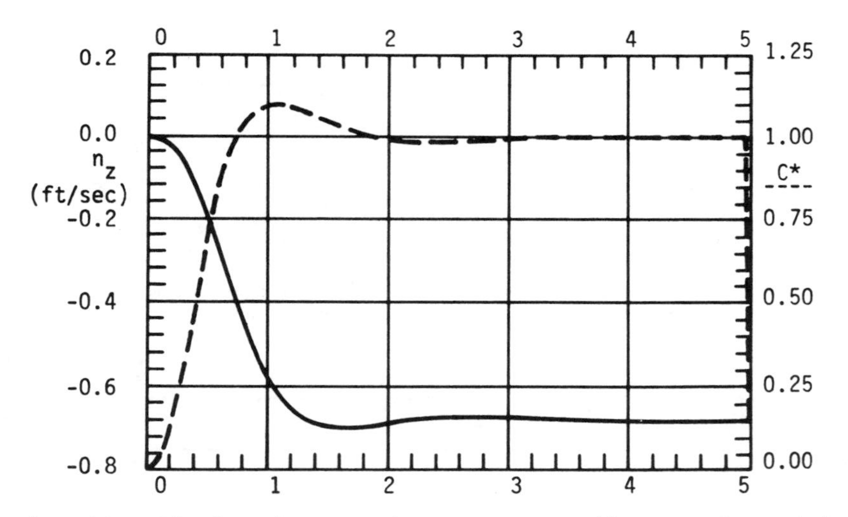

Fig. 21. Pitch axis transient response (low sample rate).

systems. Further research is required in this area, particularly in transforming it into a synthesis and design technique. Questions exist on the possible conservative nature of the stability margins computed with singular value analyses. Research is currently being done in this area. We firmly believe that stability margins may one day be specified for multivariable systems based on singular value analysis.

More complex design issues that arise when subsystems are integrated in design have not been addressed in this chapter. Tight pointing and tracking requirements for today's fighter aircraft will result in overlapping bandwidths between the flight and fire control systems. Design criteria for the two systems are not always homogenous and are often conflicting. New techniques are needed to handle multiple dissimilar design criteria.

In the singular value analysis discussion, techniques for analyzing the effect of nonlinearities were described. While these can be a powerful tool, much work still needs to be done in treating system nonlinearities.

Finally, digital control is still in its infancy. Designers are still trying to make digital systems look like their analog counterparts. Work is ongoing today in the area of finite state machines to try and discover the real power of digital structures for control application.

ACKNOWLEDGMENT

The flight control system design techniques described in this chapter were developed under U. S. Office of Naval Research Contract #N00014-75-C-0]444 and NASA Dryden Flight Research Center Contract #NAS4-2518. Honeywell's Systems and

Control Technology section, primarily G. Stein, T. Cunningham, J. Doyle, R. Stone, and G. Hartmann contributed extensively to the developments described through initial development and application of these techniques to numerous flight control design problems.

REFERENCES

1. H. H. ROSENBROCK, "Design of Multivariable Control Systems Using the Inverse Nyquist Array," *Proc. IEEE 116*, 1929-1936 (1969).
2. H. H. ROSENBROCK, "Computer-Aided Control System Design," Academic Press, London, 1974.
3. A. G. J. MacFARLANE and B. KOUVARITAKIS, "A Design Technique for Linear Multivariable Feedback Systems," *Int. J. Control 25*, 837-874 (1977).
4. A. G. J. MacFARLANE and I. POSTLETHWAITE, "Characteristic Frequency Functions and Characteristic Gain Functions," *Int. J. Control 26*, 265-278 (1977).
5. J. D. SIMON and S. K. Mitter, "A Theory of Modal Control," *Inform. Contr. 13*, 316-353 (1968).
6. B. C. MOORE, "On the Flexibility Offered by State Feedback in Multivariable Systems Beyond Closed-Loop Eigenvalue Assignment," *IEEE Trans. AC-21*, 685-692 (1976).
7. B. C. KUO, "Automatic Control Systems," Prentice-Hall, Englewood Cliffs, New Jersey, 1967.
8. J. W. BREWER, "Control Systems," Prentice-Hall, Englewood Cliffs, New Jersey, 1974.
9. I. M. HOROWITZ, "Synthesis of Feedback Systems," Academic Press, New York, 1963.
10. J. H. WILKINSON, "The Algebraic Eigenvalue Problem," Clarendon Press, London and New York, 1965.
11. G. E. FORSYTHE and C. B. MOLER, "Computer Solutions of Linear Algebraic Systems," Prentice-Hall, Englewood Cliffs, New Jersey, 1967.
12. C. A. HARVEY and J. C. DOYLE, "Optimal Linear Control (Characterization and Loop Transmission Properties of Multivariable Systems)," ONR Report CR215-238-3, 1978.
13. A. G. J. MacFARLANE and I. POSTLETHWAITE, "The Generalized Nyquist Stability Criterion and Multivariable Root Loci," *Int. J. Control 23*, No. 1, 81-128 (1977).

14. G. STEIN and A. H. HENKE, "A Design Procedure and Handling Quality Criteria for Lateral-Directional Flight Control Systems," AFFDL-TR-70-152, May 1971.

15. C. A. HARVEY and G. STEIN, "Quadratic Weights for Asymptotic Regulator Properties," *IEEE Trans. AC-23*, 378-387 (1978).

16. A. J. OSTROFF, D. R. DOWNING, and W. J. ROAD, "A technique Using a Nonlinear Helicopter Model for Determining Terms and Derivations," NASA Technical Note TN D-8159, NASA Langley Research Center, May 1976.

17. R. H. HOHEUEMSER and S. YIN, "Some Applications of the Method of Multiblade Coordinates," *J. Am. Helicopter Soc.* (July 1972).

18. M. G. SAFONOV, "Tight Bounds on the Response of Multivariable Systems with Component Uncertainty," 1978 Allerton Conference.

19. G. STEIN, "Generalized Quadratic Weights for Asymptotic Regulator Properties," *IEEE Trans. AC-24*, 559-566 (1979).

20. R. E. KALMAN, "When is a Linear System Optimal?" *J. Basic Engr. 86*, 51-60 (1964).

21. M. G. SAFONOV and M. ATHANS, "Gain and Phase Margin for Multiloop LQG Regulators," *IEEE Trans. AC-22*, 173-176 (1977).

22. J. C. DOYLE, "Guaranteed Margins for LQG Regulators," *IEEE Trans. AC-23*, 756-757 (1978).

23. J. C. DOYLE and G. STEIN, "Robustness with Observers," *IEEE Trans. AC-24*, 607-611 (1979).

24. P. H. WHITAKER, "Development of a Parameter Optimization Technique for the Design of Automatic Control Systems," NASA CR-143844, May 1977.

25. R. F. WHITBECK and L. F. Hofmann, "Analysis of Digital Flight Control Systems with Flying Qualities Applications," AFFDL-TR-78-115, September 1978.

26. U. PELED, "A Design Method with Application in Prefilter Design and Sampling-Rate Selection in Digital Flight Control Systems," Dept. of Aero/Astro, Stanford University, Stanford, California 94303, SUDAAR Rept. 512.

27. U. PELED and J. F. POWELL, "The Effect of Prefilter Design on Sample Rate Selection in Digital Flight Control Systems," AIAA Paper 78-1308 (1978).

28. T. B. CUNNINGHAM, "A Low Sample Rate Design for the F-8 Aircraft," 3rd. Digital Avionics Systems Conference, Ft. Worth, Texas, November 6-8, 1979.

SYMBOLS

A	System matrix
B	Control matrix
C	Measurement matrix
C^*	Linear combination of normal acceleration and pitch rate
c	Collective
dc	Differential collective
E	Expected value
F	Kalman filter gains
G(s)	Open-loop transfer function
gr	Rotor dynamics transfer function
H(s)	Closed-loop transfer function matrix
Hz	Hertz
I	Identity index
J	Performance index
j	$\sqrt{-1}$
k	Loop gains
K	Gain matrix
L(s)	Perturbation matrix
M	Relative maximum of closed-loop amplitude-frequency response
p_s	Stability axis roll rate
P	Riccati matrix--optimal control
q	Robust estimator design parameter
Q	Weighting matrix
r	Left singular vector
r_s	Stability axis yaw rate
R	Matrix of left singular vectors; control weighting matrix
r_{lim}	Rate limit

S	Riccati matrix--filter
s	Laplace operator
u	Control vector
U,V,G	Inverse Nyquist array matrix transformations
v	Right singular vector
V	Matrix of right singular vectors
x	System state vector
y	System measurement
$\dot{z}$	Vertical velocity
z	Scalar variable analogous to inverse Nyquist variable; discrete plane operator
Z	Transform operator
β	Intermediate variable in calculation of z
η	Nonlinearity input amplitude
δ	Control deflection
ζ	Damping ratio
λ	Eigenvalue
ϕ	Bank angle
θ	Sensor noise; pitch angle
Θ	Measurement noise covariance matrix
ω	Frequency
σ	Singular value
Σ	Diagonal matrix of singular values
ρ	Scalar weighting parameter
Λ	Diagonal matrix of eigenvalues
ξ	System noise
Ξ	System noise covariance matrix

A Discrete-Time Disturbance-Accommodating Control Theory for Digital Control of Dynamical Systems

C. D. JOHNSON

Department of Electrical Engineering
The University of Alabama in Huntsville
Huntsville, Alabama

ISBN 0-12-012718-0

I. INTRODUCTION

The assortment of inputs that act on a dynamical system can be grouped into two broad categories: the *control inputs* u, and the *disturbance inputs* w. By definition, the control inputs consist of all those inputs that "we" can manipulate. On the other hand, the disturbance inputs are defined as the set of all inputs that "we" *cannot* manipulate. The term we is used here to underscore the fact that these definitions are relative; that is, in some situations one person's control is another person's disturbance and vice versa. Such situations exist, for example, in missile pursuit problems with maneuvering targets and in similar game-type dynamical problems in engineering, economics, sociology, politics, etc. The act of manipulating the control inputs is referred to as "controlling" the system, and the logic or strategy whereby control manipulation decisions are made is called the "control policy." The assortment of hardware that actually performs the control decisions and manipulations is commonly referred to as the "controller."

In general, the overall objective of controlling a dynamical system is to modify the motions, or other behavior characteristics of the system. For instance, the given system might be inherently unstable, in which case the objective of control u(t) might be to stabilize the system in some specified sense. Alternatively, the objective of control might be to point or move the system in a certain prescribed manner.

Regardless of what the specific objectives of control might be, it is typically found that the disturbance inputs $w(t)$ acting on the system interfere with achieving those objectives. Thus, the realistic control problem becomes one of meeting the objectives of control *in the face of interference from the disturbances.*

The problem of meeting control objectives in the face of disturbances is a nontrivial task because of an air of uncertainty which almost always prevails around the nature of disturbance inputs. In particular, the arrival times and the exact description or behavior of the anticipated disturbance waveform is almost never known ahead of time. Moreover, the direct measurement of disturbances, as they arrive in real-time, is often impractical. Thus, the control objectives must be met in the face of disturbance interference that is typically not known *a priori*, not accurately predictable, and often not directly measurable in real-time. Because of this uncertainty, a control policy must possess certain adaptability features in order to effectively cope with changing disturbance patterns as they evolve in real-time.

Modern control engineering has contributed a new approach to the design of control policies that can cope with uncertain disturbances in complex, multivariable control problems. This new approach, introduced in [1-5], is called the theory of disturbance-accommodating control (DAC) and offers significant advantages over conventional design procedures [6,7,8]. The existing theory of DAC is restricted to *analog-type* control problems where sensor data are gathered and transmitted continuously and where the controller data processing is performed by conventional analog circuits. The purpose of this chapter is to develop a general form of discrete-time, sampled-data DAC theory that can be applied to the design of a wide variety of digital controllers for dynamical systems.

II. A BRIEF OVERVIEW OF DAC THEORY

Disturbance-accommodating control theory is a collection of nonstatistical modeling and controller design techniques for multivariable control problems with uncertain disturbances. In this section we briefly outline the general ideas underlying DAC theory for the benefit of those readers who may not be familiar with the subject. A more thorough discussion of these ideas may be found in [3,5].

A. *DISTURBANCE WAVEFORM MODELS*

In DAC theory the anticipated disturbances $w(t)$ are always assumed uncertain, but they *are not* represented by statistical properties such as means, covariances, etc. Rather, they are represented by a *basis-function* descriptor that displays the totality of possible waveform components which may be combined at any one moment of time to create $w(t)$. In particular, the class $\mathcal{W}$ of anticipated disturbances $w(t)$ is represented, in the scalar case, by

$$\mathcal{W} = \{w(t) \mid w = c_1 f_1(t) + c_2 f_2(t) + \cdots + c_M f_M(t)\}, \tag{1}$$

where the $f_i(t)$, $i = 1, \ldots, M$, are *completely known* linearly independent functions of time t, and the c_i are *completely unknown* "constant" weighting coefficients that may jump in value every once in a while in a *completely unknown* manner. Thus, the set of M-functions $\{f_i(t)\}$ forms a (finite) basis set for the class $\mathcal{W}$ of uncertain disturbance functions $w(t)$, where at each t the local behavior (waveform geometry) of $w(t)$

is determined by the particular values of the c_i that exist at that moment of time. In other words, the $\{f_i(t)\}$ represent the basic "modes" of $w(t)$ behavior and the weights c_i determine just how those modes are linearly combined at each moment t. In the case of vector disturbances $w = (w_1, \ldots, w_p)$, each independent component $w_i(t)$ is modeled by an expression of the form (1).

The disturbance descriptor (1) is called a "waveform model" in DAC theory and represents a time-domain version of the finite-element, or spline-function technique [9,10], which has found wide application in engineering problems involving spatially dependent phenomena (solid mechanics, fluid mechanics, electromagnetics, etc.). It should be emphasized that in DAC theory the values of the arbitrary weighting coefficients c_i in (1) are assumed piecewise-constant but otherwise completely unknown. No statistical properties or probabilistic structure is assumed about the time behavior of the c_i. Thus, for instance, the traditional properties of uncertainty such as mean and covariance of $w(t)$ are completely unknown and, in fact, are of no concern in DAC theory! This means that assumptions regarding ergodic behavior, stationary statistics, etc., are not required in the DAC approach.

In applications of control theory to realistic problems, it is found that not all disturbances $w(t)$ admit a waveform model descriptor (1). Those that *do*, are referred to as disturbances with "waveform structure." Those that do not are defined as "noise-type" disturbances. Some common examples of disturbances with waveform structure are load variations on power systems and speed regulators; winds, updrafts, and other aerodynamic loads on missiles and aircraft; flowrates and steam pressures in chemical manufacturing processes; exogenous inputs to socioeconomic models. Some common examples of noise-type disturbances are radio static, fluid turbulence, and brush-noise in electrical motors. The traditional random-process (stochastic) modeling and control policy design techniques are apparently the best approach to problems involving noise-type disturbances. On the other hand, if the disturbances have waveform structure, it is our contention that DAC techniques, using the waveform model (1), lead to more effective control policy designs.

B. DISTURBANCE STATE MODELS

The waveform model (1) is the key idea behind the DAC approach to disturbance modeling. However, the "information" reflected in the model (1) must be encoded into another format before it can be used effectively in DAC design recipes. That alternative format is called a "disturbance state model" in DAC theory and consists of a *differential equation* for which (1) is the general solution. In other words, one must solve the following inverse-problem in differential equations: Given the general solution (1), find the differential equation. In practical applications of DAC theory, this step typically leads to a *linear* differential equation which, in the case of a vector disturbance $w = (w_1, \ldots, w_p)$, can be expressed in the general form [5, pp. 402-422]:

$$w = H(t)z, \quad z = (z_1, \ldots, z_\rho), \tag{2a}$$

$$\dot{z} = D(t)z + \sigma(t), \tag{2b}$$

where $H(t)$, $D(t)$ are *completely known* matrices, $z(t)$ is a ρ vector called the "state" of the disturbance w, and $\sigma(t) = (\sigma_1(t), \ldots, \sigma_\rho(t))$ is a symbolic representation of a vector sequence of impulses with completely unknown arrival times and completely unknown randomlike intensities. Thus, the basis functions $f_i(t)$ in (1) appear as principal modes of the disturbance state model $\dot{z} = D(t)z$ in (2b).

The impulses of $\sigma(t)$ in (2) are imagined as the source (cause) of the uncertain, once-in-a-while jumps in the values of the piecewise-constant weighting coefficients c_i in (1). Thus, in DAC theory it is assumed that adjacent impulses in $\sigma(t)$ are separated by some finite time-spacing μ; i.e., $\sigma(t)$ consists of a *sparsely populated* sequence of unknown impulses. If the impulses of $\sigma(t)$ arrive "too fast," the c_i in (1) will jump in value too often and in that case we say the disturbance $w(t)$ loses its waveform structure and becomes "noise." In particular, if the impulses arrive arbitrarily close (and are totally uncorrelated), the $\sigma(t)$ sequence then behaves like a vector "white-noise" process and the DAC state model (2) then appears similar to the white-noise coloring filters traditionally used in stochastic control. However, note the subtle differences. Namely, in DAC theory the matrices $H(t)$, $D(t)$ in (2) are not chosen from statistical data, and the model (2) *is not* required to possess stability properties. In fact, it is not uncommon in DAC applications to find that many of the principal modes $f_i(t)$ of the disturbance model $\dot{z} = D(t)z$ are *unstable*, (i.e., grow with time in an unbounded fashion). For example, the waveform model (1) might appear as

$$w(t) = c_1 e^t + c_2 e^{10t}.$$

Such behavior of (2) is not permitted of the coloring filters in conventional stochastic control theories. This constitutes one of the unique features of DAC theory.

C. *THE PRINCIPLE OF OPTIMAL DISTURBANCE-ACCOMMODATION*

The presence of waveform structure in uncertain disturbances leads to an important control-information principle in DAC theory. It then turns out that, for a broad class of performance criteria, the theoretically optimal choice for the control u at time t can be expressed in terms of the current plant state $x(t)$ and the current disturbance state $z(t)$, i.e., the optimal control policy has the form

$$u(t) = \phi(x(t), z(t), t). \tag{3}$$

In other words, at each t the disturbance state $z(t)$ embodies enough information about $w(t)$ to allow a rational scientific choice for the real-time control $u(t)$, even though the future behavior of $w(t)$ is uncertain. This result, called the principle of optimal disturbance-accommodation [11], enables the control designer to formulate a deterministic control policy for accommodating the presence of uncertain disturbances. Because that control policy utilizes the real-time local behavior of $z(t)$, it automatically adapts itself to the existing real-time waveform patterns of each individual disturbance function $w(t)$.

D. *MODES OF DISTURBANCE-ACCOMMODATION*

In the theory of disturbance-accommodating control, there are three primary attitudes the control designer can take with respect to "accommodating" disturbances. Each of these attitudes leads to a different control policy and different mode of control action. First, the designer can take the attitude that disturbances are totally unwanted and therefore the best control policy is to cancel out all effects caused by disturbances. This leads to a special form of DAC called a disturbance-absorbing controller, a modern generalization of the "dynamic vibration absorbers" classically used to suppress unwanted vibrations in structures and machines [12,13]. Second, the designer can take the alternative attitude that only certain parts or aspects of disturbances are unwanted. This leads to what is called a disturbance-minimizing controller in which the controller cancels out only part of the disturbance effects. This latter form of DAC may be necessitated by physical constraints that prohibit total cancellation of disturbance effects.

The third possible attitude of disturbance accommodation is based on the fact that disturbances may sometimes produce effects that are *beneficial* to the primary control objectives. For instance, certain types of wind gusts may actually "help" the controller to steer a missile toward a specified target. In such cases the best control policy is to let the disturbances assist the controller as much as possible. This attitude leads to what is called a disturbance-utilizing controller. In some applications designers may use a combination of these three primary modes of accommodation, thus leading to a multimode disturbance-accommodating controller.

Control engineering design procedures, which accomplish the absorption, minimization, and utilization modes of disturbance accommodation described above, have been developed for a broad class of traditional analog-type control problems. Those existing results are described in the tutorial references [3,5,13]. In this chapter, we shall develop a sampled-data, discrete-time version of DAC theory, which parallels the existing analog DAC theory in [5]. Some additional topics concerning analog DAC theory are presented in [11,12,14-17], and a historical account of the subject may be found in [6].

E. *DISCRETE-TIME CONTROL POLICIES, SAMPLED-DATA, AND DIGITAL CONTROLLERS*

A discrete-time control policy is defined as a control policy in which control "decisions" are made (updated) only at certain isolated moments of time $\{t_0, t_1, t_2, \ldots, t_N\}$. Between any two consecutive decision times $(t_i \leq t < t_{i+1})$ the control action u(t) either remains constant or varies in accordance with some *a priori* prescribed interpolation rule that is initiated at t_i. In the most common case, the decision times t_i occur at equally spaced intervals $\{t_0, t_0 + T_c, t_0 + 2T_c, \ldots, t_0 + NT_c\}$ where T_c is a constant.

In feedback (closed-loop) controller designs, control decisions are made in light of real-time data provided to the controller by "sensors" which measure, for instance, the system output variables $y(t) = (y_1(t), \ldots, y_m(t))$. The term sampled-data is used to denote situations in which data provided to the controller are updated only at certain isolated moments of time. Between such moments of time, data provided to the controller are either held constant (at the last updated value) or are varied in accordance with some *a priori* prescribed interpolation rule. In the most common case, the data update times occur at equally spaced intervals $\{t_0, t_0 + T_d, t_0 + 2T_d, \ldots, t_0 + NT_d\}$, where T_d is a constant called the sampling period.

When digital devices (digital computers, microprocessors, etc.) are used to perform control decisions, the controller is referred to as a digital controller. Such controllers are, by their very nature, only capable of processing sampled-data and executing discrete-time control policies.

In this chapter, we shall consider the design of discrete-time disturbance-accommodating control policies for situations in which the system output data are sampled. Moreover, for simplicity it will be assumed that the control decision times are equally spaced and coincide with the data sample times, e.g., $T_c = T_d = T$.

III. MATHEMATICAL MODELS OF PLANTS, DISTURBANCES, AND COMMANDS FOR DISCRETE-TIME DAC

In order to develop a general-purpose theory of DAC for discrete-time control applications, it is necessary to derive appropriate mathematical models to represent the discrete-time behavior of plants, disturbances, and commands. A collection of such models is derived in this section.

A. DISCRETE-TIME MODELS OF PLANTS AND DISTURBANCES

The class of controlled and disturbed dynamical systems (plants) considered here includes those continuous-time plants which, when modeled about appropriate operating points, lead to linearized differential equation models of the form

$$\dot{x} = A(t)x + B(t)u(t) + F(t)w(t), \quad y = C(t)x, \quad (\dot{\ }) = d/dt, \tag{4}$$

where A, B, C, F are known matrices; x the plant state vector is of dimension n; u the plant control input vector is of dimension r; w the plant disturbance input vector is of dimension p; and y the plant output vector is of dimension m.

We also include consideration of plants that are intrinsically discrete-time in nature and therefore possess no underlying continuous-time model of the form (4). Plants of this latter variety arise from studies of dynamic problems in economics, business, demographics, and other socioeconomic areas. The mathematical models for such plants, and disturbances, automatically appear in a discrete-time form, which we assume has the linear structure of our final equation (16) derived in the sequel. The objective of this subsection is to develop a discrete-time model, corresponding to (4), which will describe the behavior of (4) at the isolated moments $t = t_0, t_0 + T, t_0 + 2T, \ldots$, where T is a fixed positive constant, (T = the sampling period).

It is recalled that the general solution of (4a) is given by [18]

$$x(t;\, t_0,\, u,\, w) = \Phi(t, t_0)x(t_0) + \int_{t_0}^{t} \Phi(t, \tau)B(\tau)u(\tau)\,d\tau + \int_{t_0}^{t} \Phi(t, \tau)F(\tau)w(\tau)\,d\tau, \qquad (5)$$

where $\Phi(t, t_0)$ is the state-transition matrix for A(t) in (4a). Setting $t \to (t_0 + T)$ and then $t_0 \to t$ in (5) yields

$$x(t + T;\, t,\, u,\, w) = \Phi(t + T, t)x(t) + \int_{t}^{t+T} \Phi(t + T, \tau)B(\tau)u(\tau)\,d\tau + \int_{t}^{t+T} \Phi(t + T, \tau)F(\tau)w(\tau)\,d\tau. \qquad (6)$$

In the most common case of discrete-time control, the control action u(t) is held constant between sampling times, $t_i \le t < t_i + T$. Thus, if $u(t_i)$ denotes the constant value of u over such an interval, (6) may be rewritten as

$$x(t_i + T;\ t_i,\ u,\ w) = \Phi(t_i + T,\ t_i)x(t_i) + \left[\int_{t_i}^{t_i+T} \Phi(t_i + T,\ \tau)B(\tau)d\tau\right]u(t_i) + \int_{t_i}^{t_i+T} \Phi(t_i + T,\ \tau)F(\tau)w(\tau)d\tau. \tag{7}$$

Finally, if t_i is set equal to the "previous" sample time $(t_0 + nT)$, expressions (7) and (4b) can be written as[1]

$$x((n + 1)T;\ nT,\ u,\ w) = \tilde{A}(nT)x(nT) + \tilde{B}(nT)u(nT) + \tilde{v}((n + 1)T), \tag{8a}$$

$$y(nT) = C(nT)x(nT)$$

where

$$\tilde{A}(nT) = \Phi(t_0 + (n + 1)T,\ t_0 + nT),$$

$$\tilde{B}(nT) = \int_{t_0+nT}^{t_0+(n+1)T} \Phi(t_0 + (n + 1)T,\ \tau)B(\tau)d\tau, \tag{8b}$$

$$\tilde{v}((n + 1)T) = \int_{t_0+nT}^{t_0+(n+1)T} \Phi(t_0 + (n + 1)T,\ \tau)F(\tau)w(\tau)d\tau. \tag{8c}$$

In the time-invariant case of (4), where A, B, C, F, are all *constant* matrices, the terms in (8b), (8c) simplify to

$$\tilde{A} = e^{AT}; \quad \tilde{B} = \int_0^T e^{A(T-\tau)}Bd\tau;$$

$$\tilde{v} = \int_0^T e^{A(T-\tau)}Fw(\tau + t_0 + nT)d\tau. \tag{9}$$

[1] *It is important to note that the argument symbols $(n + 1)T$ and nT in (8a) actually denote the times $t = t_0 + (n + 1)T$ and $t = t_0 + nT$, respectively. This shorthand notation will be consistently used in the sequel.*

In accordance with standard procedures in DAC theory, the uncertain disturbances $w(t)$ in (4) are assumed to be modeled by a "waveform description" that permits $w(t)$ to be viewed as the "output" of a differential equation system

$$w(t) = H(t)z(t), \quad \dot{z} = D(t)z + \sigma(t) \tag{10}$$

as already discussed in Section II.B; see remarks below (2).

The incorporation of (10) into the discrete-time plant expression [Eq. (8)] is accomplished as follows. The general solution for $z(t)$ in (10) can be expressed as

$$z(t;\, t_0,\, \sigma) = \Phi_D(t,\, t_0)z(t_0) + \int_{t_0}^{t} \Phi_D(t,\, \xi)\sigma(\xi)\,d\xi, \tag{11}$$

where $\Phi_D(t,\, t_0)$ denotes the state-transition matrix for $D(t)$ in (10). Replacing t_0 by $(t_0 + nT)$ and setting $t = \tau$, $t_0 + nT \le \tau \le t_0 + (n + 1)T$, in (11) permits the solution of (10) to be written as

$$w(\tau) = H(\tau)z(\tau),$$

$$z(\tau) = \Phi_D(\tau,\, t_0 + nT)z(t_0 + nT) + \int_{t_0+nT}^{\tau} \Phi_D(\tau,\, \xi)\sigma(\xi)\,d\xi. \tag{12}$$

Substituting (12) into (8c) yields

$$\tilde{v}((n + 1)T) = \widetilde{FH}(nT)z(nT) + \tilde{\gamma}(nT), \tag{13a}$$

where

$$\widetilde{FH}(nT) = \int_{t_0+nT}^{t_0+(n+1)T} \Phi(t_0+(n+1)T,\tau)F(\tau)H(\tau)\Phi_D(\tau,t_0+nT)\,d\tau,$$

$$\tilde{\gamma}(nT) = \int_{t_0+nT}^{t_0+(n+1)T} \times \left[\Phi(t_0+(n+1)T,\tau)F(\tau)H(\tau) \int_{t_0+nT}^{\tau} \Phi_D(\tau,\xi)\sigma(\xi)\,d\xi\right] d\tau. \tag{13b}$$

Consolidating (13) with (8) yields the discrete-time plant model as

$$\begin{aligned} x((n+1)T;\, nT,\, u,\, w) &= \tilde{A}(nT)x(nT) + \tilde{B}(nT)u(nT) \\ &\quad + \widetilde{FH}(nT)z(nT) + \tilde{\gamma}(nT), \qquad (14a) \\ y(nT) &= C(nT)x(nT). \qquad (14b) \end{aligned}$$

The reader is reminded that the arguments $(n + 1)T$, nT in (13a) and (14) actually denote the times $t = t_0 + (n + 1)T$, $t = t_0 + nT$, respectively; see previous footnote.

In the time-invariant case, $\tilde{A}$, $\tilde{B}$ in the plant model (14) simplify as indicated in (9), and $\widetilde{FH}$, $\tilde{\gamma}$ simplify to

$$\begin{aligned} \widetilde{FH} &= \int_0^T e^{A(T-\tau)} FH e^{D\tau}\, d\tau, \\ \tilde{\gamma} &= \int_0^T e^{A(T-\tau)} FH \left[\int_0^\tau e^{D(\tau-\xi)} \sigma(\xi + t_0 + nT)\, d\xi\right] d\tau. \end{aligned} \qquad (15)$$

B. A COMPOSITE PLANT/DISTURBANCE MODEL

Setting $\tau = t_0 + (n + 1)T$ in (12), the discrete-time models (12) and (14) corresponding to the continuous-time plant and disturbance equations (4), (10) can be expressed in the composite block-matrix format[2]

$$\left(\frac{x((n+1)T)}{z((n+1)T)}\right) = \left[\begin{array}{c|c} \tilde{A}(nT) & \widetilde{FH}(nT) \\ \hline O & \tilde{D}(nT) \end{array}\right] \left(\frac{x(nT)}{z(nT)}\right) + \left[\frac{\tilde{B}(nT)}{O}\right] u(nT) + \left(\frac{\tilde{\gamma}(nT)}{\tilde{\sigma}(nT)}\right), \qquad (16a)$$

$$y(nT) = [C(nT) \,|\, O] \left(\frac{x(nT)}{z(nT)}\right), \qquad (16b)$$

[2] *The format (16) may also arise from the modeling of plants and disturbances that are intrinsically discrete-time in nature; see remarks below (4).*

where

$$\begin{aligned} \widetilde{D}(nT) &= \Phi_D(t_0 + (n+1)T,\ t_0 + nT), \\ \widetilde{\sigma}(nT) &= \int_{t_0+nT}^{t_0+(n+1)T} \Phi_D(t_0 + (n+1)T,\ \xi)\sigma(\xi)d\xi. \end{aligned} \tag{17}$$

In the time-invariant case, $\widetilde{D}$, $\widetilde{\sigma}$ in (17) simplify to

$$\widetilde{D} = e^{DT}, \quad \widetilde{\sigma} = \int_0^T e^{D(T-\xi)}\sigma(\xi + t_0 + nT)d\xi. \tag{18}$$

The remainder of this chapter is concerned with methods for designing the $u(nT)$ control policy in (16) to achieve specified control objectives in the face of any admissible uncertain disturbance $w(nT) = H(nT)z(nT)$. For this purpose, the control policy for $u(nT)$ will be sought in the extended state-feedback form $u(nT) = \phi(x(nT), z(nT), nT)$.

It should be noted that the two terms $\widetilde{\gamma}(nT)$, $\widetilde{\sigma}(nT)$ in (16) represent the effects caused by the $\sigma(t)$ impulses which arrive during the interval *between* sampling instants; i.e., during $t_0 + nT < t < t_0 + (n+1)T$. Those effects, which we shall call "residuals," result in perturbations in the values of $x((n+1)T)$ and $z((n+1)T)$ in accordance with (16). Since the impulses in $\sigma(t)$ are assumed *completely unknown*, those perturbations due to residuals are likewise unknown, unpredictable, and unmeasurable. Therefore, to avoid excessive repetitions in errors associated with predictions of $x((n+1)T)$, $z((n+1)T)$ using the discrete model (16), it will be necessary to assume that the $\sigma(t)$ impulses arrive in a once-in-a-while fashion with minimal spacing μ between impulses being somewhat *larger* than the sampling period T. As a practical matter it is advisable to have $\mu \geq 5T$.

C. *SOME GENERALIZATIONS OF (16)*

The discrete-time model (16) can be generalized to include various exceptional cases. For instance, the plant output $y(t)$ in (4) may be generalized to contain additional (linear) terms involving $u(t)$ and/or $w(t)$ in which case the discrete-time expression for $y(nT)$ in (16) becomes

$$\begin{aligned} y(nT) &= C(nT)x(nT) + P(nT)u(nT) + G(nT)w(nT), \\ w(nT) &= H(nT)z(nT). \end{aligned} \tag{19}$$

Cases of the form (19) arise when, for instance, accelerometers are used to measure the "motions" of a missile.

Another generalization of (16) consists of allowing plant-state dependent terms to appear in the disturbance model (10) in which case the discrete-time model for $w(nT)$, $z(nT)$ becomes

$$w(nT) = H(nT)z(nT) + L(nT)x(nT), \tag{20a}$$

$$z((n + 1)T) = \tilde{D}(nT)z(nT) + \tilde{M}(nT)x(nT) + \tilde{\sigma}(nT). \tag{20b}$$

Situations requiring models of the type (20) are referred to as cases of "state-dependent" disturbances.

D. *A DISCRETE-TIME MODEL OF SET-POINTS AND SERVO-COMMANDS*

In this chapter, the primary objective of control is assumed to be expressible as the set-point regulation of, or servo-tracking by, certain specified plant variables $(\bar{y}_1, \bar{y}_2, \ldots, \bar{y}_{\bar{m}})$, where in general the $\bar{y}_i$ can be related to the plant state variables $(x_1, \ldots, x_n)$ by the linear expression

$$\bar{y} = \bar{C}(t)x, \quad \bar{y} = (\bar{y}_1, \ldots, \bar{y}_{\bar{m}}). \tag{21}$$

In many cases, the objective is to control the plant output $y(t)$ in (4), in which case one would choose $\bar{C} = C$ in (21). On the other hand, the desire to control the entire plant state $x(t)$ would be indicated by choosing $\bar{C} = I$ in (21).

In accordance with standard procedures in DAC [3], the *desired* (commanded) behavior of $\bar{y}(t)$ is expressed by the "set-point/servo-command" dynamical model

$$\bar{y}(t)\big|_{desired} = \bar{y}_c(t) = G(t)c(t), \tag{22a}$$

$$\dot{c} = \bar{E}(t)c + \bar{\mu}(t), \tag{22b}$$

where $\{G(t), \bar{E}(t)\}$ are determined *a priori* by appropriate "command modeling" procedures [3, p.642]; and where $c = (c_1, \ldots, c_\nu)$ represents the "state" of the command model (22). The vector $\bar{\mu} = (\bar{\mu}_1, \ldots, \bar{\mu}_\nu)$ represents a sequence of totally unknown impulses which are sparse, similar in nature to the $\sigma(t)$ impulses in (2) and (10). It is assumed that the set-point/servo-command vector $\bar{y}_c = (\bar{y}_{c_1}, \ldots, \bar{y}_{c_{\bar{m}}})$ might not be known *a priori*, but *can* be directly and accurately measured on-line in real-time.

In the case of set-point regulation problems the command $\bar{y}_c(t)$ is essentially constant, or piecewise constant, in which case $\bar{E}(t) \equiv 0$ in (22b) and one can then set $G(t) \equiv I$ in (22a), assuming the $\bar{y}_i$ in (21) are independent. In the case of servo-tracking problems, the command $\bar{y}_c(t)$ is allowed to vary continuously with time and $\bar{E}(t)$ is chosen accordingly [3, p. 642].

For purposes of designing discrete-time controllers, it is necessary to have a discrete-time version of the set-point/servo-command model (22). Following the same procedure used for (10), the discrete-time model of (22) is obtained as

$$\bar{y}_c(nT) = G(nT)c(nT), \quad c((n+1)T) = \tilde{E}(nT)c(nT) + \tilde{\mu}(nT), \tag{23}$$

where

$$\tilde{E}(nT) = \Phi_{\bar{E}}((n + 1)T, nT) = \text{transition matrix for } \bar{E}(t), \tag{24a}$$

$$\tilde{\mu}(nT) = \int_{t_0+nT}^{t_0+(n+1)T} \Phi_{\bar{E}}(t_0 + (n + 1)T, \xi)\bar{\mu}(\xi)d\xi, \tag{24b}$$

and where, as before, the arguments nT and (n + 1)T denote the actual times t_0 + nT and t_0 + (n + 1)T, respectively. In the time-invariant case of (22), expressions (24) simplify to

$$\tilde{E}(nT) = e^{\bar{E}T}, \quad \tilde{\mu}(nT) = \int_0^T e^{\bar{E}(T-\xi)}\bar{\mu}(\xi + t_0 + nT)d\xi. \tag{25}$$

The information embodied in the real-time command-state c(nT) enables the DAC controller to "accommodate" uncertain command behavior $\bar{y}_c(nT)$ similar to the way disturbances are accommodated. For this purpose, the DAC control policy (3) is generalized to the form u(nT) = ϕ(x(nT), c(nT), z(nT), nT) as will be demonstrated in Section V.C.

IV. COMPOSITE STATE-OBSERVERS FOR DISCRETE-TIME DAC

A key element in the design of disturbance-accommodating controllers is the construction of a data processing device that can produce on-line, real-time estimates of the plant state x(t), the disturbance state z(t), and the command state c(t). Such a device is called a composite state observer or state reconstructor, and general design receipes for the case of analog-type DACs are given in [3,5,14]. In this section, design procedures are given for two forms of discrete-time composite state observers. These discrete-time observer designs represent natural extensions of the analog-type DAC observers derived in [14].

A. *A FULL-DIMENSIONAL COMPOSITE STATE-OBSERVER*

Recall from (4) and (10) that x is n-dimensional and z is ρ-dimensional. The simplest form of a discrete-time composite state observer for x(nT) and z(nT) is of dimension $(n + \rho)$ and, in the case of (16), has the form[3] [note: hereafter, the shift operator notation Ex(nT), etc., is used to denote x((n + 1)T), etc.]

$$\left(\frac{E\hat{x}(nT)}{E\hat{z}(nT)}\right) = \left[\begin{array}{c|c} \tilde{A}(nT) & \tilde{FH}(nT) \\ \hline 0 & \tilde{D}(nT) \end{array}\right]\left[\frac{\hat{x}(nT)}{\hat{z}(nT)}\right] + \left[\frac{\tilde{B}(nT)}{0}\right]u(nT)$$
$$+ \left[\frac{K_{01}}{K_{02}}\right]\left[\left(C(nT) \mid 0\right)\left(\frac{\hat{x}(nT)}{\hat{z}(nT)}\right) - y(nT)\right], \tag{26}$$

where $\hat{x}$ and $\hat{z}$ denote the outputs produced by the observer and u(nT) and y(nT) denote inputs to the observer which are on-line, real-time sampled data measurements of the plant control input and plant output. The two matrices K_{01} and K_{02} in (26) are "arbitrary" gain matrices that the designer must choose in accordance with certain desired response characteristics.

If the estimation errors are denoted by $\epsilon_x = x(nT) - \hat{x}(nT)$ and $\epsilon_z = z(nT) - \hat{z}(nT)$, then it is easily shown that $\epsilon_x(nT)$ and and $\epsilon_z(nT)$ are governed by the coupled set of discrete-time equations

$$\left[\frac{E\epsilon_x(nT)}{E\epsilon_z(nT)}\right] = \left[\begin{array}{c|c} \tilde{A} + K_{01}C & \tilde{FH} \\ \hline K_{02}C & \tilde{D} \end{array}\right]\left[\frac{\epsilon_x(nT)}{\epsilon_z(nT)}\right] + \left[\frac{\tilde{\gamma}(nT)}{\tilde{\sigma}(nT)}\right], \tag{27}$$

where for notational simplicity the argument (nT) has been ommitted from the various matrices in (27). In order to

[3] *The generalization of (26) to accommodate the more general output expression (19) and/or the presence of "state-dependent" disturbances modeled by (20) is accomplished by the same procedure used for the continuous-time case [5, pp. 431-434].*

produce reliable estimates $\hat{x}(nT)$ and $\hat{z}(nT)$, the observer gain matrices K_{01} and K_{02} must be designed so that $\epsilon_x(nT)$ and $\epsilon_z(nT)$ in (27) both approach zero rapidly between impulses of $\sigma(t)$. This means that the solutions of the homogenous part of (27) must be made strongly asymptotically stable to $\epsilon_x = 0$ and $\epsilon_z = 0$. Since (27) is a *difference* equation, with possibly time-varying coefficients, it is advisable in the most general case to design K_{01} and K_{02} by means of the "discrete Riccati equation method" of optimal control theory. The details of that method may be found in [19].

If $\tilde{A}$, C, $\widetilde{FH}$, and $\tilde{D}$ in (27) are all *constant* matrices, the design of K_{01} and K_{02} is simplified. In particular, those gain matrices may then be chosen as constant matrices so as to place the eigenvalues of the (constant) characteristic matrix

$$\mathbf{A} = \left[\begin{array}{c|c} \tilde{A} & \widetilde{FH} \\ \hline 0 & \tilde{D} \end{array}\right] + \left[\begin{array}{c} K_{01} \\ \hline K_{02} \end{array}\right][C|0] = \left[\begin{array}{c|c} \tilde{A} + K_{01}C & \widetilde{FH} \\ \hline K_{02}C & \tilde{D} \end{array}\right] \tag{28}$$

at suitably damped locations within the unit circle of the complex-plane. It is remarked that one attractive choice for the eigenvalue locations of $\mathbf{A}$ is to place all λ_i at zero. This choice produces "deadbeat" response of the homogenous part of (27), which means that both $\epsilon_x(nT)$ and $\epsilon_z(nT)$ theoretically reach zero in a finite number of sample periods, assuming that the residuals $\tilde{\gamma}$ and $\tilde{\sigma}$ are "quiet" during such an interval. This level of performance is not attainable from the conventional analog-type state observer. A block-diagram of the composite observer (26) is shown in Fig. 1.

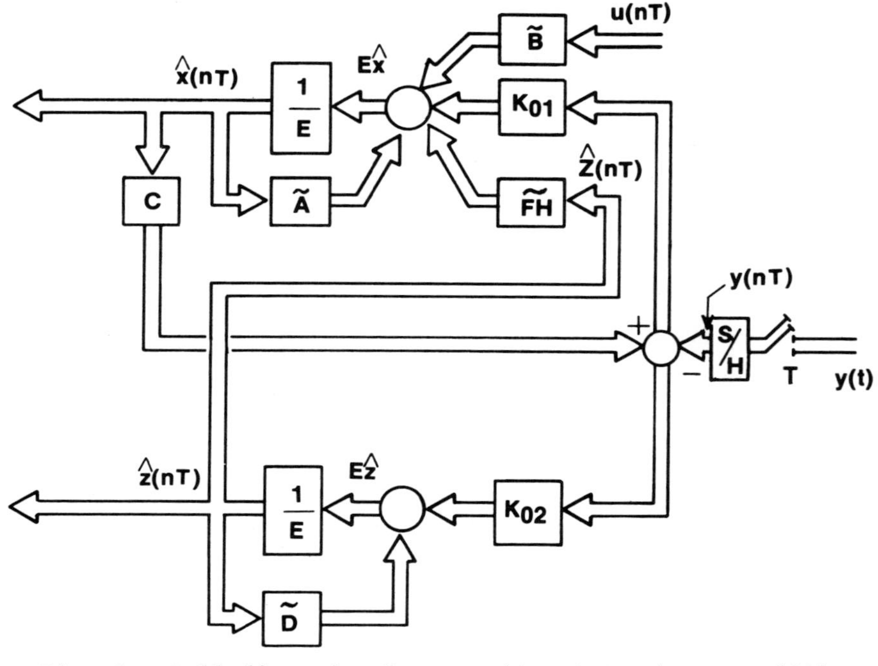

Fig. 1. Full-dimensional composite state-observer (26) for discrete-time DAC. Note: 1/E = T-unit delayor.

B. A REDUCED-DIMENSION COMPOSITE STATE-OBSERVER

The dimension of an observer indicates the number of delays (or integrators in analog systems) that the designer must use in the construction of the observer. Thus, there is an interest in seeking alternative observer designs that have a dimension less than $(n + \rho)$. One such observer, which has dimension $(n + \rho - m)$ where m, the dimension of $y(t)$, is assumed equal to the rank of C, can be constructed as follows. For simplicity, only the time-invariant case of (16) will be considered here; the time-varying case, as well as cases of (19) and/or (20), are handled just as in [5, pp. 480-487].

First, let T_{12} and T_{22} be, respectively, any $n \times (n + \rho - m)$ and $\rho \times (n + \rho - m)$ matrices that satisfy the two conditions

$$[C|0][T_{12}/T_{22}] = 0, \quad \text{rank}[T_{12}/T_{22}] = n + \rho - m. \tag{29}$$

It is remarked that such matrices always exist, are nonunique, and are easily found. Next define the two auxiliary matrices $\overline{T}_{12}$, $\overline{T}_{22}$ as

$$\overline{T}_{12} = \left(T_{12}^T T_{12} + T_{22}^T T_{22}\right)^{-1} T_{12}^T;$$

$$\overline{T}_{22} = \left(T_{12}^T T_{12} + T_{22}^T T_{22}\right)^{-1} T_{22}^T;$$

and set

$$C^{\#} = (CC^T)^{-1} C.$$

Then, a reduced-dimension composite state-observer for the time-invariant case of (16) can be designed as follows. The first section of the observer consists of an $(n + \rho - m)$-dimension *digital filter* having the vector output variable $\xi(nT)$ and obeying the difference equation

$$E\xi(nT) = (\widetilde{\mathscr{D}} + \Sigma\widetilde{\mathscr{H}})\xi(nT) + [(\overline{T}_{12} + \Sigma C)(\widetilde{A}C^{\#T}) - (\widetilde{\mathscr{D}} + \Sigma\widetilde{\mathscr{H}})\Sigma]y(nT) + (\overline{T}_{12} + \Sigma C)\widetilde{B}u(nT), \tag{30a}$$

where $y(nT)$ and $u(nT)$ are filter inputs, Σ is an "arbitrary" gain matrix, which the designer must choose to meet performance specifications, and

$$\widetilde{\mathscr{D}} = \overline{T}_{12}(\widetilde{A}T_{12} + \widetilde{F}HT_{22}) + \overline{T}_{22}\widetilde{D}T_{22}, \tag{30b}$$

$$\widetilde{\mathscr{H}} = C(\widetilde{A}T_{12} + \widetilde{F}HT_{22}). \tag{30c}$$

The second section of the observer consists of a pair of algebraic "assembly equations," which indicate just how the real-time filter output $\xi(nT)$ and the sampled-data plant output measurements $y(nT)$ are combined to form the estimates $\hat{x}(nT)$ and $\hat{z}(nT)$. Those assembly equations are

$$\hat{x}(nT) = T_{12}\xi(nT) + \left[C^{\#T} - T_{12}\Sigma\right]y(nT), \qquad (31a)$$

$$\hat{z}(nT) = T_{22}\xi(nT) - T_{22}\Sigma y(nT). \qquad (31b)$$

The design of the gain matrix Σ is guided by the dynamic behavior of the estimation errors $(x - \hat{x})$ and $(z - \hat{z})$. It is straightforward to show that

$$(x - \hat{x}) = T_{12}\epsilon(nT), \quad (z - \hat{z}) = T_{22}\epsilon(nT), \qquad (32)$$

where the error variable $\epsilon(nT)$ is governed, between impulses of $\sigma(t)$, by the homogenous difference equation

$$E\epsilon(nT) = (\tilde{\mathscr{D}} + \Sigma\tilde{\mathscr{H}})\epsilon(nT). \qquad (33)$$

Thus, in order to achieve $\hat{x}(nT) \approx x(nT)$ and $\hat{z}(nT) \approx z(nT)$ promptly, the designer should choose Σ so that (33) is strongly asymptotically stable to $\epsilon = 0$. This means that Σ should be

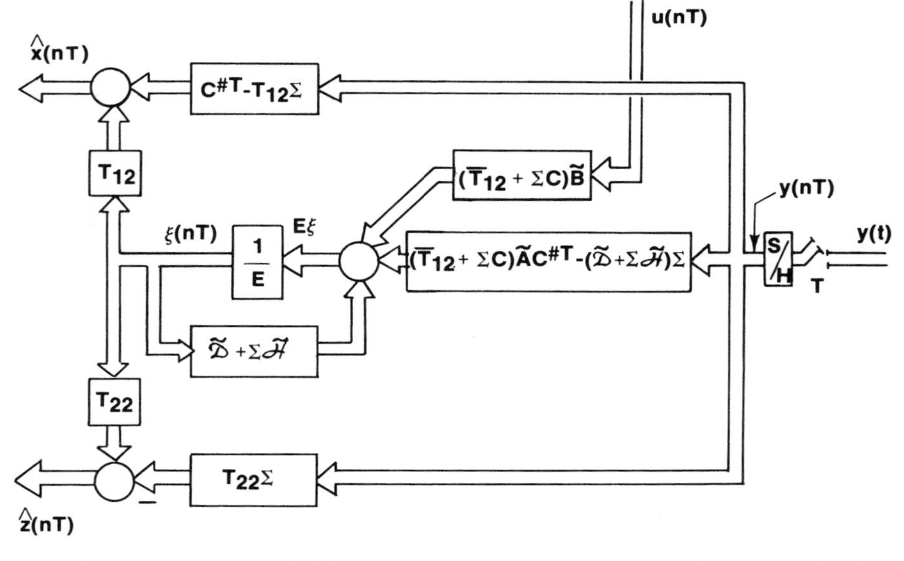

Fig. 2. Reduced-dimension composite state-observer (30), (31) for discrete-time DAC. Note: 1/E = T-unit delayor.

chosen to place the eigenvalues of $(\tilde{\mathscr{D}} + \Sigma\tilde{\mathscr{H}})$ at appropriately damped locations within the interior of the unit circle in the complex plane. Therefore, all the remarks previously stated concerning the design of K_{01} and K_{02} in (28) apply also to the design of Σ in (33).

A block diagram of the reduced-dimension observer (30), (31) is shown in Fig. 2.

Comparison of Figs. 1 and 2 shows that the full-dimensional observer (26) produces on-line, real-time estimates of *both* $\{x(nT), z(nT)\}$ and $\{x((n + 1)T), z((n + 1)T)\}$, whereas the reduced-dimension observer (30), (31) produces on-line, real-time estimates of only the current states $\{x(nT), z(nT)\}$. This "one-step-ahead" prediction capability of the full-dimensional observer (26) is an important advantage in certain DAC applications. Two such applications are described in Sections V.D-4; F-2 of this chapter.

C. A DISTURBANCE STATE-OBSERVER FOR DIRECTLY MEASURABLE DISTURBANCES

In some applications, particularly in chemical process control problems, it turns out that some of the disturbance inputs $w(t)$ can be directly measured on-line in real-time. We denote the subset of such measurable disturbances by $w^{(m)} = (w_{m1}, \ldots, w_{ms})$. For such cases, one can employ a separate observer to estimate the state $z^{(m)}(nT)$ associated with the measurable disturbance vector $w^{(m)}(t)$ as follows. A full-dimensional separate observer for $z^{(m)}(nT)$ can be realized by using a minor variation of (26):

$$E\hat{z}^{(m)} = \tilde{D}_m\hat{z}^{(m)} + K_m\left[H_m\hat{z}^{(m)} - w^{(m)}\right], \tag{34}$$

where $w^{(m)}$ denotes the on-line, real-time sampled-data disturbance measurements and it is assumed, as in (10), that the $w^{(m)}(t)$ is modeled by[4] $w^{(m)} = H_m z^{(m)}$ and $\dot{z}^{(m)} = D_m z^{(m)} + \sigma^{(m)}(t)$. The matrix $\tilde{D}_m(nT)$ is defined in terms of $D_m(t)$ by expressions analogus to (17) and (18). The estimation error $\epsilon_z^{(m)} = z^{(m)} - \hat{z}^{(m)}$ associated with (34) is governed by $E\epsilon_z^{(m)} = (\tilde{D}_m + K_m H_m)\epsilon_z^{(m)} + \tilde{\sigma}^{(m)}$ and thus the gain matrix K_m can be designed by the procedures described below (27) and (28).

A reduced-order version of (34) can be developed by mimicking the structure of (30) and (31); e.g., treat $z^{(m)}$ as x in (16), (30), and (31), setting z = 0. If the behavior of $w^{(m)}(t)$ is somehow coupled with the dynamics of the *un*measurable disturbances, then one must employ a modified version of (26) [or (30), (31)] rather than (34). The continuous-time (analog) versions of such composite observers for partially measurable coupled disturbances is presented in [5, pp. 431-434]. The structures of the discrete-time counterparts follow by obvious modifications analogous to those represented by (26), (30), and (31). The latter results are not presented here.

D. A COMMAND STATE-OBSERVER

The anticipated servo-commands $\bar{y}_c(t)$ in discrete-time DAC are assumed to be modeled by (22), which leads to the discrete-time expressions (23) and (24). Since the DAC requires on-line real-time knowledge of the command *state* c(nT) rather than just the command value $\bar{y}_c(nT)$, see remarks below (25), it is necessary to employ a separate on-line servo-command observer to estimate c(nT) from the on-line measurements $\bar{y}_c(nT)$.

[4] *The allowance for state-dependent disturbances (20) is handled just as in [5, pp. 431-434].*

For this purpose, a full-dimensional command state-observer can be realized in a form analogous to (34):

$$E\hat{c} = \tilde{E}\hat{c} + K_c[G\hat{c} - \bar{y}_c], \tag{35}$$

where the design of the gain matrix K_c is treated just as for K_m in (34). A reduced-dimension version of (35) can also be realized following the procedure already described.

This completes our discussion of state observers for discrete-time DAC applications. We now turn to the central issue in discrete-time DAC: how to design the feedback control function (policy) $u(nT) = \phi(x(nT), c(nT), z(nT), nT)$.

V. DESIGN OF DISCRETE-TIME DAC CONTROL POLICIES

In this section we develop design algorithms for synthesizing discrete-time DAC control laws (policies) which accomplish the absorption, minimization, and utilization modes of disturbance accommodation. The various discrete-time models of plants, disturbances, and commands, upon which these design algorithms are based, have been derived in Section III and are summarized here for convenience.

The plant state x and output y are assumed to be modeled by the discrete-time expressions [see (14)]

$$Ex(nT) = \tilde{A}(nT)x(nT) + \tilde{B}(nT)u(nT) + \widetilde{FH}(nT)z(nT) + \tilde{\gamma}(nT), \tag{36a}$$

$$y(nT) = C(nT)x(nT), \quad Ex(nT) \triangleq x((n+1)T). \tag{36b}$$

The disturbance w is likewise modeled by [see (12) and (16)]

$$w(nT) = H(nT)z(nT), \quad Ez(nT) = \tilde{D}(nT)z(nT) + \tilde{\sigma}(nT), \tag{37}$$

and the plant variables $\bar{y} = (\bar{y}_1, \bar{y}_2, \ldots, \bar{y}_{\bar{m}})$ to be controlled are assumed expressible as [see (21)]

$$\bar{y}(nT) = \bar{C}(nT)x(nT). \tag{38}$$

Finally, the commanded (desired) behavior of $\bar{y}(nT)$ is modeled by [see (23) and (24)]

$$\bar{y}_c(nT) = G(nT)c(nT), \quad Ec(nT) = \tilde{E}(nT)c(nT) + \tilde{\mu}(nT). \tag{39}$$

The design algorithms derived here can be easily extended to accommodate the more general model (19) of (36b) and/or the more general model (20) of (37).

A. DESIGN FOR THE DISTURBANCE-ABSORPTION MODE

The disturbance-absorption mode of accommodation is concerned with manipulating the control u so as to absorb (cancel-out) the effects of disturbances on the plant response. To accomplish that task and at the same time achieve the primary control task of regulation, servo-tracking, etc., it is standard procedure in DAC theory to allocate the control effort into two parts as follows:

$$u(nT) = u_p(nT) + u_d(nT), \tag{40}$$

where the part u_d is responsible for absorbing the disturbance effects, and the part u_p is responsible for achieving the primary control task. In many cases one can design the parts u_d and u_p independently.

Substitution of (40) into (36a) yields

$$Ex(nT) = \tilde{A}(nT)x(nT) + \tilde{B}(nT)u_p(nT) + \tilde{B}(nT)u_d(nT) + \widetilde{FH}(nT)z(nT) + \tilde{\gamma}(nT). \tag{41}$$

We shall now describe a design procedure for synthesizing u_d in (41).

1. Complete Absorption of Disturbances

Disturbances w(t) typically act on a plant in a continuously varying manner. On the other hand, the control action in a discrete-time control problem is typically held *constant* between sampling times t_i. In such cases it is usually impossible to absorb completely all disturbance effects for *all* $t_i \le t \le (t_i + T)$, unless the disturbance itself happens to remain constant between sampling times. Moreover, the presence of the totally uncertain "residual" term $\tilde{\gamma}(nT)$ in (41) imposes a further limit on the degree of complete absorption achievable by discrete-time control. Thus, in discrete-time DAC design the concept of "complete absorption" is understood to mean the ideal cancellation of the disturbance effects $\widetilde{FH}(nT)z(nT)$ as they appear at the isolated[5] sample-times t_i. It is recalled that we are using the shorthand notation $t_i = t_n = nT$, $n = 0, 1, 2, \ldots$ to denote the actual sample times $t_i = t_n = t_0 + nT$.

It follows that complete disturbance absorption for the discrete-time model (41) implies that $u_d(nT)$ should be manipulated so as to achieve the condition

$$\tilde{B}(nT)u_d(nT) + \widetilde{FH}(nT)z(nT) = 0; \quad n = 0, 1, 2, \ldots \tag{42}$$

for all $z(nT)$. The necessary and sufficient condition for existence of a control $u_d(nT)$ satisfying (42) is given by:

Theorem. Existence condition for complete disturbance absorption in discrete-time: A control $u_d(nT)$ satisfying (42) exists if, and only if $\text{rank}[\tilde{B}|\widetilde{FH}] = \text{rank}[\tilde{B}]$; that is, if and only if

$$\widetilde{FH}(nT) = \tilde{B}(nT)\tilde{\Gamma}(nT); \quad n = 0, 1, 2, \ldots \tag{43}$$

for some matrix $\tilde{\Gamma}(nT)$. ▮

[5]*The possibility of achieving disturbance absorption* between *sample-times, using a disturbance interpolater, is discussed in Section V.F.*

The condition (43) is called the "complete absorbability" condition for discrete-time DAC. It is remarked that condition (43) is the discrete-time version of the condition $FH = B\Gamma$ associated with continuous-time DAC theory [5, p. 439]. In this regard, it should be noted that condition (43) may *fail* to be satisfied even though $FH = B\Gamma$ *is* satisfied. Thus, the property of "complete absorbability" may be lost as a consequence of converting from continuous-time to discrete-time control. When (43) does fail, the designer can pursue several alternative options as discussed in Section V.B.

Assuming condition (43) is satisfied, the control $u_d(nT)$ satisfying (42) may be chosen ideally as

$$u_d(nT) = -\widetilde{\Gamma}(nT)z(nT). \tag{44}$$

However, the ideal expression (44) is not physically realizable because $z(nT)$ cannot be directly measured. Thus, as a physically realizable approximation to (44), we shall design u_d as

$$u_d(nT) = -\widetilde{\Gamma}(nT)\hat{z}(nT), \tag{45}$$

where $\hat{z}(nT)$ is the on-line, real-time estimate of $z(nT)$ produced by a disturbance-state observer as described in Section IV. Substituting (45) into (41), using (43) and the observer error expression $\epsilon_z = z - \hat{z}$, yields the "closed-loop" plant dynamics

$$\begin{aligned} Ex(nT) = {} & \widetilde{A}(nT)x(nT) + \widetilde{B}(nT)u_p(nT) \\ & + \widetilde{B}(nT)\widetilde{\Gamma}(nT)\epsilon_z(nT) + \widetilde{\gamma}(nT). \end{aligned} \tag{46}$$

Expression (46) shows that the effect of disturbances on (41) has been reduced to the residual $\widetilde{\gamma}(nT)$ and the term $\widetilde{B}(nT)\widetilde{\Gamma}(nT)\epsilon_z(nT)$, where supposedly $\epsilon_z(nT) \to 0$ rapidly by virtue of the observer design. Since the $\sigma(t)$ impulses that produce the residual term $\tilde{\gamma}(nT)$ are assumed completely unknown, unpredictable, and unmeasurable, there is no scientific way to absorb the remaining effects in (46). In other words, the DAC policy (45) absorbs the disturbance effects in (41) as thoroughly as possible.

The primary control term $u_p(nT)$ in (46) can now be designed by well-known conventional techniques [19], where the remaining disturbance-related terms $\tilde{B}\tilde{\Gamma}\epsilon_z(nT) + \tilde{\gamma}(nT)$ in (46) are ignored. For instance, in the case of stabilization of $x(nT)$ to zero, the term $u_p(nT)$ typically has the form $u_p(nT) = K_p(nT)x(nT)$. The design of DACs for primary control tasks of the set-point or servo-tracking type is outlined in Section V.C.

A complete block-diagram of the original continuous-time plant model (4), original continuous-time disturbance model (10), and the proposed disturbance-absorbing digital controller (45) [using the full-dimensional discrete-time composite state-observer (26) and $u_p = K_p x$] is shown in Fig. 3.

The typical response of the plant state $x(t)$, under the ideal discrete-time disturbance-absorption control policy (44), is shown in Fig. 4. At the isolated sample times $t_n = nT$, $n = 0, 1, 2$, the disturbed response $x(t)$ ideally coincides with the response of the undisturbed system [assuming that $\tilde{\gamma}(nT)$ is quiet]. Between the sample times t_n, the disturbed response $x(t)$ deviates from the undisturbed response due to variations in $w(t)$. These latter deviations in $x(t)$ are referred to as "intersample ripple" and their extent is one factor in determining the appropriate choice for the controller

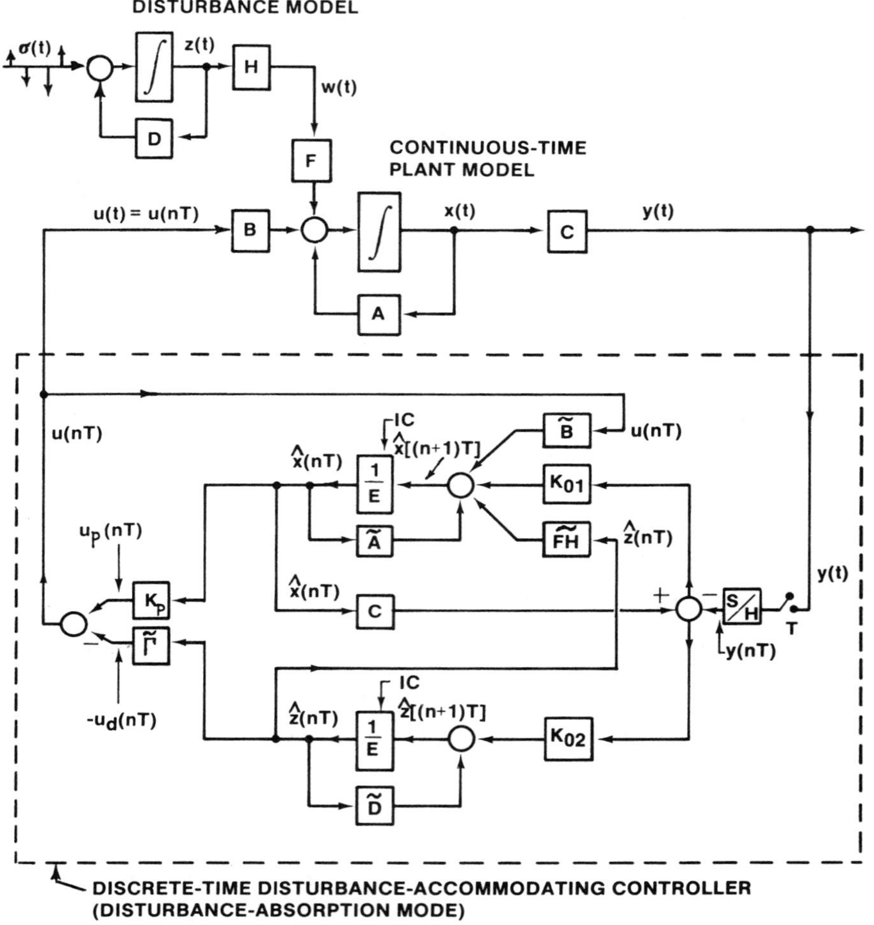

Fig. 3. General structure of discrete-time disturbance-absorbing controller. Note: $1/E \triangleq T$-unit delayor.

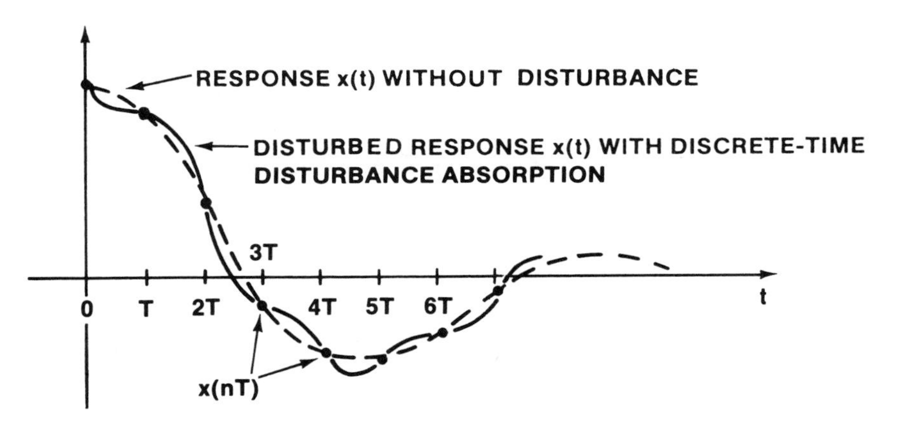

Fig. 4. Effect of discrete-time disturbance-absorption control policy (44) on motion of plant state x(t) in (4) [intersample ripple].

sample period T. In particular, as T is made smaller, the extent of intersample ripple tends to diminish, in general.

2. *Example Design of a Discrete-Time Disturbance-Absorbing Controller*

The general design algorithm outlined in the preceding section is illustrated by the following detailed example.

Suppose the continuous-time plant (4) is of first order ($n = 1$) with known equation of motion

$$\dot{x} = ax + bu + fw(t); \quad x = \text{scalar}; \tag{47a}$$

$$y = cx; \quad y = \text{scalar}, \quad c \neq 0. \tag{47b}$$

The plant coefficients a, b, c, and f in (47) are arbitrary real constants, assumed known. Suppose further that the scalar disturbance $w(t)$ has been experimentally modeled and found to obey a first-order continuous equation (10) of the form

$$w = z(t), \quad \text{i.e.,} \ h = 1 \tag{48a}$$

$$\dot{z} = dz + \sigma(t), \tag{48b}$$

where the (real) coefficient d in (48b) is an arbitrary constant assumed known. For instance, by setting $d = 0$ in (48b) the case of piecewise "constant" disturbances is obtained; setting $d = -2$ yields the case of piecewise exponentially decaying disturbances $w(t) = Ce^{-2t}$, etc.

The first step in the design is to create appropriate discrete-time models to represent the given continuous-time models (47) and (48). In other words, the designer must calculate the discrete-time parameters $\tilde{A}$, $\tilde{B}$, $\widetilde{FH}$, $\tilde{D}$ for the discrete-time models (36) and (37). Since (47) and (48) are both time-invariant, the appropriate equations are (6), (12), and (15), which lead to the following results [$\tilde{\gamma}$ and $\tilde{\sigma}$ are not needed in the design algorithm and therefore need not be calculated]:

$$\tilde{A} = e^{AT} = e^{aT}, \tag{49a}$$

$$\tilde{B} = \int_0^T e^{A(T-\tau)} B\,d\tau = \frac{b}{a}(e^{aT} - 1), \quad [= bT \quad \text{if } a = 0], \tag{49b}$$

$$\tilde{D} = e^{DT} = e^{dT}, \tag{49c}$$

$$\widetilde{FH} = \int_0^T e^{A(T-\tau)} FHe^{D\tau} d\tau = \left(\frac{f}{d - a}\right)[e^{dT} - e^{aT}],$$

$$[= fTe^{aT} \quad \text{if } a = d]. \tag{49d}$$

Thus the discrete-time models (36) and (37) of the original equations (47) and (48) are

$$Ex = e^{aT}x(nT) + \frac{b}{a}(e^{aT} - 1)u(nT)$$

$$+ \left(\frac{f}{d - a}\right)[e^{dT} - e^{aT}]z(nT) + \tilde{\gamma}(nT), \tag{50a}$$

$$Ez = e^{dT}z(nT) + \tilde{\sigma}(nT). \tag{50b}$$

Next, the designer agrees to split (allocate) the total control effort $u(t)$ into the two parts (40), thereby bringing (50a) to the form

$$Ex = e^{aT}x(nT) + \frac{b}{a}(e^{aT} - 1)u_p + \frac{b}{a}(e^{aT} - 1)u_d$$
$$+ \left(\frac{f}{d - a}\right)[e^{dT} - e^{aT}]z(nT). \tag{51}$$

The design of u_d in (51) begins by first checking for satisfaction of the "complete absorbability" condition (43), which, for the present example, reduces to the question

$$\left(\frac{f}{d - a}\right)[e^{dT} - e^{aT}] \overset{?}{=} \frac{b}{a}(e^{aT} - 1)\tilde{\Gamma}, \quad \text{for some } \tilde{\Gamma}. \tag{52}$$

It is obvious that (52) *can* be satisfied by some $\tilde{\Gamma}$. In fact, $\tilde{\Gamma}$ is given explicitly by

$$\tilde{\Gamma} = \frac{af}{b(d - a)}\left[\frac{e^{dT} - e^{aT}}{e^{aT} - 1}\right]. \tag{53}$$

The specific expression for $u_d(nT)$ is now given by the algorithm (45)

$$u_d(nT) = \left(\frac{af}{b(a - d)}\right)\left[\frac{e^{dT} - e^{aT}}{e^{aT} - 1}\right]\hat{z}(nT), \tag{54}$$

where $\hat{z}(nT)$ must be obtained from a composite state observer. We shall design such an observer in a moment.

Substituting (54) into (51), and setting $\hat{z} = z - \epsilon_z$, the discrete-time plant equation becomes

$$Ex = e^{aT}x(nT) + \frac{b}{a}(e^{aT} - 1)u_p$$
$$+ \frac{f}{(d - a)}[e^{dT} - e^{aT}]\epsilon_z(nT) + \tilde{\gamma}(nT). \tag{55}$$

The primary control term u_p can now be designed by ignoring the disturbance-related terms in (55) and using conventional digital control design procedures on the undisturbed plant equation

$$Ex = e^{aT}x(nT) + \frac{b}{a}(e^{aT} - 1)u_p(nT). \tag{56}$$

Suppose, for instance, that the primary control objective is to stabilize the original plant (47) to the equilibrium state $x = 0$. In that case, the digital control $u_p(nT)$ in (56) can be chosen as the linear, state-feedback expression

$$u_p(nT) = k\hat{x}(nT), \tag{57}$$

where $\hat{x}(nT)$ is to be obtained from a composite state observer and where the gain constant k is chosen to place the closed-loop root $\tilde{\lambda}$ of (56) at an appropriately damped location inside the "unit-circle" [interval(-1, +1)] of the real-line. In other words, k is designed subject to the constraint

$$|\tilde{\lambda}| = |e^{aT} + (b/a)(e^{aT} - 1)k| < 1. \tag{58}$$

Using (57), the closed-loop dynamics of (56) becomes

$$Ex = [e^{aT} + (b/a)(e^{aT} - 1)k]x(nT) + (b/a)(1 - e^{aT})k\epsilon_x(nT), \tag{59}$$

where the expression $\hat{x} = x - \epsilon_x$ has been incorporated in (59).

One attractive choice for k in (59) is the value that ideally produces "deadbeat response" of $x(nT)$. In that case, k is chosen to place the characteristic root $\tilde{\lambda}$ of (59) at zero, which means

$$k = k_{db} = ae^{aT}/b(1 - e^{aT}). \tag{60}$$

The corresponding closed-loop equation (59) then becomes

$$Ex = e^{aT}\epsilon_x(nT), \tag{61}$$

which indicates that $x(nT) \to 0$ in one sample period *after* the approximation $\epsilon_x \cong 0$ is reached. This suggests that deadbeat behavior of $\epsilon_x(nT)$ would be an attractive design option in the design of the observer.

In summary, the final design of the digital control $u(nT)$ is given by (54) and (57) as

$$u(nT) = u_p(nT) + u_d(nT)$$

$$= k\hat{x}(nT) + \left(\frac{af}{b(a - d)}\right)\left[\frac{e^{dT} - e^{aT}}{e^{aT} - 1}\right]\hat{z}(nT). \tag{62}$$

The next step is to design an appropriate composite observer. For this example, a reduced-order observer of the type (30) and (31) will be designed. To proceed, the matrices T_{12} and T_{22} for this example are both scalars and can be chosen to satisfy the conditions (29) by picking

$$T_{12} = 0, \quad T_{22} = 1. \tag{63a}$$

As a consequence of (63a), $\bar{T}_{12}$ and $\bar{T}_{22}$ are given by [see (29)]

$$\bar{T}_{12} = 0, \quad \bar{T}_{22} = 1 \tag{63b}$$

and

$$C^{\#} = \frac{1}{c}, \quad c = \text{given by (47)}. \tag{63c}$$

The digital filter parameters $\tilde{\mathscr{D}}$ and $\tilde{\mathscr{H}}$ of the observer are now easily calculated from (30b,c) as

$$\tilde{\mathscr{D}} = e^{dT}, \quad \tilde{\mathscr{H}} = \frac{cf}{d - a}[e^{dT} - e^{aT}]. \tag{64}$$

Substituting these calculated parameters into the observer filter equation (30a) and assembly equations (31), the final form of the observer for this example is obtained as the recursive digital filter

$$E\xi = \left[e^{dT} + \frac{\Sigma fc}{d - a}(e^{dT} - e^{aT})\right]\xi(nT)$$
$$+ \left[\Sigma(e^{aT} - e^{dT})\left(1 + \frac{\Sigma fc}{d - a}\right)\right]y(nT) \tag{65a}$$
$$+ \left[\frac{\Sigma cb}{a}(e^{aT} - 1)\right]u(nT),$$

$$\hat{x} = \left(\frac{1}{c}\right)y(nT) = x, \tag{65b}$$

$$\hat{z} = \xi(nT) - \Sigma y(nT). \tag{65c}$$

The observer gain parameter Σ in (65) is designed by considering the dynamical equation (33) of the estimation errors, which for this example is given by

$$E\epsilon(nT) = \left[e^{dT} + \Sigma\left(\frac{cf}{d - a}\right)(e^{dT} - e^{aT})\right]\epsilon(nT). \tag{66}$$

If $\tilde{\lambda}_0$ denotes the desired root for (66) $[|\tilde{\lambda}_0| < 1]$, the corresponding value of Σ is thereby specified as

$$\Sigma = \left(\tilde{\lambda}_0 - e^{dT}\right)(d - a)/cf(e^{dT} - e^{aT}). \tag{67}$$

For instance, if the designer decides to achieve deadbeat response for (66), then $\tilde{\lambda}_0 = 0$ and therefore Σ should be designed as

$$\Sigma = \Sigma_{db} = e^{dT}(a - d)/cf(e^{dT} - e^{aT}). \tag{68}$$

In the special case (68), the observer filter (65a) reduces to the nonrecursive digital filter

$$E\xi = \left[\frac{(a - d)e^{(a+d)T}}{cf(e^{dT} - e^{aT})}\right]y(nT)$$
$$+ \left[\frac{b(a - d)e^{dT}(e^{aT} - 1)}{af(e^{dT} - e^{aT})}\right]u(nT). \tag{69}$$

A complete block-diagram of the plant, disturbance, and controller for this example is shown in Fig. 5.

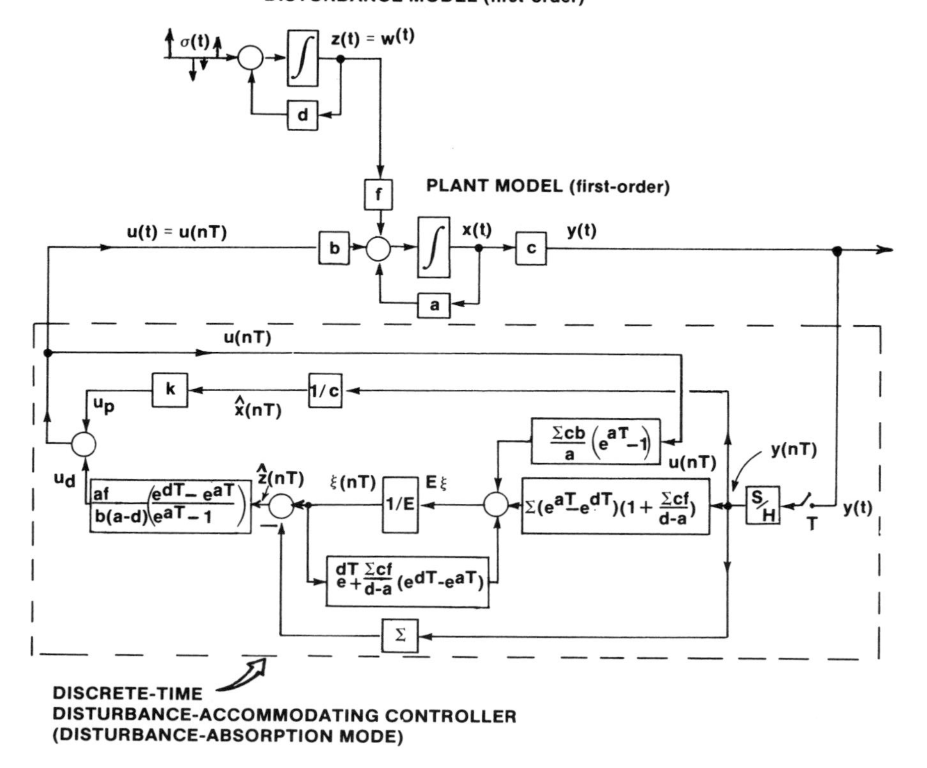

Fig. 5. Block diagram of plant, disturbance, and disturbance-absorbing digital controller for the example of Section V.A-2. Note: 1/E = T-unit delayor.

The disturbance-absorbing digital controller (62) and (65) represents a general stabilizing control policy for the first-order plant (47) with general first-order disturbance model (48). It is interesting to consider the structure of that controller for the special case of piecewise constant disturbances $w(t) = \text{constant} = C_i$ and deadbeat observer gain (68). For that purpose, it is only necessary to set $d = 0$ in (48) in which case the controller (62) and (69) reduces to the

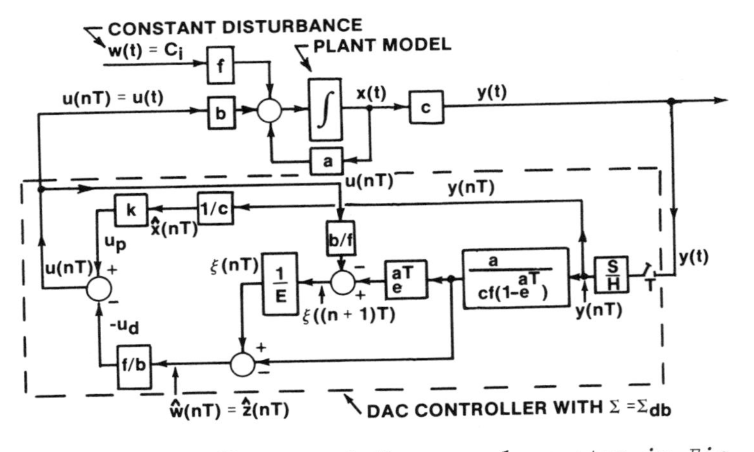

Fig. 6. Block diagram of the example system in Fig. 5 with constant disturbance and disturbance-absorbing digital controller (70). Note: To achieve plant deadbeat response, design k as $k_{db} = ae^{aT}/b(1 - e^{aT})$.

special form

$$u(nT) = k\hat{x}(nT) - (f/b)\hat{z}(nT), \tag{70a}$$

$$E\xi(nT) = \left[\frac{ae^{aT}}{cf(1 - e^{aT})}\right]y(nT) - \left(\frac{b}{f}\right)u(nT), \tag{70b}$$

$$\hat{x} = \left(\frac{1}{c}\right)y(nT) = x, \tag{70c}$$

$$\hat{z} = \xi(nT) - \left(\frac{a}{cf[1 - e^{aT}]}\right)y(nT). \tag{70d}$$

A block diagram of the system (47) with disturbance-absorbing digital controller (70) is shown in Fig. 6. This particular case of DAC is comparable to the classical controller scheme known as (digital) "integral feedback." Note that the u_d loop cancels the disturbance quickly even before $x(nT) \to 0$. Moreover, as $x(nT) \to 0$ in Fig. 6, the $u_d(nT)$ section of the controller becomes a T-unit delayor (1/E) with unity positive feedback and zero input. This means that the control u_d then "settles-down" to hold the constant value (-wf/b) thereby continuing to cancel the constant external disturbance term fw.

When the disturbance changes its "constant" value, the resulting perturbation in $y(t)$ drives the delayor loop to update the steady value of $u_d(nT)$.

3. Disturbance Absorption for Critical Variables

In some applications, it is only the effect of disturbances on certain "critical variables" of the plant that is important. For such cases, it may be more economical to design u_d to accomplish disturbance absorption *only* with respect to the plant critical variables rather than complete disturbance absorption.

To pursue this idea, suppose the plant variables $\bar{y} = (\bar{y}_1, \bar{y}_2, \ldots, \bar{y}_{\bar{m}})$ introduced in (21) and (38) are considered as the set of "critical variables." The design of u_d to absorb the effects of disturbances $w(t)$ on the behavior of the $\bar{y}_j(nT)$, $j = 1, 2, \ldots, \bar{m}$, can be accomplished by the following systematic procedure. First, one must derive a special auxiliary set of difference equations which govern the discrete-time motions $\bar{y}_j(nT)$. This auxiliary set of equations will, in general, contain "forcing terms" involving $u(nT)$, $u[(n + 1)T]$, $u[(n + 2)T], \ldots$, $z(nT)$, $z[(n + 1)T]$, $z[(n + 2)T]$, $\ldots$, and must be *homogenous* in the critical variables $\bar{y}_1$, $\bar{y}_2, \ldots, \bar{y}_{\bar{m}}$. Recall from (38) that $\bar{y}$ is defined by $\bar{y} = \bar{C}x$, which can be written as

$$\bar{y} = \bar{C}x = \begin{bmatrix} \bar{c}_1 \\ \bar{c}_2 \\ \cdot \\ \cdot \\ \cdot \\ \bar{c}_{\bar{m}} \end{bmatrix} (x) \tag{71}$$

where $\bar{c}_j$ represents the jth row of $\bar{C}$. Therefore, the $\bar{y}_j$ can be expressed as

$$\bar{y}_j = \bar{c}_j x, \quad j = 1, 2, \ldots, \bar{m}. \tag{72}$$

For simplicity, we shall assume in this section that $\bar{C}$ in (71) and $\tilde{A}$, $\tilde{B}$, $\widetilde{FH}$, $\tilde{D}$ in (36) and (37) are all constant.

The minimal-order difference equations governing each of the $\bar{y}_j(nT)$, subject to the required homogeneity condition just cited, are found by taking successively higher (forward) time-shifts $E(\cdot)$ of (72), using (36), (37), and (71). For that purpose, the residuals $\tilde{\gamma}(nT)$ and $\tilde{\sigma}(nT)$ in (36) and (37) can be disregarded since they are sparse and totally unknown [see remarks following (18)]. Starting with $\bar{y}_1$, the first time-shift of (72) yields

$$\begin{aligned} E\bar{y}_1(nT) &\triangleq \bar{y}_1((n+1)T) = \bar{c}_1 Ex(nT) \\ &= \bar{c}_1 [\tilde{A}x(nT) + \tilde{B}u(nT) + \widetilde{FH}z(nT)]. \end{aligned} \tag{73}$$

We now ask the question: Can $\bar{y}_1((n+1)T)$ in (73) be expressed as some weighted linear combination of the variables $\bar{y}_1(nT), \bar{y}_2(nT), \ldots, \bar{y}_{\bar{m}}(nT)$, $u(nT)$, and $z(nT)$? In other words, can one express the right side of (73) as

$$\bar{y}_1((n+1)T) = \beta_1 \bar{C}x(nT) + \bar{c}_1 \tilde{B}u(nT) + \bar{c}_1 \widetilde{FH}z(nT) \tag{74}$$

for some row vector $\beta_1 = (\beta_{11}, \beta_{12}, \ldots, \beta_{1\bar{m}})$? Clearly, (73) *can* be written in the form of (74) if and only if

$$\bar{c}_1 \tilde{A}x = \beta_1 \bar{C}x, \quad \text{for all } x, \tag{75}$$

which implies

$$\bar{c}_1 \tilde{A} = \beta_1 \bar{C}. \tag{76}$$

The necessary and sufficient condition for existence of a vector β_1 satisfying (76) is

$$\text{rank}\left[\bar{C}^T | \tilde{A}^T \bar{c}_1^T\right] = \text{rank}\,[\bar{C}^T] \tag{77}$$

or, stated equivalently,

$$\mathscr{R}\left[\tilde{A}^T \bar{c}_1^T\right] \subseteq \mathscr{R}[\bar{C}^T], \; \mathscr{R}[\cdot] = \text{col. range space}. \tag{78}$$

If (77) is satisfied, and rank $[\bar{C}] = \bar{m}$, the vector β_1 in (76) is unique, otherwise β_1 is nonunique (see [5, pp. 439, 440]).

Assuming (77) is satisfied, it follows from (74) that $\bar{y}_1((n + 1)T)$ may be written as

$$\bar{y}_1((n + 1)T) = \beta_1 \bar{y}(nT) + \bar{c}_1 \tilde{B} u(nT) + \bar{c}_1 F\tilde{H} z(nT), \tag{79}$$

which shows that $\bar{y}_1(nT)$ obeys a first-order difference equation homogenous in $\bar{y}_1, \bar{y}_2, \ldots, \bar{y}_{\bar{m}}$ and forced by $u(nT)$ and $z(nT)$. This is the *kind* of equation we seek for each of the critical variables $\bar{y}_j$, although those equations will not necessarily be first-order like (79) nor will they all be of the same order, in general.

Suppose condition (77) fails. Then $\bar{y}_1$ does *not* satisfy a first-order difference equation of the form (79) and we must therefore search for a higher-order equation. For that purpose, the second time-shift $E^2\bar{y}_1(nT)$ is computed from (73), (36), and (37) as

$$\begin{aligned} E^2\bar{y}_1(nT) &= \bar{y}_1((n + 2)T) = \bar{c}_1 \tilde{A} E x(nT) + \bar{c}_1 \tilde{B} E u(nT) + \bar{c}_1 F\tilde{H} E z(nT) \\ &= \bar{c}_1 \tilde{A}^2 x(nT) + \bar{c}_1 \tilde{A}\tilde{B} u(nT) + \bar{c}_1 \tilde{A} F\tilde{H} z(nT) \\ &\quad + \bar{c}_1 \tilde{B} E u(nT) + \bar{c}_1 F\tilde{H} E z(nT). \end{aligned} \tag{80}$$

We now ask: Can $\bar{y}_1((n + 2)T)$ in (80) be expressed as some weighted linear combination of $\bar{y}_1, \bar{y}_2, \ldots, \bar{y}_{\bar{m}}$;

$E\bar{y}_1, E\bar{y}_2, \ldots, E\bar{y}_{\bar{m}}$; u, z; Eu, Ez? In other words, can (80) be expressed in the form

$$\bar{y}_1((n+2)T) = \beta_1\bar{y} + \beta_2 E\bar{y} + \theta_1 u + \alpha_1 z + \bar{c}_1\tilde{B}Eu + \bar{c}_1\tilde{F}\tilde{H}Ez \tag{81}$$

for some row vectors θ_1, α_1, β_1, and β_2? To explore this question we substitute (71) and the expression

$$E\bar{y} = \bar{C}Ex = \bar{C}\tilde{A}x(nT) + \bar{C}\tilde{B}u(nT) + \bar{C}\tilde{F}\tilde{H}z(nT) \tag{82}$$

into (81) to obtain

$$\bar{y}_1((n+2)T) = \beta_1\bar{C}x + \beta_2\bar{C}\tilde{A}x + \beta_2\bar{C}\tilde{B}u + \beta_2\bar{C}\tilde{F}\tilde{H}z + \theta_1 u + \alpha_1 z + \bar{c}_1\tilde{B}Eu + \bar{c}_1\tilde{F}\tilde{H}Ez. \tag{83}$$

Comparison of (83) with (80) shows that (81) exists if and only if

$$\bar{c}_1\tilde{A}^2 = [\beta_1|\beta_2][\bar{C}/\bar{C}\tilde{A}], \quad \text{for some } \beta_1, \beta_2 \tag{84}$$

in which case θ_1 and α_1 in (81) are automatically defined as

$$\theta_1 = (-\beta_2\bar{C} + \bar{c}_1\tilde{A})\tilde{B}, \quad \alpha_1 = (-\beta_2\bar{C} + \bar{c}_1\tilde{A})\tilde{F}\tilde{H}. \tag{85}$$

The necessary and sufficient condition for existence of β_1 and β_2 satisfying (84) is

$$\text{rank}\left[\bar{C}^T|\tilde{A}^T\bar{C}^T|\tilde{A}^{T2}\bar{c}_1^T\right] = \text{rank}[\bar{C}^T|\tilde{A}^T\bar{C}^T] \tag{86}$$

or, stated equivalently,

$$\mathscr{R}\left[\tilde{A}^{T2}\bar{c}_1^T\right] \subseteq \mathscr{R}[\bar{C}^T|\tilde{A}^T\bar{C}^T]. \tag{87}$$

The results (86) and (87) are seen to be natural generalizations of the previous results (77) and (78). Thus, if (86) is satisfied, we have proven that $\bar{y}_1$ satisfies a second-order difference equation given by (81) and (85).

If condition (86) fails to be satisfied, we compute the third time-shift of $\bar{y}_1(nT)$ and repeat the preceding procedure. This process is continued until the first (i.e., *lowest*) index v_1 is found such that

$$\text{rank}\left[\bar{C}^T|\tilde{A}^T\bar{C}^T|\cdots|\tilde{A}^{T(v_1-1)}\bar{C}^T|\tilde{A}^{T(v_1)}\bar{c}_1^T\right]$$

$$= \text{rank}\left[\bar{C}^T|\tilde{A}^T\bar{C}^T|\cdots|\tilde{A}^{T(v_1-1)}\bar{C}^T\right]. \tag{88}$$

This latter event determines that the lowest-order difference equation satisfied by $\bar{y}_1(nT)$ is a v_1th order equation of the form

$$E^{(v_1)}\bar{y}_1(nT) = \beta^{1,1}\bar{y}(nT) + \cdots + \beta^{1,v_1}E^{(v_1-1)}\bar{y}(nT)$$
$$+ \theta^{1,1}u + \theta^{1,2}Eu + \cdots + \alpha^{1,1}z$$
$$+ \alpha^{1,2}Ez + \cdots, \tag{89}$$

where $\beta^{1,k}$, $\theta^{1,k}$, and $\alpha^{1,k}$ denote row-vectors.

This same procedure is now repeated for each of the remaining critical variables $\bar{y}_2, \bar{y}_3, \ldots, \bar{y}_{\bar{m}}$ to obtain a set of simultaneous difference equations of the general form (89). For each $\bar{y}_j$, the corresponding test for the lowest order v_j has the form (88), where $\bar{c}_1^T$ in (88) is replaced by $\bar{c}_j^T$. That set of simultaneous difference equations may finally be expressed in the compact form

$$\begin{pmatrix} E^{(v_1)}\bar{y}_1 \\ E^{(v_2)}\bar{y}_2 \\ \vdots \\ E^{(v_{\bar{m}})}\bar{y}_{\bar{m}} \end{pmatrix} = \begin{bmatrix} \beta^{11} & \beta^{12} & \cdots & \beta^{1s} \\ \beta^{21} & \beta^{22} & & \beta^{2s} \\ \vdots & & & \\ \beta^{\bar{m}1} & \beta^{\bar{m}2} & \cdots & \beta^{\bar{m}s} \end{bmatrix} \begin{pmatrix} \bar{y} \\ E\bar{y} \\ \vdots \\ E^{(s-1)}\bar{y} \end{pmatrix}$$

$$+ \begin{bmatrix} \theta^{11} & \cdots & \theta^{1x} \\ \theta^{21} & & \theta^{2x} \\ \vdots & & \\ \theta^{\bar{m}1} & \cdots & \theta^{\bar{m}x} \end{bmatrix} \begin{pmatrix} u \\ Eu \\ \vdots \\ E^{(x-1)}u \end{pmatrix} + \begin{bmatrix} \alpha^{11} & \cdots & \alpha^{1t} \\ \alpha^{21} & & \alpha^{2t} \\ \vdots & & \\ \alpha^{\bar{m}1} & \cdots & \alpha^{\bar{m}t} \end{bmatrix} \begin{pmatrix} z \\ Ez \\ \vdots \\ E^{(t-1)}z \end{pmatrix}, \tag{90}$$

where the index notations $(s - 1)$, $(x - 1)$, and $(t - 1)$ denote the highest powers of the shift operator E, which appear with $\bar{y}$, u, and z, respectively, on the right side of (90). Note that the auxiliary set of equations (90) is homogenous with respect to the critical variables $\bar{y} = \bar{y}_1, \ldots, \bar{y}_{\bar{m}})$ as we require, and the individual orders v_j of the equations may be different.

Now, one can allocate (split) the control u in (90) in the usual DAC manner:

$$u = u_p + u_d, \tag{91}$$

where u_d is assigned the task of completely absorbing (counteracting) the discrete-time disturbance terms in (90) ["complete" disturbance absorption with respect to the critical variables], and u_p is then assigned the task of accomplishing the primary control objective (i.e., stabilization, set-point regulation, servo-tracking, etc.) for the resulting *undisturbed* system[6] (90). Thus, u_d should be designed ideally to satisfy the condition

$$\begin{bmatrix} \theta^{11} & \cdots & \theta^{1x} \\ \theta^{21} & & \theta^{2x} \\ \vdots & & \vdots \\ \theta^{\bar{m}1} & \cdots & \theta^{\bar{m}x} \end{bmatrix} \begin{pmatrix} u_d \\ Eu_d \\ \vdots \\ E^{(x-1)}u_d \end{pmatrix} = - \begin{bmatrix} \alpha^{11} & \cdots & \alpha^{1t} \\ \alpha^{21} & & \alpha^{2t} \\ \vdots & & \vdots \\ \alpha^{\bar{m}1} & \cdots & \alpha^{\bar{m}t} \end{bmatrix} \begin{pmatrix} z \\ Ez \\ \vdots \\ E^{(t-1)}z \end{pmatrix}. \qquad (92)$$

Expression (92) is recognized as a set of implicit simultaneous vector difference equations for iteratively generating the discrete-time control u_d, provided (92) is consistent. In other words, the control u_d is generated by a discrete-time dynamical "compensator" [digital filter] (92) where ideally the "inputs" to that compensator consist of the past and present measurements of $z(nT)$. For implementation purposes, of course, one would use the observer-produced estimates $\hat{z}(nT)$ as inputs to the compensator (92).

Since the disturbance state $z(nT)$ is arbitrary, it is clear that (92) is consistent [i.e., it can be explicitly solved for the highest shift-orders $E^{(\cdot)}u_{d,i}$ of each *individual* control element $u_{d,i}$ in $u_d = (u_{d,1}, u_{d,2}, \ldots, u_{d,r})$] if and only if (92) admits a factorization similar to that associated with (42) and (43). This latter condition, derived by setting $E^k z = \tilde{D}^k z$, $k = 1, 2, \ldots$ in (92), constitutes the existence condition for a critical-variable disturbance-absorbing controller u_d. If (92) turns out to be *in*consistent, it is not

[6] *In critical-variable disturbance-absorption problems, the primary control objective implicitly includes also the "proper" control of all noncritical plant variables. For instance, the noncritical variables* ($\triangleq$ *the set of states* x *orthogonal to* $\mathcal{R}[\bar{C}^T]$) *are typically required to remain bounded or be asymptotically stable.*

possible to accomplish complete disturbance absorption with respect to all of the critical variables $(\bar{y}_1, \ldots, \bar{y}_{\bar{m}})$.

Assuming that u_d is designed to achieve the disturbance-absorption condition (92), the design of u_p to accomplish the primary control objective can proceed by considering the *un*disturbed version of (90). This procedure typically leads to another discrete-time dynamical compensator (digital filter) for the on-line generation of u_p. For example, if the primary control objective is to stabilize the critical variables $\bar{y} = (\bar{y}_1, \ldots, \bar{y}_{\bar{m}})$ to zero, it follows from (90) that u_p can be sought in the idealized form

$$\begin{bmatrix} \theta^{11} & \cdots & \theta^{1k} \\ \theta^{21} & & \\ \vdots & & \\ \theta^{\bar{m}1} & & \end{bmatrix} \begin{pmatrix} u_p \\ Eu_p \\ \vdots \\ E^{(x-1)} u_p \end{pmatrix} = \begin{bmatrix} \\ \mathscr{K}_1 \\ \\ \end{bmatrix} (\bar{y}) + \begin{bmatrix} \\ \mathscr{K}_2 \\ \\ \end{bmatrix} (E\bar{y}) + \cdots + \begin{bmatrix} \\ \mathscr{K}_{v_{\bar{m}}} \\ \\ \end{bmatrix} \left(E^{(v_1-1)} \bar{y}\right), \tag{93}$$

where the "gain matrices" $\mathscr{K}_i$ are designed by Routh-Hurwitz considerations, etc., to make the undisturbed part of (90) asymptotically stable to $\bar{y} = 0$. Expression (93) is recognized as an implicit discrete-time feedback compensator for generating u_p, where the inputs to that compensator consist of the past and present measurements of the critical-variable $\bar{y}(nT)$. The questions of consistency and implementation of (93) are handled just as for (92), where it may be necessary to use observer-produced estimates $\hat{\bar{y}}(nT) = \bar{C}\hat{x}(nT)$ in place of $\bar{y}$ in (93).

In the case of primary control objectives of the nonzero set-point regulation or servo-tracking type, the preceding design algorithm for u_p is modified as follows. Using (39), the control "error variables" $\bar{\epsilon}_j = (\bar{y}_{c,j} - \bar{y}_j)$, $j = 1,\ldots, \bar{m}$, are introduced so that the primary control objective now becomes the prompt regulation of $\bar{\epsilon}_j(nT) \to 0$, for all j. Using (39), together with the undisturbed part of (90) with $u \to u_p$, a set of simultaneous, higher-order difference equations for the motions of $\bar{\epsilon}_j(nT)$ can then be derived [comparable to the undisturbed part of (90)], which are "forced" by $u_p(nT)$, $u_p[(n + 1)T]$, $u_p[(n + 2)T]$, etc., and are otherwise homogenous in the error variables $\bar{\epsilon}_j$. The latter derivation proceeds in a manner similar to (73) - (89). Finally, u_p is designed to satisfy a controller expression similar to (93), with $\bar{y}$ replaced by $\bar{\epsilon}$, where the gain matrices $\mathscr{K}$ are designed to stabilize the $\bar{\epsilon}_j(nT)$ to zero.

4. *Example Design of a Discrete-Time Disturbance-Absorbing Controller for Critical Variables*

The critical-variable disturbance-absorber design procedure outlined in the preceding section is illustrated by the following specific example. Suppose the plant to be controlled is second-order and modeled by the continuous-time equations

$$\begin{pmatrix} \dot{x}_1 \\ \dot{x}_2 \end{pmatrix} = \begin{bmatrix} 0 & 1 \\ 0 & 0 \end{bmatrix} \begin{pmatrix} x_1 \\ x_2 \end{pmatrix} + \begin{pmatrix} 0 \\ 1 \end{pmatrix} u + \begin{pmatrix} 0 \\ 1 \end{pmatrix} w \tag{94}$$

with a general type of exponential disturbance $w(t)$ governed by

$$\begin{aligned} w &= z, \\ \dot{z} &= \alpha z + \sigma(t), \quad \alpha = \text{real, scalar constant.} \end{aligned} \tag{95}$$

The discrete-time models (36) and (37) corresponding to (94) and (95) are found to be

$$\begin{pmatrix} Ex_1(nT) \\ Ex_2(nT) \end{pmatrix} = \begin{bmatrix} 1 & T \\ 0 & 1 \end{bmatrix} \begin{pmatrix} x_1(nT) \\ x_2(nT) \end{pmatrix} + \begin{pmatrix} \frac{T^2}{2} \\ T \end{pmatrix} u(nT) + \begin{pmatrix} \frac{e^{\alpha T} - 1 - \alpha T}{\alpha^2} \\ \frac{e^{\alpha T} - 1}{\alpha} \end{pmatrix} z(nT) + \left(\tilde{\gamma}(nT) \right), \tag{96}$$

$$Ez(nT) = e^{\alpha T} z(nT) + \tilde{\sigma}(nT). \tag{97}$$

If one now applies the complete absorbability test (43) to (96), the result is

$$\begin{pmatrix} \frac{e^{\alpha T} - 1 - \alpha T}{\alpha^2} \\ \frac{e^{\alpha T} - 1}{\alpha} \end{pmatrix} \overset{?}{=} \begin{pmatrix} \frac{T^2}{2} \\ T \end{pmatrix} \tilde{\Gamma}, \text{ for some scalar } \tilde{\Gamma}. \tag{98}$$

It is easy to see that (98) *cannot* be satisfied unless[7] $\alpha = 0$, which means unless $w(t)$ is piecewise constant. Thus, if $\alpha \neq 0$, one cannot achieve complete disturbance absorption for *both* plant states $x_1(nT)$ and $x_2(nT)$ in (96).

To apply the critical-variable approach to disturbance-absorption, we must first decide which plant state will be called "critical." For this example, it will be assumed that

[7] *The left side of (98) is evaluated for* $\alpha = 0$ *by first expanding* $e^{\alpha T}$ *in a power series and then letting* $\alpha \to 0$.

the state variable $x_1(t)$ in (94) and (96) is the critical variable. Therefore, we shall attempt to design $u_d(\cdot)$ to absorb completely the discrete-time effects of $w(t)$ on the one state-variable $x_1(t)$. Moreover, it will be assumed that the primary control objective is to regulate promptly $\bar{y} = x_1(nT)$ to the given set-point value $\bar{y}_c = c$, c = constant. Thus, for this example expression (38) becomes

$$\bar{y} = \bar{C}x = \bar{y}_1 = (1, 0)\begin{pmatrix} x_1 \\ x_2 \end{pmatrix}, \tag{99}$$

and we therefore have only the one row-vector

$$\bar{C} = \bar{c}_1 = (1, 0) \tag{100}$$

in (71). The first step is to apply the series of rank tests (88) using $v_1 = 1$, $v_1 = 2, \ldots$. Using $v_1 = 1$, the test is

$$\text{rank}\left[\bar{C}^T \mid \tilde{A}^T \bar{c}_1^T\right] \overset{?}{=} \text{rank}\,[\bar{C}^T], \tag{101}$$

which for this example becomes the specific test

$$\text{rank}\begin{bmatrix} 1 & 1 \\ 0 & T \end{bmatrix} \overset{?}{=} \text{rank}\begin{bmatrix} 1 \\ 0 \end{bmatrix}. \tag{102}$$

Clearly (102) *is not* satisfied. Proceeding next to the case $v_1 = 2$, we have the test

$$\text{rank}\left[\bar{C}^T \mid \tilde{A}^T \bar{C}^T \mid \tilde{A}^{T^2} \bar{c}_1^T\right] \overset{?}{=} \text{rank}\,[\bar{C}^T \mid \tilde{A}^T \bar{C}^T], \tag{103}$$

which, for the example at hand, becomes the specific test

$$\text{rank}\begin{bmatrix} 1 & 1 & 1 \\ 0 & T & 2T \end{bmatrix} \overset{?}{=} \text{rank}\begin{bmatrix} 1 & 1 \\ 0 & T \end{bmatrix}. \tag{104}$$

It is clear that (104) *is* satisfied, and therefore there exist scalar constants β_1 and β_2 such that $\bar{y}_1$ in (99) satisfies a second-order difference equation of the form (81) and (85).

The constants β_1 and β_2 are determined from (84) which for this example becomes

$$(1,\ 2T) = (\beta_1,\ \beta_2)\begin{bmatrix} 1 & 0 \\ 1 & T \end{bmatrix}. \tag{105}$$

Solving (105) for β_1 and β_2 yields

$$(\beta_1,\ \beta_2) = (1,\ 2T)\begin{bmatrix} 1 & 0 \\ 1 & T \end{bmatrix}^{-1} = (-1,\ 2). \tag{106}$$

Substituting (106) [and the numerical expressions for $\tilde{A}$, $\tilde{B}$, $\widetilde{FH}$, $\bar{C}$ as read-off from (96) and (100)] into (81) and (85), the auxiliary difference equation (81) for the critical variable $\bar{y}_1 = x_1$ is finally obtained as

$$\begin{aligned} E^2x_1(nT) &= -x_1(nT) + 2Ex_1(nT) \\ &\quad + \frac{T^2}{2}\,u(nT) + \left[\frac{e^{\alpha T}(\alpha T - 1) + 1}{\alpha^2}\right]z(nT) \\ &\quad + \frac{T^2}{2}\,Eu(nT) + \left[\frac{e^{\alpha T} - 1 - \alpha T}{\alpha^2}\right]Ez(nT). \end{aligned} \tag{107}$$

Since we are disregarding the uncertain residual terms $\tilde{\gamma}(nT)$ and $\tilde{\sigma}(nT)$ in (96) and (97), expression (107) can be further simplified by incorporating the relation $Ez = e^{\alpha T}z$ from (97) to obtain

$$\begin{aligned} E^2x_1(nT) &= -x_1(nT) + 2Ex_1(nT) \\ &\quad + \frac{T^2}{2}[u(nT) + Eu(nT)] + \left[\frac{e^{\alpha T} - 1}{\alpha}\right]^2 z(nT). \end{aligned} \tag{108}$$

Now, substitution of the standard allocation rule $u = u_p + u_d$ into (108) yields

$$E^2x_1 = -x_1 + 2Ex_1 + \frac{T^2}{2}[u_p + Eu_p] + \frac{T^2}{2}[u_d + Eu_d] + \left[\frac{e^{\alpha T} - 1}{\alpha}\right]^2 z. \qquad (109)$$

The design of u_d in (109) to absorb completely the discrete-time disturbance effects on the critical-variable motion $x_1(nT)$ is accomplished by setting

$$\frac{T^2}{2}[u_d + Eu_d] \equiv -\left[\frac{e^{\alpha T} - 1}{\alpha}\right]^2 z, \qquad (110)$$

which corresponds to controller expression (92) for this example. Since (110) is a *single* difference equation, there is no question of consistency in (110). Thus one is assured that (110) can be solved for a well-defined control expression, namely

$$Eu_d(nT) = -u_d(nT) - \frac{2}{T^2}\left[\frac{e^{\alpha T} - 1}{\alpha}\right]^2 z(nT). \qquad (111)$$

In practice, one would use an on-line observer-produced estimate $\hat{z}(nT)$ for $z(nT)$ in (111). Note that (111) represents a recursive digital filter for generation of $u_d(nT)$, where z (or $\hat{z}$) is the "input" to that filter. A block-diagram of such a digital filter is shown in Fig. 7. This completes the design of the u_d term in (109).

The design of the u_p term in (109) is guided by the desired behavior of the critical variable, which in this example is simply prompt set-point regulation of $\bar{y} = x_1(nT)$ to $\bar{y}_c = c$ = constant. Thus, assuming that u_d in (111) is "installed and working" in (109), the task of u_p is to stabilize promptly the

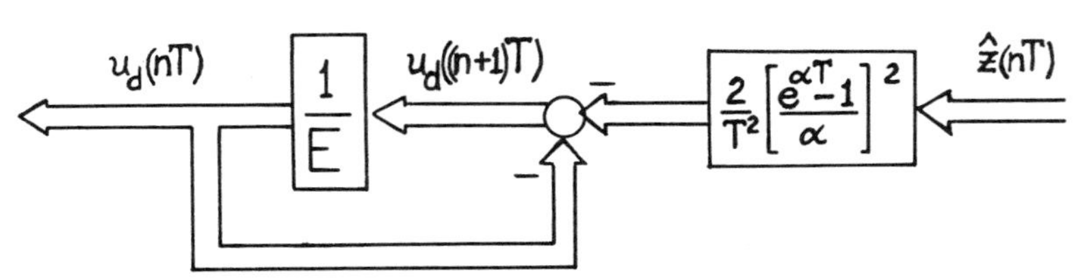

Fig. 7. Block diagram of the digital filter (111) for generation of $u_d(nT)$.

control error $\bar{\epsilon} = \bar{y}_c - \bar{y}$ to zero for the *un*disturbed plant model

$$E^2 x_1(nT) = -x_1(nT) + 2Ex_1(nT) + \frac{T^2}{2}(u_p + Eu_p). \tag{112}$$

To accomplish the latter task, it is first necessary to derive the auxiliary difference equation governing $\bar{\epsilon}(nT)$. Thus, using (112) and the set-point command model [see (39)]

$$\bar{y}_c = c, \quad Ec = Ic + \tilde{\mu}(nT), \tag{113}$$

we compute

$$E\bar{\epsilon} = E\bar{y}_c - E\bar{y} = c - Ex_1, \tag{114a}$$

$$E^2\bar{\epsilon} = Ec - E^2 x_1 = c + x_1 - 2Ex_1 - \frac{T^2}{2}(u_p + Eu_p). \tag{114b}$$

Note that the residual $\tilde{\mu}(nT)$ in (113) has been disregarded in the calculation of (114); see remarks after (72). Since $x_1 = c - \bar{\epsilon}$ and $Ex_1 = c - E\bar{\epsilon}$, it follows that (114b) can be expressed in the homogenous form

$$E^2\bar{\epsilon} = -\bar{\epsilon} + 2E\bar{\epsilon} - \frac{1}{2}T^2(u_p + Eu_p). \tag{115}$$

Now, one simply designs u_p in (115) to make $\bar{\epsilon}(nT) \to 0$ promptly. For this purpose, the required net contribution of u_p to the right side of (115) can be expressed generally as a weighted linear combination of $\bar{\epsilon}(nT)$ and $E\bar{\epsilon}(nT)$ of the form $k_1\bar{\epsilon}(nT) + k_2E\bar{\epsilon}(nT)$, where k_1 and k_2 are gain matrices to be designed.

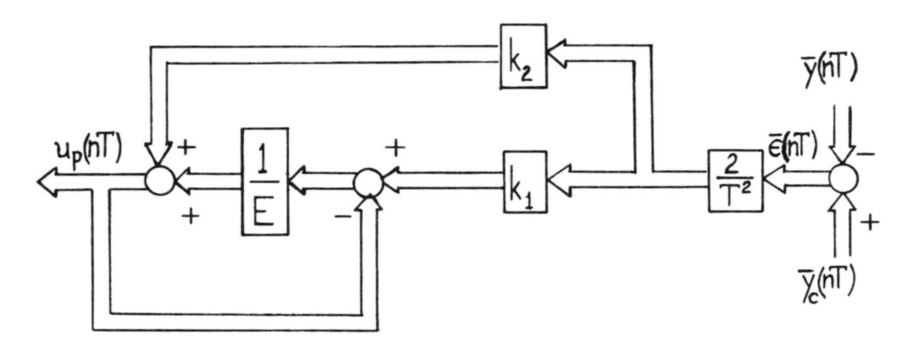

Fig. 8. Block diagram of the digital filter (117) for generation of $u_p(nT)$.

Thus, to design the u_p controller, we set

$$\frac{1}{2} T^2(u_p + Eu_p) = k_1\bar{\epsilon} + k_2E\bar{\epsilon}, \tag{116}$$

which is comparable to (93). The controller expression (116) is recognized as the recursive digital filter (difference equation)

$$Eu_p = -u_p + \frac{2}{T^2}[k_1\bar{\epsilon} + k_2E\bar{\epsilon}], \tag{117}$$

where the on-line measurements $\bar{\epsilon}(nT)$ are the inputs to that filter. Thus, (117) represents the discrete-time dynamical compensator for generation of $u_p(nT)$. One possible implementation of (117) is shown in Fig. 8.

The design of (k_1, k_2) in (117) is accomplished by substitution of (117) into (115) to obtain the "closed-loop" error equation

$$E^2\bar{\epsilon} + (k_2 - 2)E\bar{\epsilon} + (k_1 + 1)\bar{\epsilon} = 0. \tag{118}$$

Now, the designer can pick the characteristic roots $(\tilde{\lambda}_1, \tilde{\lambda}_2)$ *desired* for (118) and then choose k_1 and k_2 to yield those two roots. It is recalled that for asymptotic stability of (118), the two roots $(\tilde{\lambda}_1, \tilde{\lambda}_2)$ must lie *interior to* the unit circle in the complex plane. Assuming $\tilde{\lambda}_1$ and $\tilde{\lambda}_2$ have been appropriately chosen (to give satisfactory damping, settling-time, etc.), the corresponding "desired" characteristic equation $\mathscr{P}(\tilde{\lambda})$ is simply

$$\mathscr{P}_{\text{des.}}(\tilde{\lambda}) = (\tilde{\lambda} - \tilde{\lambda}_1)(\tilde{\lambda} - \tilde{\lambda}_2) = \tilde{\lambda}^2 - (\tilde{\lambda}_1 + \tilde{\lambda}_2)\tilde{\lambda} + \tilde{\lambda}_1\tilde{\lambda}_2 = 0. \tag{119}$$

The *actual* characteristic equation of (118) is

$$\mathscr{P}_{\text{act.}}(\tilde{\lambda}) = \tilde{\lambda}^2 + (k_2 - 2)\tilde{\lambda} + (k_1 + 1) = 0 \tag{120}$$

and, therefore, by equating (119) and (120) one obtains the required k_1 and k_2 immediately as

$$k_1 = \tilde{\lambda}_1\tilde{\lambda}_2 - 1, \tag{121a}$$

$$k_2 = 2 - (\tilde{\lambda}_1 + \tilde{\lambda}_2). \tag{121b}$$

It is remarked that a convenient choice for $\tilde{\lambda}_1$ and $\tilde{\lambda}_2$ is

$$\tilde{\lambda}_1 = \tilde{\lambda}_2 = 0, \tag{122}$$

which produces "deadbeat" response of (118); i.e., the error $\bar{\epsilon}(nT)$ theoretically reaches zero in a finite number of sample periods T. The k_1 and k_2 required for deadbeat response of (118) are obtained from (121) as

$$k_1 = -1, \quad k_2 = 2. \tag{123}$$

This completes the design of $u_p(nT)$ for this example.

In summary, the discrete-time controller for critical-variable disturbance-absorption and set-point regulation in this example is given by $u = u_p + u_d$, where u_p and u_d are generated by the pair of parallel-processing digital filters associated with Eqs. (111) and (117) as shown in Figs. 7 and 8. It is remarked that those two separate digital filters could be combined into *one* digital filter, having *only one delayor*, by writing

$$Eu = Eu_p + Eu_d = -u_p + \frac{2}{T^2}[k_1\bar{\epsilon} + k_2E\bar{\epsilon}] - u_d - \frac{2}{T^2}\left[\frac{e^{\alpha T} - 1}{\alpha}\right]^2 z, \qquad (124)$$

and then dividing by E to obtain

$$u = \frac{1}{E}\left[-(u_p + u_d) + \frac{2}{T^2}\left(k_1\bar{\epsilon} - \left[\frac{e^{\alpha T} - 1}{\alpha}\right]^2 z\right)\right] + \frac{2}{T^2} k_2\bar{\epsilon}. \qquad (125)$$

The implementation of (125) will involve only *one* delayor and the two "inputs" $\bar{\epsilon}(nT)$ and $z(nT)$, since $u_p + u_d = u$. The use of (125) saves one delayor, but also puts the designer in the risky position of assigning all controller processing to one central "filter" having one delayor. If that one delayor fails, both the u_p and u_d loops are disabled simultaneously. Thus, there may be some "fail-safe" reasons for NOT combining the u_p and u_d filters into the one filter expressed by (125).

It should be noted that the DAC controller (111) and (117) absorbs the discrete-time disturbance effects of $w(t)$ on the critical-variable $x_1(nT)$ and simultaneously regulates $x_1(nT) \to c$, *even though* complete disturbance absorption for both $x_1(nT)$ and $x_2(nT)$ is generally impossible [i.e., (98) fails unless $\alpha = 0$].

B. DESIGN FOR THE DISTURBANCE-MINIMIZATION MODE

The complete absorbability condition (43) must be satisfied in order to achieve total cancellation of the disturbance term $\widetilde{FH}z(nT)$, which appears in (41). In some cases it turns out that condition (43) cannot be satisfied owing to the structure of the given matrices $\tilde{A}$, $\tilde{B}$, $\widetilde{FH}$, and $\tilde{D}$. For instance, when the disturbance $w(t)$ is not an essentially piecewise-constant function of time, no constant control $u(nT)$ of dimension $r < n$ can completely cancel all effects of $w(t)$ on the n-dimensional state vector $x[(n + 1)T]$, in general.[8] In such cases the designer can take the alternative approach of designing $u_d(nT)$ in (41) to "best" approximate the ideal absorption condition (42) in some specified sense. Such DACs are called disturbance-minimizing controllers.

There are literally hundreds of different ways to specify the "sense" in which a disturbance-minimizing controller is to approximate the ideal condition (42). In fact, once the designer realizes that (42) cannot be achieved owing to the failure of (43), it is possible that the best alternative approach will then not involve an attempt to approximate (42) but rather to take a whole new approach to the approximate disturbance-absorption strategy. These various issues are discussed at some length in [5]. In this section, some representative cases are considered as a means of demonstrating the general ideas involved in designing a disturbance-minimizing controller for discrete-time applications.

1. The Norm Minimization Method

The simplest approach to approximating the ideal disturbance-absorption condition (42) is to choose $u_d(nT)$ to minimize the vector "norm" of the left side of (42)

$$\min_{u_d} \|\tilde{B}u_d(nT) + \tilde{F}Hz(nT)\|. \tag{126}$$

Condition (126) is a natural generalization of (42) because if condition (43) happens to be satisfied, the control $u_d(nT)$

[8] *If $r = n$, and rank $\tilde{B} = n$, such cancellation is possible; i.e., (43) is always satisfied.*

which minimizes (126) will automatically satisfy (42); i.e., the norm (126) will then be made zero. The control $u_d(nT)$ which minimizes (126) is not unique, in general. However the control $u_d^0(nT)$ which minimizes (126) and which *itself* has minimum norm $\|u_d(nT)\|$ *is* unique and is given by

$$u_d^0(nT) = -\tilde{B}^{\dagger}\tilde{F}Hz(nT), \tag{127}$$

where $\tilde{B}^{\dagger}$ denotes the Moore-Penrose "generalized inverse" of $\tilde{B}$. If $\tilde{B}$ has rank = r, then $\tilde{B}^{\dagger} = (\tilde{B}^T\tilde{B})^{-1}\tilde{B}^T$. Otherwise $\tilde{B}^{\dagger}$ can be computed as described in [20]. For physical realization, the control in (127) would be implemented as

$$u_d^0(nT) = -\tilde{B}^{\dagger}\tilde{F}H\hat{z}(nT). \tag{128}$$

The remaining steps in the design and implementation of $u = u_p + u_d$ for disturbance-minimization would be exactly the same as outlined in Section V.A for disturbance-absorption.

Expression (126) can be generalized by introducing $\epsilon_d = \tilde{B}u_d + \tilde{F}Hz$ and choosing u_d to minimize the positive definite quadratic form $\epsilon_d^T Q \epsilon_d$, $Q = Q^T > 0$.

2. *The Critical Disturbance Method*

In some applications, the plant disruptions caused by certain components of the disturbance vector $w = (w_1, w_2, \ldots, w_p)$ are more critical (significant) than the others. Thus, if complete disturbance absorption (42) cannot be achieved, one can attempt instead to absorb the critical components w_i^* of w. For this purpose, the expressions Fw, w = Hz, in (4) and (7) are written as

$$Fw = [f^1 | f^2 | \cdots | f^p] \begin{pmatrix} w_1 \\ w_2 \\ \vdots \\ w_p \end{pmatrix} = w_1 f^1 + \cdots + w_p f^p;$$

$$\begin{pmatrix} w_1 \\ w_2 \\ \vdots \\ w_p \end{pmatrix} = \begin{bmatrix} h^1 \\ h^2 \\ \vdots \\ h^p \end{bmatrix} \begin{pmatrix} z \end{pmatrix}, \tag{129}$$

so that the disturbance term Fw in (4) can be written in the component form

$$Fw = f^1 h^1 z + f^2 h^2 z + \cdots + f^p h^p z. \tag{130}$$

The term $\tilde{F}\tilde{H}z$ in the corresponding discrete-time plant model (14a) then has the component form

$$\tilde{F}\tilde{H}z = \widetilde{f^1 h^1} z + \cdots + \widetilde{f^p h^p} z, \tag{131}$$

where [see (13b)]

$$\widetilde{f^i h^i} = \int_{t_0+nT}^{t_0+(n+1)T} \Phi(t_0+(n+1)T, \tau) f^i h^i \Phi_D(\tau, t_0+nT)\, d\tau. \tag{132}$$

Thus, if $w_i^* = h_*^i z$ denotes a critical component of the disturbance w, one can completely absorb that particular disturbance effect if u_d can be designed to satisfy

$$\tilde{B}u_d + \widetilde{f_*^i h_*^i} z = 0. \tag{133}$$

If several components of w are critical, one would replace (133) by

$$\tilde{B}u_d + \sum \widetilde{f_*^i h_*^i} z = 0. \tag{134}$$

The necessary and sufficient condition for existence of a solution u_d of (133) and (134) is given by an expression analogous to (43).

3. The Absorbable Subspace Method

If (43) fails, it implies that $\mathscr{R}[\widetilde{FH}] \not\subseteq \mathscr{R}[\tilde{B}]$, where $\mathscr{R}[\cdot]$ denotes column range space. In that event one can consider disturbance-absorption within the largest *subspace* $\tilde{\mathscr{S}}$ of $\mathscr{R}[\widetilde{FH}]$, which *is* contained in $\mathscr{R}[\tilde{B}]$. We call $\tilde{\mathscr{S}}$ the *absorbable subspace*, and it is clear that

$$\tilde{\mathscr{S}} = \mathscr{R}[\widetilde{FH}] \cap \mathscr{R}[\tilde{B}]. \tag{135}$$

Thus, we may write

$$\widetilde{FH} = \tilde{B}\tilde{\Gamma}_1 + \tilde{\Omega}\tilde{\Gamma}_2, \tag{136}$$

where $\tilde{\mathscr{S}} = \mathscr{R}[\tilde{B}\tilde{\Gamma}_1]$ and $\tilde{\Omega}$ is chosen with a minimal number of columns such that $\mathscr{R}[\widetilde{FH}] = \mathscr{R}[\tilde{B}|\tilde{\Omega}]$. Now, u_d can be chosen as $u_d = -\tilde{\Gamma}_1 z$ to absorb completely that "part" of $\widetilde{FH}z$ which lies in $\tilde{\mathscr{S}}$. Of course, this approach yields a larger norm (126) than does (127), in general, and is ineffective if $\tilde{\mathscr{S}}$ has dimension zero.

4. The Critical-Variable Method

When (43) fails, it is inevitable that the disturbance will affect at least some of the plant variables $x_i(nT)$, $i = 1, 2, \ldots, n$. In that case one can attempt the absorption of disturbances for that subset of plant variables which is considered most critical. This method has already been described and illustrated in Sections V.A.3,4.

C. *DESIGN FOR COMMAND "DISTURBANCES" IN SET-POINT AND SERVO-TRACKING PROBLEMS*

The disturbance-absorbing design procedure outlined in Section V.A can also be used as a tool for designing effective set-point and servo-tracking controllers, where the "commands" create disturbance-like terms in the control error equations. For this purpose, one can proceed as follows.

1. *Design for Set-Point Commands*

Suppose the objective of control is to regulate the plant state-vector $x(nT)$ to the given set-point x_c, where x_c is a known constant n-vector. In that case, the control error is defined as

$$\epsilon = x_c - x, \tag{137}$$

and it is easy to verify that $\epsilon(nT)$ obeys the difference equation

$$E\epsilon = \tilde{A}\epsilon - \tilde{B}u_p - \tilde{B}u_d - \widetilde{FH}z - \tilde{A}x_c - \tilde{\gamma}. \tag{138}$$

From the view point of the error dynamics (138), the primary control task, for u_p, is to regulate promptly $\epsilon(nT)$ to the value $\epsilon = 0$, whereas the task for u_d is to absorb (counteract) the collection of "disturbance" terms $-\widetilde{FH}z(nT) - \tilde{A}x_c$ in (138). Thus, u_d should ideally be chosen to satisfy the condition

$$\tilde{B}u_d + \widetilde{FH}z + \tilde{A}x_c = 0. \tag{139}$$

The necessary and sufficient condition for existence of a $u_d(nT)$ satisfying (139) is

$$[\widetilde{FH} | \tilde{A}x_c] = \tilde{B}\tilde{\Gamma}_c, \text{ for all admissible } x_c, \tag{140a}$$

where $\tilde{\Gamma}_c$ is typically a function of x_c having the form

$$\tilde{\Gamma}_c = [\tilde{\Gamma}_1 | \tilde{\Gamma}_2 x_c], \quad \tilde{B}\tilde{\Gamma}_1 = \widetilde{FH}, \quad \tilde{B}\tilde{\Gamma}_2 x_c = \tilde{A}x_c. \tag{140b}$$

If the set-point x_c is an *arbitrary* vector in n-dimensional space, then it is necessary to further specify $\tilde{B}\tilde{\Gamma}_2 = \tilde{A}$ in (140b). Assuming (140) is satisfied, the designer can synthesize $u_d(nT)$ (ideally) as

$$u_d(nT) = -\tilde{\Gamma}_1 z(nT) - \tilde{\Gamma}_2 x_c, \tag{141}$$

which will then leave (138) in the idealized form

$$E\epsilon = \tilde{A}\epsilon - \tilde{B}u_p, \tag{142}$$

where $\tilde{\gamma}(nT)$ in (138) has been disregarded. The remaining control term $u_p(nT)$ in (142) can now be designed to stabilize $\epsilon(nT) \to 0$ by choosing the structure

$$u_p(nT) = -\tilde{K}\epsilon(nT), \tag{143}$$

where $\tilde{K}$ is designed to place all eigenvalues of $(\tilde{A} + \tilde{B}\tilde{K})$ *interior to* the unit circle in the complex plane. We therefore say a specific set-point x_c is "regulatable" if and only if $(\tilde{A}, \tilde{B})$ is stabilizable *and* (140) is satisfied.[9] Finally, the total DAC control $u = u_p + u_d$ in (141) and (143) is physically realized as

$$u = u_p + u_d = -\tilde{K}(x_c - \hat{x}) - \tilde{\Gamma}_1\hat{z} - \tilde{\Gamma}_2 x_c. \tag{144}$$

The more general case (38) and (39) of a $\bar{y}(nT)$ set-point $\bar{y}_{desired} = \bar{y}_c$, $\bar{y}_c$ = constant $\in \mathscr{R}[\bar{C}]$, can be handled as follows. Since $\bar{y} = \bar{C}x$, the condition $\bar{y}(nT) \to \bar{y}_c$ implies that $x(nT) \to$ a value x^0 such that $\bar{C}x^0 = \bar{y}_c$. The set of all such x^0 is denoted by $\mathscr{X}^0(\bar{y}_c)$ and, assuming rank $\bar{C} = \bar{m}$, it is easy to show that $\mathscr{X}^0 = \left\{x^0 \,|\, x^0 = \bar{C}^T(\bar{C}\bar{C}^T)^{-1}\bar{y}_c + N\zeta\right\}$, where $\mathscr{R}[N]$ forms a basis for the null-space of $\bar{C}$ and ζ is an arbitrary $(n - \bar{m})$-vector. Thus, to achieve $\bar{y}(nT) \to \bar{y}_c$, it is only necessary to regulate $x(nT)$ to any convenient "regulatable" state $x^0 \in \mathscr{X}^0(\bar{y}_c)$. This

[9] *The plant (36) cannot be stabilized to the desired set-point x_c if x_c fails to be regulatable.*

leads to a state set-point regulator problem of the type already discussed above. It is remarked that not every state $x^0 \in \mathscr{X}^0(\bar{y}_c)$ is regulatable, in general. Moreover, the state x^0 having minimum-norm corresponds to the choice $\zeta = 0$; however, there is no guarantee that such x^0 will be regulatable for any given $\bar{y}_c$. For another approach see (154) with $u \to u_p$.

2. *Design for Servo-Tracking Commands (see also Section V.D.2)*

The design of a discrete-time DAC to achieve servo-control of $\bar{y}(nT)$ for a plant of the form (36)-(38) can be accomplished as follows. Assume, for instance, that the servo-commands $\bar{y}_c(nT)$ are modeled by the difference equation (39) and introduce the servo-error

$$\epsilon_s = (\epsilon_{s,1}, \epsilon_{s,2}, \ldots, \epsilon_{s,\bar{m}}) = \bar{y}_c(nT) - \bar{y}(nT). \tag{145}$$

Now, one can take successively higher shifts $E\epsilon_{s,i}(nT)$, $E^2\epsilon_{s,i}(nT)$, etc., of each component $\epsilon_{s,i}$ in (145), using (36)-(39), to develop a set of auxiliary higher-order difference equations for the $\epsilon_{s,i}(nT)$, which are forced by $u(nT)$, $u((n + 1)T), \ldots, z(nT), z((n + 1)T), \ldots$, and are otherwise *homogenous* in $\epsilon_s(nT)$. In other words, one uses the same procedure as (73)-(89) to develop a set of auxiliary equations for the components $\epsilon_{s,i}(nT)$ having the same structure as (90).

Then, proceeding as in (90)-(93), one can design $u_d(nT)$ to absorb all the "disturbance" terms involving z and c and then design $u_p(nT)$ to regulate $\epsilon_s(nT)$ to zero. This procedure leads to discrete-time dynamical compensators for both $u_d(nT)$ and $u_p(nT)$, similar to (92) and (93). The resulting DAC servo-controller $u = u_p + u_d$ will steer $\bar{y}(nT)$ so as to acquire quickly and track $\bar{y}_c(nT)$ for all commands generated by (39). However, before such a design is finalized, one should investigate the corresponding behavior of $x(nT)$ to make sure that $\|x(nT)\|$ remains well-behaved; see footnote 6 and the discussion of this point in [4]. Note that the on-line, real-time command-state data $c(nT)$ required by the DAC servo-controller can be generated by the command state-observer (35).

D. *DESIGN FOR THE DISTURBANCE-UTILIZATION MODE*

The modes of disturbance accommodation considered so far are designed to cope with disturbances by counteracting (absorbing) their effects. This design attitude reflects the traditional view of disturbances as causing only unwanted, disruptive effects on the plant behavior. However, there are realistic situations in which disturbances are capable of producing *desirable* effects on the plant behavior. In particular, it is possible that at least some of the action of disturbances can be constructively used to assist the controller in accomplishing the primary control task. The trick, of course, is to know just how to manipulate the control $u(nT)$, in real-time, so as to harness and exploit any useful effects inherent in the (uncertain) disturbance actions.

The systematic design of continuous-time controllers to utilize optimally the action of uncertain disturbances was first introduced in [11], and has since been refined and applied [3,5,7,15,21]. In this section, we shall derive discrete-time versions of "disturbance-utilizing" controller design procedures which parallel the results in [3], [5], and [15].

1. *The Choice of a Performance Index $\tilde{J}$ in Disturbance-Utilizing Control Problems*

The objective of disturbance-utilizing control is to make maximum (optimal) use of the disturbance $w(t)$ as an aid in accomplishing the primary control task. For instance, if the primary control task is to achieve set-point regulation or servo-tracking with minimal expenditure of control resources (fuel, energy, etc.), it is conceivable that the action of disturbances $w(t)$ might be able to reduce the drain on control energy and/or achieve "better" set-point regulation or servo-tracking, if $u(nT)$ is manipulated properly. On the other hand, if the disturbance actions are such that they are totally counter-productive to the primary control task, the use of an optimal disturbance-utilizing controller will serve to *minimize* the inevitable loss of performance contributed by the disturbance.

The optimal utilization of disturbances is achieved by application of optimal control theory, where the performance-index functional J is structured such that the minimization of J by $u(nT)$ achieves the primary control task while simultaneously making maximum "use" of $w(t)$. In the continuous-time version of disturbance-utilizing control theory [5], the most common choice of performance index J for set-point and servo-tracking problems is the classical error/control quadratic functional

$$J = \frac{1}{2}\epsilon^T(T_f)S\epsilon(T_f) + \frac{1}{2}\int_{t_0}^{T_f} [\epsilon^T(t)Q(t)\epsilon(t) + u^T(t)R(t)u(t)]dt, \tag{146}$$

where $\epsilon(t)$ denotes the instantaneous "control error," i.e., the error between desired response $\bar{y}_c(t)$ and actual response $\bar{y}(t)$; and S, Q, and R are positive definite symmetric matrices chosen by the designer. The design of u(t) to minimize (146) automatically achieves the primary control task of $\|\epsilon(T_f)\|$ = "small" [and $\|\epsilon(t)\|$ = "small"], while simultaneously letting w(t) "assist" in that task and/or in (possibly) reducing control resource consumption as measured by the time-integral of $u^T(t)R(t)u(t)$. If the disturbance w(t) is such that it cannot "assist" in reducing J in (146), the control u(t) which minimizes (146) will then automatically minimize any performance deterioration (increase in J) which w(t) contributes.

In discrete-time optimal control theory, the most common discrete-time version of (146) is expressed as

$$\tilde{J} = \frac{1}{2}\epsilon^T(NT)\tilde{S}\epsilon(NT) + \frac{1}{2}\sum_{n=0}^{n=(N-1)} [\epsilon^T(nT)\tilde{Q}(nT)\epsilon(nT) + u^T(nT)\tilde{R}(nT)u(nT)], \tag{147}$$

where the interval of control $[t_0, T_f]$ is divided into N equal segments $t_n = t_0 + nT$; $n = 0, 1, 2, \ldots, N$. Actually, if one evaluates the original continuous-time performance index (146) over each segment $t_0 + nT \leq t \leq t_0 + (n + 1)T$, using the known solution expression for (4), (21), and (22), it can be shown [19] that (146) may finally be expressed in the form[10] (147) with the exception that there is then an additional term $2u^T(nT)\tilde{M}(nT)\epsilon(nT)$ in the summation on the right side of (147). In that case, the matrices $\tilde{Q}$, $\tilde{M}$, and $\tilde{R}$ are related to Q, R, B,

[10] *Note that the arguments NT and nT in (147) actually represent the times $t = t_0 + NT$ and $t = t_0 + nT$. This short-hand notation is consistent with that used in (36)-(39), etc.; see footnote 1.*

and A in (146) and (4) through some rather involved integrals. In practical applications of discrete-time optimal control, it is generally preferable to adopt the format (147) as the *starting point* for structuring the performance index $\tilde{J}$ and then design the weighting matrices $\tilde{S}$, $\tilde{Q}$, and $\tilde{R}$ in (147) to attach proper emphasis on the minimization of $\epsilon(nT)$, $u(nT)$; $n = 0, 1, 2, \ldots, N$. For this reason we hereafter adopt (147) as the basic performance index $\tilde{J}$ for the design of discrete-time disturbance-utilizing controllers for set-point and servo-tracking problems.

2. *Formulation of a General Class of Discrete-Time Disturbance-Utilizing Control Problems*

The systematic design of disturbance-utilizing controllers can be achieved by formulating the problem as a conventional (undisturbed) linear-quadratic discrete-time control problem for which solution algorithms are known. For this purpose, the discrete-time models (36)-(39) are consolidated into one composite "plant" model and written as

$$\begin{pmatrix} Ex(nT) \\ Ec(nT) \\ Ez(nT) \end{pmatrix} = \left[\begin{array}{c|c|c} \tilde{A} & O & \tilde{F}\tilde{H} \\ \hline O & \tilde{E} & O \\ \hline O & O & \tilde{D} \end{array}\right] \begin{pmatrix} x(nT) \\ c(nT) \\ z(nT) \end{pmatrix} + \left[\begin{array}{c} \tilde{B} \\ \hline O \\ \hline O \end{array}\right] u(nT) + \begin{pmatrix} \tilde{\gamma} \\ \hline \tilde{\mu} \\ \hline \tilde{\sigma} \end{pmatrix}, \tag{148a}$$

$$y(nT) = [C(nT) \,|\, O \,|\, O] \begin{pmatrix} x(nT) \\ c(nT) \\ z(nT) \end{pmatrix}. \tag{148b}$$

For simplicity, the model (148a) is written in the more compact form

$$E\tilde{x} = \bar{A}(nT)\tilde{x}(nT) + \bar{B}(nT)u(nT) + \bar{\delta}(nT), \quad \tilde{x} = (x|c|z)^T, \tag{149}$$

where the meanings of $\bar{A}$, $\bar{B}$, and $\bar{\delta}$, are clear from examination of (148a).

The instantaneous control error $\epsilon(t)$ in (146) is the difference between the desired response and the actual response. Since $\bar{y}_c(t)$ and $\bar{y}(t)$ represent those two responses, we write $\epsilon(t)$ as

$$\epsilon(t) = \bar{y}_c(t) - \bar{y}(t), \tag{150}$$

or, in terms of discrete-time $t = nT$, as

$$\epsilon(nT) = \bar{y}_c(nT) - \bar{y}(nT). \tag{151}$$

Using (38) and (39), expression (151) may be expressed in terms of $\tilde{x}$ as

$$\epsilon = [-\bar{C}|G|O]\tilde{x} = \hat{C}\tilde{x}, \quad \hat{C} = [-\bar{C}|G|O]. \tag{152}$$

Now, the quadratic forms in the discrete-time performance-index $\tilde{J}$ in (147) may be expressed as

$$\epsilon^T\tilde{S}\epsilon = \tilde{x}^T\hat{C}^T\tilde{S}\hat{C}\tilde{x} = \tilde{x}^T\hat{S}\tilde{x}, \quad \hat{S} = \hat{C}^T\tilde{S}\hat{C}, \tag{153a}$$

$$\epsilon^T\tilde{Q}\epsilon = \tilde{x}^T\hat{C}^T\tilde{Q}\hat{C}\tilde{x} = \tilde{x}^T\hat{Q}\tilde{x}, \quad \hat{Q} = \hat{C}^T\tilde{Q}\hat{C}. \tag{153b}$$

Using (153), $\tilde{J}$ in (147) may finally be expressed in terms of the composite state $\tilde{x}$ as follows

$$\tilde{J} = \frac{1}{2}\tilde{x}^T(NT)\hat{S}\tilde{x}(NT) + \frac{1}{2}\sum_{n=0}^{n=(N-1)} [\tilde{x}^T(nT)\hat{Q}(nT)\tilde{x}(nT) + u^T(nT)\tilde{R}(nT)u(nT)]. \tag{154}$$

The optimal disturbance-utilizing control problem for discrete-time set-point regulation and servo-tracking may now be expressed precisely as follows. Find the control sequence $u(nT) = u^0(nT)$, $n = 0, 1, 2, \ldots, (N - 1)$, which minimizes the performance index (154) subject to the difference equation constraint (148) and (149) and for arbitrary initial conditions $\{x(0), c(0), z(0)\} = \tilde{x}(0)$. Since the $\sigma(t)$ and $\mu(t)$ impulses that create the terms $\tilde{\gamma}$, $\tilde{\mu}$, $\tilde{\sigma}$ in (148a) are completely unknown and *sparse*, we shall follow standard procedure in DAC theory and disregard the presence of those terms in (148a); note remarks in [3, p. 639].

3. *Solution of the Discrete-Time Disturbance-Utilizing Control Problem*

The minimization of (154) subject to (148) and (149) has the form of the conventional (undisturbed) discrete-time linear quadratic regulator problem which has already been solved; see for instance [19]. That known solution, when applied to the specific plant (148) and (149) and performance index (154) leads to the following expressions for the optimal discrete-time disturbance-utilizing control $u^0(nT)$, $n = 0, 1, \ldots, (N - 1)$. The optimal control $u^0(nT)$ is given by

$$u^0(nT) = -\left[\tilde{R}(nT) + \bar{B}^T(nT)\bar{P}[(n + 1)T]\bar{B}(nT)\right]^{-1} \times \left[\bar{B}^T(nT)\bar{P}[(n + 1)T]\bar{A}(nT)\right]\tilde{x}(nT) \tag{155}$$

where the matrix $\bar{P}(\cdot)$ is symmetric, positive definite, and governed by the Riccati difference equation

$$\bar{P}(nT) = \left[\bar{A}^T(nT)\bar{P}[(n + 1)T]\bar{A}(nT) + \hat{Q}(nT)\right] - U^T\left[\tilde{R}(nT) + \bar{B}^T(nT)\bar{P}[(n + 1)T]\bar{B}(nT)\right]^{-1}U, \tag{156a}$$

$$U = \bar{B}^T(nT)\bar{P}[(n + 1)T]\bar{A}(nT),$$

with the boundary condition

$$\bar{P}(NT) = \hat{S}. \tag{156b}$$

Note that the Riccati difference equation (156) is automatically set-up for backward-time solution, "starting" at $t = T_f = t_0 + NT$ and progressing backward: $t = t_0 + (N - 1)T$, $t = t_0 + (N - 2)T, \ldots, t = t_0 + T$, $t = t_0$. In other words, one successively sets $n = (N - 1), (N - 2), (N - 3), \ldots, 1,$

0 in (156a). The resulting sequence of values $\overline{P}(nT)$ is then stored for future playback in the forward-time control law expression (155). In particular, at each time $t = t_0 + nT$ the "current" values of $\widetilde{R}(nT)$, $\overline{B}(nT)$, and $\overline{A}(nT)$ are substituted into (155), together with the "one-step-ahead" value of $\overline{P}[(n + 1)T]$, to compute the overall state-feedback gain matrix $\mathscr{K}[nT, (n + 1)T]$:

$$\mathscr{K} = -\left[\widetilde{R}(nT) + \overline{B}^T(nT)\overline{P}[(n + 1)T]\overline{B}(nT)\right]^{-1} \times \left[\overline{B}^T(nT)\overline{P}[(n + 1)T]\overline{A}(nT)\right] \tag{157a}$$

such that

$$u^0(nT) = \mathscr{K}[nT, (n + 1)T]\widetilde{x}(nT). \tag{157b}$$

Note that the stored values $\overline{P}[(n + 1)T]$, $n = 0, 1, 2, \ldots, (N - 1)$, must be used in two locations within the $\mathscr{K}$-expression, one of which involves a complicated composite-matrix inverse operation. This characteristic feature consitutes one of the traditionally undesirable aspects of the optimal discrete-time linear regulator controller. Fortunately, the computation of $\mathscr{K}(\cdot)$ can be done off-line [it does not depend on knowledge of $\widetilde{x}(nT)$] and therefore the time required for accurate calculation of $\mathscr{K}$ does not impact on the real-time performance of the controller. In the next section, we shall derive a novel alternative expression for $u^0(nT)$ in (157), which is computationally attractive because it *does not* require calculation of the composite inverse in (157a).

The forms of (155), (156), and (157) do not yield much insight into the fine structure of the optimal disturbance-utilizing control $u^0(nT)$. To see that fine structure, it is necessary to decompose $\overline{P}(\cdot)$ into smaller blocks corresponding to the block structure of $\overline{A}$ in (148) and (149). For that purpose, we set

$$\overline{P} = \left[\begin{array}{c|c|c} K_x & K_{xc} & K_{xz} \\ \hline K_{xc}^T & K_c & K_{cz} \\ \hline K_{xz}^T & K_{cz}^T & K_z \end{array}\right], \quad \begin{array}{l} K_x = n \times n,\ K_{xc} = n \times \nu,\ K_{xz} = n \times \rho \\ K_c = \nu \times \nu,\ K_{cz} = \nu \times \rho,\ K_z = \rho \times \rho \end{array} \tag{158}$$

and substitute (158), together with the expressions for $\widetilde{R}$, $\overline{B}$, $\overline{A}$, $\hat{Q}$, and $\hat{S}$, into (155), (156), and (157) to obtain the optimal control $u^0(nT)$ expressed equivalently as

$$\begin{aligned} u^0(nT) = -\Big[\widetilde{R}(nT) &+ \widetilde{B}^T(nT)K_x[(n+1)T]\widetilde{B}(nT)\Big]^{-1}\widetilde{B}^T(nT) \\ &\times \Big[K_x[(n+1)T]\widetilde{A}(nT)x(nT) \\ &+ K_{xc}[(n+1)T]\widetilde{E}(nT)c(nT) \\ &+ \Big\{K_x[(n+1)T]\widetilde{FH}(nT) \\ &+ K_{xz}[(n+1)T]\widetilde{D}(nT)\Big\}z(nT)\Big] \end{aligned} \tag{159}$$

where the six block matrices comprising (158) obey the following set of coupled matrix difference equations

$$\begin{aligned} K_x(nT) = \Big[&\widetilde{A}(nT) - \widetilde{B}(nT)\Big[\widetilde{R} + \widetilde{B}^T(nT)K_x[(n+1)T]\widetilde{B}(nT)\Big]^{-1} \\ &\times \widetilde{B}^T(nT)K_x[(n+1)T]\widetilde{A}(nT)\Big]^T K_x[(n+1)T]\widetilde{A}(nT) \\ &+ \overline{C}^T(nT)\widetilde{Q}(nT)\overline{C}(nT); \quad K_x(NT) = \overline{C}^T(NT)\widetilde{S}\overline{C}(NT), \end{aligned} \tag{160a}$$

$$\begin{aligned} K_{xc}(nT) = \Big[&\widetilde{A}(nT) - \widetilde{B}(nT)\Big[\widetilde{R}(nT) + \widetilde{B}^T(nT)K_x[(n+1)T]\widetilde{B}(nT)\Big]^{-1} \\ &\times \widetilde{B}^T(nT)K_x[(n+1)T]\widetilde{A}(nT)\Big]^T K_{xc}[(n+1)T]\widetilde{E}(nT) \\ &- \overline{C}^T(nT)\widetilde{Q}(nT)G(nT); \quad K_{xc}(NT) = -\overline{C}^T(NT)\widetilde{S}G(NT), \end{aligned} \tag{160b}$$

$$K_{xz}(nT) = \Big[\widetilde{A}(nT)-\widetilde{B}(nT)\Big[\widetilde{R}(nT)+\widetilde{B}^T(nT)K_x[(n+1)T]\widetilde{B}(nT)\Big]^{-1}$$

$$\times\widetilde{B}^T(nT)K_x[(n+1)T]\widetilde{A}(nT)\Big]^T\cdot\Big[K_{xz}[(n+1)T]\widetilde{D}(nT)$$

$$+K_x[(n+1)T]\widetilde{FH}(nT)\Big];\ K_{xz}(NT) = 0, \qquad (160c)$$

$$K_c(nT) = \widetilde{E}^T(nT)\Big[K_c[(n+1)T]-K_{xc}^T[(n+1)T]\widetilde{B}(nT)\Big[\widetilde{R}(nT)$$

$$+\widetilde{B}^T(nT)K_x[(n+1)T]\widetilde{B}(nT)\Big]^{-1}\widetilde{B}^T(nT)K_{xc}[(n+1)T]\Big]\widetilde{E}(nT)$$

$$+G^T(nT)\widetilde{Q}(nT)G(nT);\ K_c(NT) = G^T(NT)\widetilde{S}G(NT), \qquad (160d)$$

$$K_{cz}(nT) = \widetilde{E}(nT)\Big[K_{xc}^T[(n+1)T]\widetilde{FH}(nT)+K_{cz}[(n+1)T]\widetilde{D}(nT)$$

$$-K_{xc}^T[(n+1)T]\widetilde{B}(nT)\Big[\widetilde{R}(nT)+\widetilde{B}^T(nT)K_x[(n+1)T]\widetilde{B}(nT)\Big]^{-1}$$

$$\times\widetilde{B}^T(nT)\Big[K_x[(n+1)T]\widetilde{FH}(nT)$$

$$+K_{xz}[(n+1)T]\widetilde{D}(nT)\Big]\Big];\ K_{cz}(NT) = 0, \qquad (160e)$$

$$K_z(nT) = \widetilde{FH}^T(nT)\Big[K_{xz}[(n+1)T]\widetilde{D}(nT)+K_x[(n+1)T]\widetilde{FH}(nT)\Big]$$

$$+\widetilde{D}^T(nT)\Big[K_{xz}^T[(n+1)T]\widetilde{FH}(nT)+K_z[(n+1)T]\widetilde{D}(nT)\Big]$$

$$-\Big[K_x[(n+1)T]\widetilde{FH}(nT)+K_{xz}[(n+1)T]\widetilde{D}(nT)\Big]^T$$

$$\times\widetilde{B}(nT)\Big[\widetilde{R}(nT)+\widetilde{B}^T(nT)K_x[(n+1)T]\widetilde{B}(nT)\Big]^{-1}$$

$$\times\widetilde{B}^T(nT)\Big[K_x[(n+1)T]\widetilde{FH}(nT)$$

$$+K_{xz}[(n+1)T]\widetilde{D}(nT)\Big];\ K_z(NT) = 0. \qquad (160f)$$

Thus, to implement the disturbance-utilizing optimal control $u^0(nT)$ in (159) one must first solve for $K_x(nT)$, $K_{xc}(nT)$, and $K_{xz}(nT)$ by solving (160a), (160b), and (160c) in backward time $n = (N - 1), (N - 2), \ldots, 2, 1, 0$, using the indicated "initial conditions" at $t = t_0 + NT$. As already mentioned, this step can be carried out off-line (ahead of time) and the computed values stored for future use. Note that at each time $t = t_0 + nT$, the real-time disturbance-utilizing control (159) depends on the values of $\tilde{R}$, $\tilde{B}$, $\tilde{A}$, $\tilde{E}$, $\widetilde{FH}$, and $\tilde{D}$, and x, c, and z *evaluated at* $t = t_0 + nT$, and the values of K_x, K_{xc}, and K_{xz} *evaluated at the "one-step-ahead" time* $t = t_0 + (n + 1)T$.

4. A "Better" Expression for the Disturbance-Utilizing Control Law

The necessity of computing the composite-matrix inverse $[\tilde{R}(nT) + \bar{B}^T(nT)\bar{P}((n + 1)T)\bar{B}(nT)]^{-1}$ in (155) and (159) is a traditionally undesirable feature of the conventional solution of the discrete-time optimal linear regulator problem [19,22]. In this section we derive a mathematically equivalent alternative expression for u^0 in (155) and (159), which does not require computation of that composite-matrix inverse.

The alternative expression for $u^0(nT)$ is derived by first multiplying both sides of (155) by $[\tilde{R} + \bar{B}^T\overline{PB}]$ to obtain

$$\tilde{R}u^0 + \bar{B}^T\overline{PB}u^0 = -\bar{B}^T\overline{PA}\tilde{x}. \tag{161}$$

Next, recall from (149) that [disregarding $\bar{\delta}(nT)$]

$$\bar{B}u^0 = \tilde{x}[(n + 1)T] - \bar{A}\tilde{x}(nT). \tag{162}$$

Now, if one substitutes (162) into (161), there obtains

$$\tilde{R}(nT)u^0(nT) = -\bar{B}^T(nT)\bar{P}[(n + 1)T]\tilde{x}[(n + 1)T], \tag{163}$$

which, assuming that $\tilde{R}^{-1}$ exists for all n, implies that $u^\circ(nT)$ can be written in the alternative format

$$u^0(nT) = -[\tilde{R}(nT)]^{-1}\bar{B}^T(nT)\bar{P}[(n + 1)T]\tilde{x}[(n + 1)T]. \tag{164}$$

Comparison of (164) with (155) shows that (164) requires considerably fewer computations (matrix multiplications, additions, etc.) and involves only the one simple matrix inverse $[\tilde{R}(nT)]^{-1}$. In fact, the simple structure of (164) is seen to be identical with that of the expression $u^0(t)$ for the *continuous-time* optimal linear-quadratic regulator problem [23]. This latter observation suggests that (164) is the "natural" discrete-time counterpart to the continuous-time linear-quadratic optimal control $u^0(t)$.

The price one pays for the computational simplicity of (164) is the necessity of using the "one-step-ahead" state value $\tilde{x}((n + 1)T)$ rather than the current state value $\tilde{x}(nT)$ in the on-line, real-time generation of $u^0(nT)$. However, this requirement is easily met if one employs *full-dimensional* composite state observers (26) and (35) to generate $\{\hat{x}(nT), \hat{c}(nT), \hat{z}(nT)\}$, because at each time $t = t_0 + nT$ such state observers automatically generate also the "one-step-ahead" estimates $\{\hat{x}[(n + 1)T], \hat{c}[(n + 1)T], \hat{z}[(n + 1)T]\}$ (see Fig. 1). Thus, the "one-step-ahead" data estimate $\hat{\tilde{x}}[(n + 1)T]$ required to implement (164) is readily available from the same full-dimensional observer which produces $\hat{\tilde{x}}(nT)$ for the implementation of (155). Note that reduced-dimension observers of the types (30) and (31) *do not* produce "one-step-ahead" estimates of the *entire* state-vector and therefore they are not generally suited for implementation of (164). This points out one of the intrinsic advantages of using full-dimensional observers to implement discrete-time control laws.

If the uncertain, unpredictable residual term $\bar{\delta}(nT)$ in (149) is not "quiet" between adjacent sample times $(t_0 + nT) < t < [t_0 + (n + 1)T]$, then expression (162) involves some degree of error and, in that case, one can argue that (164) is not precisely equivalent to (155). However, it should be recalled that the derivation of (155) also disregarded the residual $\bar{\delta}(nT)$. Thus at those moments where $\bar{\delta}(nT)$ is not quiet, both (155) and (164) involve some degree of inevitable error due to the inability to know the correct value of $\bar{\delta}(\cdot)$. Note that these errors *are not* cumulative because at each sample time $t_0 + nT$, the measured value of $y(nT)$ reflects the effects of all residual activity that has occurred *since* the last sample[11] $y[(n-1)T]$.

The simplification of (159) resulting from use of the alternative control expression (164) is seen by substituting (158) into (164) to obtain the disturbance-utilizing control law $u^0(nT)$ in the form

$$u^0(nT) = -[\tilde{R}(nT)]^{-1}\tilde{B}^T(nT)\Big[K_x[(n+1)T]x[(n+1)T] + K_{xc}[(n+1)T]c[(n+1)T] + K_{xz}[(n+1)T]z[(n+1)T]\Big] \tag{165}$$

where, as before, the matrices $K_x(\cdot)$, $K_{xc}(\cdot)$, and $K_{xz}(\cdot)$ are governed by (160). Comparison of (165) with (159) shows a significant reduction in the amount of matrix computations required to generate $u^0(nT)$.

5. *The Notions of Fixed Cost, Assistance, Burden, and Utility in Disturbance-Utilizing Control Problems*

The optimal disturbance-utilizing control $u^0(nT)$ in (159) [or (165)] achieves the minimum possible value of $\tilde{J}$ in (147). That minimum value of $\tilde{J}$ will be denoted by the scalar function $\mathcal{V} = \mathcal{V}[x(nT), c(nT), z(nT), (n+1)T]$, where $x(nT)$, $c(nT)$, and $z(nT)$ denote arbitrary "initial conditions" in (x, c, z)-space.

[11] *There is a subtlety associated with discrete-time observers that should be remembered here. Namely, the information contained in the real-time measurement $y(nT)$ is used by the observer to generate the "one-step-ahead" estimate $\hat{x}[(n+1)T]$, rather than to generate the "current" estimate $\hat{x}(nT)$; (the current estimate $\hat{x}(nT)$ was itself determined earlier by the previous measurement $y[(n-1)T]$).*

It can be shown [19] that the function $\mathscr{V}[x, c, z, (n+1)T]$ for the disturbance-utilization problem (147)-(149) has the explicit form

$$\mathscr{V} = \frac{1}{2}\,\tilde{x}^T(nT)\bar{P}[(n+1)T]\tilde{x}(nT) = \frac{1}{2}(x|c|z)^T \times \left[\begin{array}{c|c|c} K_x & K_{xc} & K_{xz} \\ \hline K_{xc}^T & K_c & K_{cz} \\ \hline K_{xz}^T & K_{cz}^T & K_z \end{array}\right]\left(\begin{array}{c} x \\ \hline c \\ \hline z \end{array}\right), \tag{166}$$

which can be expanded to yield, (note: all K-expressions in (166) and (167) are evaluated at $t = (n+1)T$, whereas x, c, and z are evaluated at $t = nT$.)

$$\mathscr{V} = \underbrace{\frac{1}{2}\left(x^T K_x x + c^T K_c c + 2x^T K_{xc} c\right)}_{\mathscr{F}\,=\,\text{Fixed cost}} + \underbrace{\left(x^T K_{xz} + c^T K_{cz}\right) z}_{-\mathscr{A}\,=\,-\text{Assistance}} + \underbrace{\frac{1}{2}\, z^T K_z z}_{\mathscr{B}\,=\,\text{Burden}}. \tag{167}$$

The role of the disturbance w(t) in reducing the minimum possible value of $\tilde{J}$ can now be clearly seen in (167). Namely, the impact of w(t) on $\mathscr{V} = \min \tilde{J}$ is reflected in the z-related terms in (167). If the collection of terms labeled $\mathscr{A}$ (assistance) is greater than the burden term $\mathscr{B}$, then, and only then, will min $\tilde{J}$ be further reduced by the action of w(t). Thus, following the ideas in [15] we define the utility $\mathscr{U}$ of the disturbance w(t) as

$$\mathscr{U} = \text{Assistance-Burden} = -\left(x^T K_{xz} + c^T K_{cz}\right) z - \frac{1}{2}\, z^T K_z z. \tag{168}$$

The condition $\mathscr{U} > 0$ indicates that the current behavior of w(t) is such that it can help in reducing min $\tilde{J}$. On the other hand, the condition $\mathscr{U} < 0$ indicates that the current behavior of w(t) is such that w(t) can only aggravate (increase) the value of min $\tilde{J}$. The collection of terms in (167) that do not involve z is referred to as the "fixed-cost" $\mathscr{F}$ because that contribution to $\mathscr{V} = \min \tilde{J}$ is *invariant* with respect to the behavior of disturbances w(t).

The disturbance utility function $\mathscr{U}$ defined by (168) can be studied in the (x, c, z, t)-space to identify the domains of positive and negative utility; the details are outlined in [15] and some specific examples, from continuous-time disturbance-utilizing DAC theory, are presented in [21] and [24]. Note that as time progresses, n = 0, 1, 2,..., the sign of $\mathscr{U}$ can change back and forth.

6. *The Effectiveness $\mathscr{E}$ of Optimal Disturbance-Utilizing Control*

The linear-quadratic regulator theory is widely used to design feedback control laws of the form $u(\cdot) = \mathscr{K}(\cdot)x(\cdot)$. Traditionally, such applications have ignored the presence of persistent disturbances w(t) and therefore, when confronted with actual, real-life disturbances in the field, such control laws do not yield "optimal" performance. Thus, it is interesting to study how much better the disturbance-utilizing control law performs, compared to the conventional linear-quadratic control law, when the two closed-loop systems are subjected to the same typical realistic disturbances w(t). To quantify such a comparison, Kelly [21] has proposed the concept of "effectiveness" $\mathscr{E}$ defined for the discrete-time case as

$$\mathscr{E} = \frac{\tilde{J}_{LQ} - \tilde{J}_{DUC}}{\tilde{J}_{LQ}} \times 100\%, \tag{169}$$

where $\tilde{J}_{LQ}$ is the value of (147) obtained by using the conventional (undisturbed) discrete-time linear-quadratic control law, and $\tilde{J}_{DUC} = \mathscr{V}$ is the value (167) of (147) obtained by using the optimal disturbance-utilizing control law (159) [or (165)] — in *both* cases the plant (36) is subjected to the *same* disturbance w(t) [as generated by the assumed disturbance model (37)]. Thus, if the disturbance-utilizing controller (159) is a better performer (as it should be), one should find that $\tilde{J}_{DUC}$ is *less* than $\tilde{J}_{LQ}$ and therefore $\mathscr{E}$ is *positive*. The maximum possible value of $\mathscr{E}$ is 100%, which would correspond to the (unlikely) case of $\tilde{J}_{DUC} = 0$. Thus, the closer $\mathscr{E}$ is to 100% the greater is the effectiveness of (159) compared to the conventional linear-quadratic control law.

It is interesting to note that the conventional linear-quadratic control law used in such comparisons can be obtained directly from (159) [or (165)] by simply setting $z(nT) = 0, \forall n$, [and also $c(nT) = 0, \forall n$, if $\bar{y}(T_f) = 0$ is the desired response]. This observation shows that the disturbance-utilizing control law (159) [or (165)] automatically reduces to the conventional linear-quadratic control law whenever the disturbance w(t) vanishes. In other words, the matrix $K_x(\cdot)$ in (159) [or (165)] and (160a) is precisely the *same* matrix used in the conventional (undisturbed) linear-quadratic regulator control law.

7. *Example Design of a Discrete-Time Disturbance-Utilizing Controller*

To demonstrate application of the controller design algorithm for discrete-time disturbance-utilization, we shall consider a rather general version of a first-order plant with first-order disturbance. The plant discrete-time model (36) and disturbance discrete-time model (37) are expressed as

$$Ex = \tilde{a}x(nT) + \tilde{b}u(nT) + \widetilde{fh}z(nT) + \tilde{\gamma}(nT), \tag{170a}$$

$$y = cx(nT), \tag{170b}$$

$$Ez = \tilde{d}z(nT) + \tilde{\sigma}(nT), \tag{170c}$$

where x, u, and z are scalars. We shall assume that the desired behavior is set-point regulation to $x(nT) = 0 \; \forall \; n$; therefore $\bar{y} = x$ in (38) and $\bar{y}_c \equiv 0$ in (39). Thus, we may set $G(t) \equiv 0$ in (39). The parameters $\tilde{a}$, $\tilde{b}$, $\widetilde{fh}$, and $\tilde{d}$ may be time-varying.

The discrete-time performance index $\tilde{J}$ in (147) is expressed as

$$\tilde{J} = \frac{1}{2} x^T(NT)\tilde{s}x(NT) + \frac{1}{2} \sum_{0}^{(N-1)} [x^T(nT)\tilde{q}x(nT)$$

$$+ u^T(nT)\tilde{R}u(nT)]; \quad \tilde{s} > 0, \; \tilde{q} > 0, \; \tilde{R} > 0, \tag{171}$$

where, in this example, $\tilde{s}$, $\tilde{q}$, and $\tilde{R}$ are arbitrary positive scalars and $\epsilon = -x$.

The optimal disturbance-utilizing control (159) for this example has the specific form

$$u^0(nT) = -\left[\frac{\tilde{b}}{\tilde{R} + \tilde{b}^2 k_x[(n+1)T]}\right]$$

$$\times \left[\tilde{a}k_x[(n+1)T]x(nT) + \left\{\widetilde{fh}k_x[(n+1)T] + \tilde{d}k_{xz}[(n+1)T]\right\}z(nT)\right], \tag{172a}$$

which can be expressed also in the alternative form (165) as

$$u^0(nT) = -\tilde{R}^{-1}\tilde{b}\left\{k_x[(n+1)T]x[(n+1)T] + k_{xz}[(n+1)T]z[(n+1)T]\right\}, \tag{172b}$$

where we have set $k_{xc} \equiv 0$ because $G(t) \equiv 0$ (set-point $\bar{y}_c$ is zero). The time-varying gains $k_x(\cdot)$ and $k_{xz}(\cdot)$ associated with the control (172) are computed from the difference equations (160a,c) which for this example reduce to

$$k_x(nT) = \frac{\tilde{R}\tilde{a}^2 k_x[(n+1)T]}{\tilde{R} + \tilde{b}^2 k_x[(n+1)T]} + \tilde{q},$$

$$\bar{c} = +1, \quad k_x(NT) = \tilde{s} > 0 \tag{173a}$$

$$k_{xz}(nT) = \left(\frac{\tilde{R}\tilde{a}}{\tilde{R} + \tilde{b}^2 k_x[(n+1)T]}\right)\Big[k_{xz}[(n+1)T]\tilde{d} + k_x[(n+1)T]\widetilde{fh}\Big], \quad k_{xz}(NT) = 0. \tag{173b}$$

One can now compute the successive values of k_x and k_{xz} for $n = (N-1), (N-2), (N-3), \ldots, 2, 1, 0$ using (173) and the indicated "starting" conditions at $t = NT$. Those computed values are then stored and used later in the real-time computation of (172).

Consideration of the time-invariant case of the example. If one assumes that the plant and disturbance models (170) came from a *time-invariant* continuous-time plant and disturbance model, the preceding results (172) and (173) can be expressed in more explicit form. In particular, if (170) are assumed to derive from the continuous-time models

$$\dot{x} = ax + bu + fw, \tag{174a}$$

$$\dot{z} = dz + \sigma(t), \quad w = hz, \tag{174b}$$

where a, b, f, h, and d are *constant* scalars, then

$$\tilde{a} = e^{aT}, \quad \tilde{b} = \int_0^T e^{a(T-\tau)} b\, d\tau = \frac{b}{a}(e^{aT} - 1), \quad (= bT \text{ if } a = 0)$$

$$\widetilde{fh} = \int_0^T e^{a(T-\tau)} fhe^{d\tau} d\tau = \left(\frac{f}{d-a}\right) \times [e^{dT} - e^{aT}], \quad (= fTe^{aT} \text{ if } a = d) \tag{175}$$

$$\tilde{d} = e^{dT}.$$

Using (175) in (172) and (173) leads to the following explicit expressions (shown for the case $a \neq 0$, $d \neq a$). The conventional form (172a) of the control law becomes

$$u^0(nT) = -\left[\frac{\frac{b}{a}(e^{aT} - 1)}{\tilde{R} + \left(\frac{b}{a}\right)^2 (e^{aT} - 1)^2 k_x[(n + 1)T]}\right]$$

$$\times \Big[e^{aT} k_x[(n + 1)T]x(nT) + \Big\{\left(\frac{f}{d - a}\right)(e^{dT} - e^{aT})$$

$$\times k_x[(n + 1)T] + e^{dT} k_{xz}[(n + 1)T]\Big\} z(nT)\Big] \tag{176a}$$

and the alternative form (172b) becomes

$$u^0(nT) = -\frac{b}{Ra}(e^{aT} - 1)\Big\{k_x[(n + 1)T]x[(n + 1)T]$$

$$+ k_{xz}[(n + 1)T]z[(n + 1)T]\Big\}. \tag{176b}$$

Expressions (173) for the gain matrices become

$$k_x(nT) = \frac{\tilde{R}e^{2aT} k_x[(n + 1)T]}{\tilde{R} + \left(\frac{b}{a}\right)^2 (e^{aT} - 1)^2 k_x[(n + 1)T]} + \tilde{q},$$

$$k_x(NT) = \tilde{s}, \tag{177a}$$

$$k_{xz}(nT) = \left(\frac{\tilde{R}e^{aT}}{\tilde{R} + \left(\frac{b}{a}\right)^2 (e^{aT} - 1)^2 k_x[(n + 1)T]}\right)$$

$$\times \Big[e^{dT} k_{xz}[(n + 1)T] + \left(\frac{f}{d - a}\right)(e^{dT} - e^{aT})$$

$$\times k_x[(n + 1)T]\Big], \quad k_{xz}(NT) = 0. \tag{177b}$$

8. *Other Classes of Disturbance-Utilizing Controllers*

The disturbance-utilizing controllers described so far have been derived for the conventional error/control quadratic performance index $\tilde{J}$ in (147). Although (147) has a broad range of applications, there are practical situations for which other performance indexes $\tilde{J}$ are more appropriate. For example, in the case of disturbance-utilization for time-optimal control, one would replace (147) by

$$\tilde{J} = T_f - t_0 = NT; \quad T = \text{fixed (sample period)}, \quad N = \text{unspecified positive integer to be minimized} \tag{178}$$

with specified terminal conditions on $x(T_f)$ and appropriate constraints imposed on the admissible values of $u(t)$. In the case of time-optimal control with energy or fuel minimization, the optimal utilization of disturbances might be approached by considering the performance index

$$\tilde{J} = NT + \sum_{n=0}^{n=(N-1)} \beta[u(nT)], \tag{179}$$

where $\beta[u(\cdot)]$ is an appropriately weighted measure of energy or fuel consumption.

The design of disturbance-utilizing controllers for non-quadratic performance indexes such as (178), (179), etc., proceeds just as in (148)-(154). In particular, the composite "plant" models (148) and (149) are first used to reduce the problem to an equivalent undisturbed problem, and then conventional optimal control theory is used to derive the optimal feedback control in the form $u^0 = u^0[\tilde{x}(nT), nT, (n + 1)T]$. In this way, disturbance-utilizing controllers can be designed for any of the performance indexes $\tilde{J}$ which are amenable to conventional optimal control theory for undisturbed plants [22].

E. DESIGN FOR MULTIMODE DISTURBANCE-ACCOMMODATION

In some practical applications, the appropriate mode of disturbance-accommodation varies with the condition of the plant state, the value of time, the nature of the disturbance behavior, etc. For such cases, one can use the foregoing design procedures to design a multimode disturbance-accommodating controller that automatically changes "mode" in accordance with changes in the environment. As an example, the disturbance-utilization mode might be used when the utility $\mathscr{E}$ is positive, with the mode automatically changing to disturbance-absorption whenever utility becomes negative.

F. DESIGN OF DISCRETE-TIME DACs WITH INTERSAMPLE DISTURBANCE INTERPOLATION

In the most common applications of discrete-time controllers, the control action u(t) is held *constant* between sample-times. The DAC design procedures described in the previous sections of this chapter have all been tailored for such applications. As mentioned in Section V.A, the use of a constant control action u(nT) to "accommodate" a varying disturbance action w(t) leads to certain compromises. For instance, it is then not generally possible to absorb completely all disturbance effects on the plant state motion x(t). As a consequence, the disturbance w(t) typically produces intersample ripple behavior in x(t) as depicted in Fig. 4. If the extent of that ripple is judged to be excessive, one must either reduce the controller sample-period T or seek some means for accomplishing a higher degree of disturbance-absorption between sample times.

An effective means for improving the degree of disturbance-absorption *between* controller sample-times is to allow the control action u(•) to vary between sample-times (open-loop fashion) in accordance with some prescribed interpolation rule. Such an interpolation is initiated at each sample-time t_i, and guided by the information available at that time. In this section we describe two approaches to the design of discrete-time DACs with intersample interpolation capability. The first of these approaches requires a modified form of composite state-observer that we call a *hybrid observer*.

1. *A Hybrid Composite State-Observer*

When the control action $u(\cdot)$ is allowed to vary between sample-times, in a discrete-time control problem, the structure of the discrete-time state-observer must be modified accordingly. For our purposes, we shall only be concerned with modifying the full-dimensional composite state-observer (26) since only that observer produces the one-step-ahead estimates $x[(n + 1)T]$ and $z[(n + 1)T]$, which we shall need for the control interpolation rule.

In the most general case, the total control action $u(\cdot)$ in a discrete-time control problem will consist of a constant part $u_c(\cdot)$ of the usual type and a time-varying interpolating part $u_t(\cdot)$. Thus, for generality we shall write the total discrete-time control $u(\cdot)$ as

$$u(\tau)\big|_{nT \le \tau < (n+1)T} = u_c(nT) + u_t(\tau); \quad u_c(\cdot) = \text{const.} \tag{180}$$

The presence of the continuously varying control term $u_t(\cdot)$ in (180) makes it necessary to modify the structure of the observer (26) as follows. Recall from (6) that the contribution of $u(\cdot)$ to the discrete-time motion of $x(t)$ is expressed by the convolution integral

$$\int_t^{t+T} \Phi(t + T, \tau)B(\tau)u(\tau)\,d\tau. \tag{181}$$

Substituting (180) into (181), and setting $t = t_i = t_0 + nT$, expression (181) can be written as

$$\int_{t_0+nT}^{t_0+(n+1)T} \Phi[t_0 + (n + 1)T, \tau]B(\tau)[u_c(nT) + u_t(\tau)]d\tau$$

$$= \tilde{B}(nT)u_c(nT) + \int_{t_0+nT}^{t_0+(n+1)T} \Phi[t_0 + (n + 1)T, \tau] \times B(\tau)u_t(\tau)d\tau, \tag{182}$$

where $\tilde{B}(nT)$ is defined by (8b). As a consequence of (182), the discrete-time plant model (8a) must be written in the generalized form

$$x[(n + 1)T; nT, u, w] = \tilde{A}(nT)x(nT) + \tilde{B}(nT)u_c(nT) + \psi(u_t) + \tilde{v}[(n + 1)T], \tag{183a}$$

where

$$\psi(u_t) = \int_{t_0+nT}^{t_0+(n+1)T} \Phi[t_0 + (n + 1)T, \tau] \times B(\tau)u_t(\tau)d\tau. \tag{183b}$$

The term $\tilde{v}[(n + 1)T]$ in (183a) can now be evaluated exactly as in (12) and (13) to obtain finally the modified plant/disturbance composite model (16) in the form

$$\left(\frac{x[(n + 1)T]}{z[(n + 1)T]}\right) = \left[\begin{array}{c|c} \tilde{A}(nT) & \widetilde{FH}(nT) \\ \hline O & \tilde{D}(nT) \end{array}\right] \left(\frac{x(nT)}{z(nT)}\right) + \left[\frac{\tilde{B}(nT)}{O}\right] u_c(nT) + \left(\frac{\psi(u_t)}{O}\right) + \left(\frac{\tilde{\gamma}(nT)}{\tilde{\sigma}(nT)}\right). \tag{184}$$

The generalization of the full-dimensional observer (26) corresponding to the modified composite model (184) is now obtained by simply adding the control-related term $[\psi(u_t)|O]^T$ to the right side of (26) — and setting $u \to u_c$ in (26). We call such an observer a *hybrid* composite state-observer. The corresponding estimation error dynamics (27) are *unchanged* by this modification because the term $\psi(u_t)$ can presumably be directly and accurately measured in real-time. Thus, all the comments following (27) and (28) apply also to the modified (hybrid) version of (26).

2. *Design of a Linear-Interpolating Discrete-Time DAC for Improved Disturbance-Absorption*

The simplest way to improve disturbance-absorption performance between sample-times is to perform a linear interpolation of the behavior of the disturbance state $z(t)$ between nT and $(n + 1)T$. Referring to Fig. 9, it can be seen that if $\hat{z}(nT)$ and $\hat{z}((n + 1)T)$ are both known at time nT, then the linear interpolation $\tilde{z}(\tau)$ of the actual behavior of $z(\tau)$, $nT \leq \tau < (n + 1)T$, can be expressed as[12]

$$\tilde{z}(\tau) \cong \hat{z}(nT) + \left(\frac{\tau - nT}{T}\right)[\hat{z}((n + 1)T) - \hat{z}(nT)]. \qquad (185)$$

Thus, instead of designing a constant u_d as in (42), one can attempt to choose $u_d = u_t$ so as to accomplish *continuous-time* disturbance absorption of the linearly interpolated disturbance term

$$F(\tau)w(\tau) \approx F(\tau)H(\tau)\tilde{z}(\tau), \quad \tilde{z} \text{ given by (185)}, \qquad (186)$$

between sample times $nT \leq \tau < (n + 1)T$. For this purpose, we return to (4) and note that, with $u = u_p(nT) + u_d(t)$ and $u_d = u_t$, the condition of complete-absorption of the disturbance approximation (186) (in continuous-time) implies that

[12]*The reader is reminded of our shorthand notation whereby* nT, τ *actually represent the times* $t = t_0 + nT$, *and* $t = t_0 + \tau$ *respectively.*

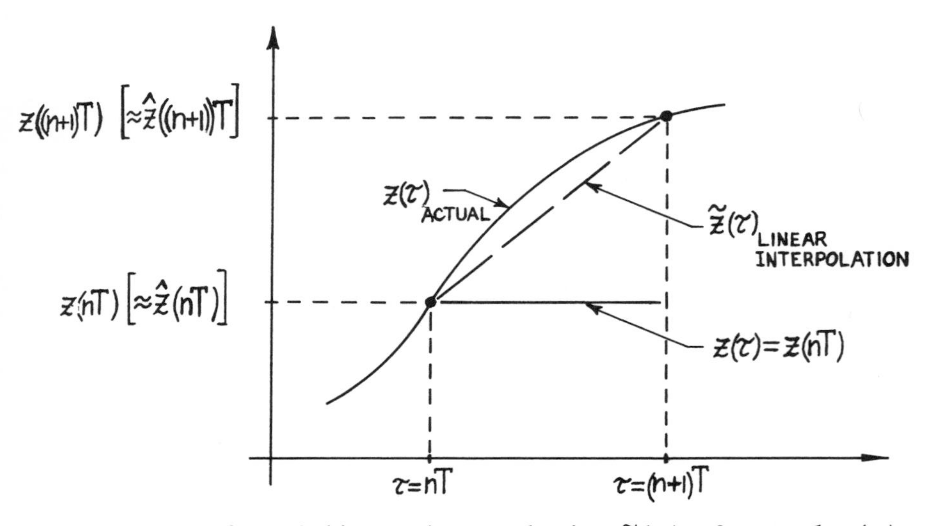

Fig. 9. Idea of linear interpolation $\tilde{z}(\tau)$ of actual $z(\tau)$ over $nT \leq \tau < (n+1)T$.

$u_t(\cdot)$ must be chosen to satisfy

$$B(\tau)u_t(\tau) + F(\tau)H(\tau)\tilde{z}(\tau) \equiv 0, \quad nT \leq \tau < (n+1)T. \tag{187}$$

Assuming $\tilde{z}(\tau)$ is "arbitrary," the necessary and sufficient condition for existence of a $u_t(\tau)$ satisfying (187) is

$$F(\tau)H(\tau) \equiv B(\tau)\Gamma(\tau), \quad \text{for some } \Gamma(\tau), \tag{188}$$

which is simply the complete-absorbability condition for continuous-time DAC; see remarks after (43). Assuming (188) is satisfied, the control term $u_t(\tau)$ in (187) can be designed as

$$\begin{aligned} u_t(\tau) &= -\Gamma(\tau)\tilde{z}(\tau) \\ &= -\Gamma(\tau)\left\{\hat{z}(nT) + \left(\frac{\tau - nT}{T}\right)\left[\hat{z}[(n+1)T] - \hat{z}(nT)\right]\right\}; \quad nT \leq \tau < (n+1)T. \end{aligned} \tag{189}$$

Note that the control interpolation rule (189) only requires knowledge of the two values $\hat{z}(nT)$ and $\hat{z}[(n+1)T]$, and these two values are directly available in real-time (at each time nT) from the hybrid composite state-observer described in Section V.F.1. Thus, at each sample time nT the interpolation rule (189) is initialized by setting $\tau = nT$ and inserting the "current" estimates $\hat{z}(nT)$ and $\hat{z}[(n+1)T]$ as obtained from the hybrid observer. Then, τ in (189) is varied over the interval $nT \leq \tau < (n+1)T$ (by a clock, analog integrator, or other time-base generator) to generate the desired interpolating control $u_d = u_t(\tau)$. Note that the control $u_t(\tau)$ must be fed back into the hybrid observer through the term (183b). The discrete-time primary control term $u_p(nT)$ can now be designed by the methods described in the earlier sections of this chapter.

3. *Design of an Exact-Interpolating Discrete-Time DAC for Improved Disturbance-Absorption*

The linear interpolating control rule (189) can be generalized to higher-order polynomials in time by replacing (185) with an appropriate polynomial fit to the data values $\hat{z}[(n+1)T]$, $\hat{z}(nT)$, $\hat{z}[(n-1)T], \ldots, \hat{z}[(n-k)T]$. In this way, an improvement in the degree of disturbance-absorption over that obtained with (189) should be realized — at the expense of a more elaborate interpolation rule.

Alternatively, one can replace the class of polynomial approximations of $z(\tau)$ by an approximation scheme that employs the actual (natural) waveform modes of $z(\tau)$ as reflected in the model (10). For this purpose, one can use the *known* matrix $D(t)$ in (10) to approximate $z(\tau)$ over $nT \leq \tau < (n+1)T$ by the "exact" expression

$$\tilde{z}(\tau) = \Phi_D(\tau, nT)\hat{z}(nT), \tag{190}$$

where $\hat{z}(nT)$ is obtained from a hybrid observer. If (190) is now used in place of (185), the interpolating control rule $u_t(\tau)$ in (189) becomes the "exact" expression

$$u_t(\tau) = -\Gamma(\tau)\Phi_D(\tau, nT)\hat{z}(nT). \tag{191}$$

The practical implementation of (191) is simplified if one makes use of the fact that $\tilde{z}(\tau)$ in (190) satisfies the differential equation (10). In this way the implementation of (191) can be realized as $u(\tau) = -\Gamma(\tau)\tilde{z}(\tau)$ using an analog circuit (continuous-time realization) of $\dot{\tilde{z}} = D(t)\tilde{z}$ to generate $\tilde{z}(\tau)$ with the "initial condition" $\tilde{z}(nT) = \hat{z}(nT)$ *reset* at each sample time. Such an implementation, along with the

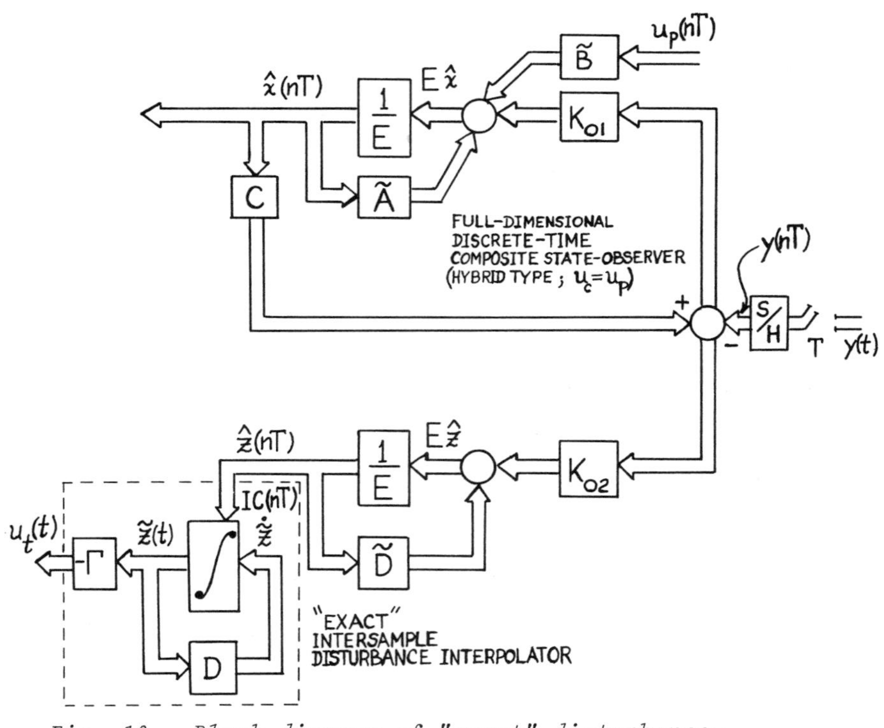

Fig. 10. Block diagram of "exact" disturbance-interpolating controller (191). Note: 1/E = T-unit delayor.

associated hybrid observer, is shown in Fig. 10. Note that when the "exact" interpolating control (191) is fed back into the hybrid observer, through the term (183b), the condition

$$\widetilde{FH}(nT)\hat{z}(nT) + \psi(u_t) = 0 \qquad (192)$$

is automatically realized, provided (188) is satisfied. This explains why the terms involving u_t and $\widetilde{FH}$ *do not* appear in the specialized hybrid observer in Fig. 10.

VI. CONCLUSIONS

The discrete-time disturbance-accommodating control theory presented in this chapter provides a systematic procedure for designing digital controllers that can maintain system performance in the face of a broad range of uncertain disturbances. Such controllers can accommodate disturbances in basically three modes: disturbance-absorption (cancellation, rejection), disturbance-minimization, and disturbance-utilization. Controller designs for each of these modes have been illustrated by worked-out examples.

The material presented in this chapter is the discrete-time counterpart to the (continuous-time) disturbance-accommodating control theory published in [5].

REFERENCES

1. C. D. JOHNSON, "Optimal Control of the Linear Regulator with Constant Disturbances," *IEEE Trans. Autom. Control AC-13*, No. 4, 416-421 (1968).

2. C. D. JOHNSON, "Further Study of the Linear Regulator with Disturbances: The Case of Vector Disturbances Satisfying a Linear Differential Equation," *IEEE Trans. Autom. Control AC-15*, No. 2, 222-228 (1970).

3. C. D. JOHNSON, "Accommodation of Disturbances in Linear Regulator and Servo-Mechanism Problems," *IEEE Trans. Autom. Control* (special issue on the "Linear-Quadratic-Gaussian Problem") *AC-16*, No. 6 (1971).

4. C. D. JOHNSON, "Algebraic Solution of the Servo-Mechanism Problem with External Disturbances," *ASME Trans., J. Dyn. Syst. Meas. Control 96*, Series G, No. 1, 25-35 (1974); see also *97*, No. 2, 161 (1975).

5. C. D. JOHNSON, "Theory of Disturbance Accommodating Controllers," Chapter 7 in "Advances in Control and Dynamics Systems," Vol. 12, (C. T. Leondes, ed.), Academic Press, New York, 1976.

6. C. D. JOHNSON, "Disturbance-Accommodating Control; A History of Its Development," Proceedings of the 15th Annual Meeting of the Society for Engineering Science, Gainesville, Florida, p. 331, December 1978.

7. C. D. JOHNSON and R. E. SKELTON, "Optimal Desaturation of Momentum Exchange Control System," The *11th Joint Automatic Control Conference*, Atlanta, Georgia, June 1970; also, *J. AIAA 9*, No. 1, 12-22 (1971).

8. C. D. JOHNSON and G. A. MILLER, "Design of a Disturbance-Accommodating Controller for an Airborne Pointing Device," *Proc. 1976 Conf. Decision Control*, Clearwater Beach, Florida, 1171-1179 (1976).

9. J. T. ODEN, "Finite Elements of Nonlinear Continua," McGraw-Hill, New York, 1971.

10. T. J. CHUNG, "Finite Element Analysis in Fluid Dynamics," McGraw-Hill, New York, 1978.

11. C. D. JOHNSON, "Accommodation of Disturbances in Optimal Control Problems," *Int. J. Control 15*, No. 2, 209-231 (1972). Also *Proc. Third Southeastern Symposium on System Theory*, Atlanta, Georgia, April 1971.

12. C. D. JOHNSON, "Comments on 'Optimal Control of the Linear Regulator with Constant Disturbances,'" *IEEE Trans. Autom. Control AC-15*, No. 4, 516-518 (1970).

13. C. D. JOHNSON, "Control of Dynamical Systems in the Face of Uncertain Disturbances," "Stochastic Problems in Mechanics," S. M. Series, University of Waterloo Press, Waterloo, Ontario, Canada, June 1974.

14. C. D. JOHNSON, "On Observers for Systems with Unknown Inaccessible Inputs," *Int. J. Control 21*, No. 5, 825-831 (1975).

15. C. D. JOHNSON, "Utility of Disturbances in Disturbance-Accommodating Control Problems," *Proc. 15th Annu. Meeting Soc. Eng. Sci.*, Gainesville, Florida, p. 347 (1978).

16. C. D. JOHNSON, "Design of Disturbance-Accommodating Controllers by the Algebraic/Stabilization Method," *Proc. First Int. Symp. Policy Analysis Inform. Syst.*, Session No. SA-VI, Duke University, June 1979; also in *Int. J. Policy Inform. 4*, No. 1 (1980).

17. C. D. JOHNSON, "Disturbance-Accommodating Control; An Overview of the Subject," *J. Interdisc. Modeling Simulation* (special issue on Disturbance-Accommodating Control), *3*, No. 1 (1980).

18. L. ZADEH and C. DESOER, "Linear Systems Theory," McGraw-Hill, New York, 1963.

19. P. DORATO and A. H. LEVIS, "Optimal Linear Regulators: The Discrete-Time Case," *IEEE Trans. Autom. Control AC-16*, No. 6, 613-620 (1971).

20. R. PENROSE, "A Generalized Inverse for Matrices," *Proc. Cambridge Phil. Soc. 51*, 3, 406-413 (1955).

21. W. C. KELLY, "Homing Missile Guidance with Disturbance-Utilizing Control," *Proc. 12th Southeastern Symp. Syst. Theory*, 260 (1980).

22. G. F. FRANKLIN and J. D. POWELL, "Digital Control of Dynamic Systems," Addison-Wesley, Reading, Massachusetts, 1980.

23. A. P. SAGE, "Optimum Systems Control," Prentice-Hall, Englewood Cliffs, New Jersey, 1968.

24. W. C. KELLY, "Theory of Disturbance-Utilizing Control with Applications to Missile Intercept Problems," Ph.D. Dissertation, 1979, The University of Alabama in Huntsville, Huntsville, Alabama.

Ship Propulsion Dynamics Simulation

C. JOSEPH RUBIS

THURBER R. HARPER

Propulsion Dynamics, Inc.
Annapolis, Maryland

I. INTRODUCTION

The mathematical modeling and simulation of ship propulsion systems has become an accepted practice worldwide as part of the ship design process. Such simulations are based on fundamental principles, but depend heavily on experimental hydrodynamic, thermodynamic, machinery, and control system

ISBN 0-12-012718-0

data derived from model scale and full scale tests. Engineers may conduct total propulsion plant and ship motion simulations involving up to six degrees-of-freedom and perhaps 200 computed parameters involving ship motion and propulsion system variables all influenced to a greater or lesser degree by the control system design. A successful approach to simulations of such complexity requires considerable simplification and decoupling with a continual appraisal of assumptions and conformance to physical reality. This chapter presents an overview of the overall approach to ship propulsion systems simulations with discussions of methods and limitations rather than modeling details. A gas turbine, controllable pitch propeller ship propulsion system is used throughout as an example. Such power plants are in wide use today in the Navies of the world for powering gunboats, frigates, and destroyers.

II. THE SHIP AND ITS ENVIRONMENT

The marine environment as it affects ship performance is unique in several respects. Some of these unique aspects are the long mission times and physical isolation together with the possibility of extreme loads imposed by the sea and by maneuvering transients. The long mission times require a propulsion plant capable of continuous rated power operation for perhaps weeks at a time. The sea-imposed loads demand structures and machines of the utmost endurance. Primarily for these two reasons, i.e., long sustained duty and the ultimate in safety, marine systems have evolved to be among the most conservative, highest safety factor designs in engineering. In contrast, airplanes while subject to large loads and an even greater demand for reliable machinery and structures have very short mission times, e.g., hours as compared to weeks or months for ships. The ship in a seaway is subjected to potentially enormous forces that can lead to large ship motions and varying loads on the propulsion system. In a heavy sea, the ship resistance changes greatly with wave action while the propeller loading also varies as the ship pitches and heaves. Moreover, in a storm such conditions can persist unabated for days at a time subjecting the hull structure and propulsion plant to forces that at best strain or wear systems and at worst sink the ship. Ship maneuvering transients, e.g., in various "crash" maneuvers, or high-speed turns can cause extreme loads on the propulsion system. Propulsion power levels may be required to vary back and forth between idle and full power within a matter of seconds, whereas land-based power plants generally operate at relatively unvarying power levels. Such large and frequent power demands are the norm in Naval ships engaged in continual training and other exercises. Merchant ship power plants, on the other hand, are more akin to shore-based plants operating at constant power.

Ship dynamics simulations are conducted for many different reasons and points of view. As an example, dynamics simulations are conducted to study ship motions in a seaway to understand structural loading and improve seakeeping qualities, while others may be used to evaluate ship maneuvering and stability. Propulsion machinery system simulations are conducted to design propulsion systems that must operate in the sea-imposed environment and respond consistently and safely to maneuvering commands. This chapter is primarily intended to illustrate the development of a ship propulsion system dynamic simulation. The reasons for ship propulsion dynamics and control simulations are the following:

Feasibility and tradeoff evaluations — selection of processes, machinery, engines, combined plants, controls, automation, and subsystem configurations.

Prediction of steady-state and transient conditions — loads, stresses, temperatures, pressures, currents, voltages, flows, energy levels, ship control, and machinery responses.

Machinery and control system design — optimization, failure analysis, hardware selections, performance criteria, and specifications.

System performance evaluation and correction — qualitative and quantitative evaluation of complex sequences (cause-effect relationships), identify and correct dynamics-control problems, baseline predictions for tests and trials.

III. SHIP'S MOTION EQUATIONS

Each of the six degrees-of-freedom of ship's motion affect the propulsion system to a varying degree. But, as in virtually all complex, multidimensional problems where solutions must be developed, the typical engineering approach is to decouple the equations and simplify the system complexity whenever possible. The six degrees-of-freedom of ship motion are shown in Fig. 1. The arrangement of the six motion components in the two columns is an aid to their understanding. Surge caused by propeller thrust and yaw by rudder action are the primary ship maneuvering control parameters. Heave and pitch are similar because they are the up and down motions, whereas

Translational motions	*Rotational motions*
Surge	*Yaw*
Heave	*Pitch*
Sway	*Roll*

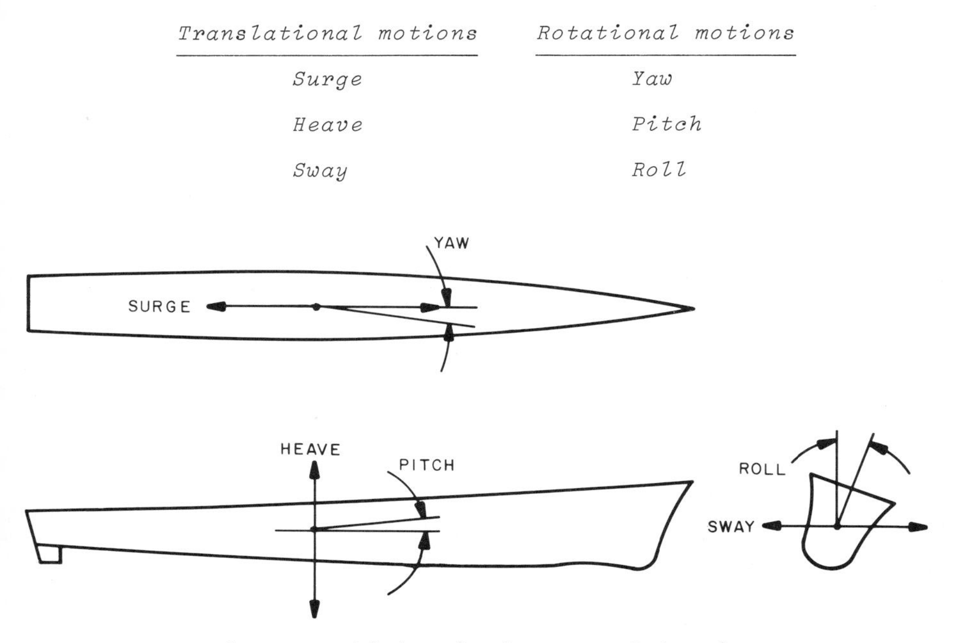

Fig. 1. Ship's six degrees-of-freedom.

sway and roll are the sideways motions. In some ships roll is countered by means of automatic roll stabilization systems.

A typical set of six simultaneous nonlinear first-order differential equations representing ship response in six degrees-of-freedom has more than 100 terms. The computations, interrelationships, and the physical understanding of such a complex system renders solutions of these equations in all six degrees-of-freedom extremely difficult even with the most sophisticated computer techniques. "In the simulation of any system, experience has shown that unless some simplifications can be made, the amount of computation involved increases at least as fast as the square of the number of equations" [1]. Restricting the analysis to four of these six ship motion equations by excluding pitch and heave results in the equations shown below. These are typical of ship's calm water maneuvering simulations where pitch and heave have been ignored.

Surge:

$$(m' - X'_{\dot{u}})\dot{u}' = F_1(u', v', r', \delta, \phi),$$

Sway:

$$(m' - Y'_{\dot{v}})\dot{v}' + (m'x'_G - Y'_{\dot{r}})\dot{r}' - m'z'_G\dot{p}' = F_2(u', v', r', \delta, \phi),$$

Yaw:

$$(m'x'_G - N'_{\dot{v}})\dot{v}' + (I'_z - N'_{\dot{r}})\dot{r}' - m'x'_G z'_G\dot{p}' = F_3(u', v', r', \delta, \phi),$$

Roll:

$$-m'z'_G\dot{v}' - m'z'_G x'_G\dot{r}' + (I'_x - K'_{\dot{p}})\dot{p}' = F_4(u', v', r', \delta, \phi),$$

where u' is the nondimensional change in surge velocity; v', nondimensional sway velocity; r', nondimensional turning rate; δ, rudder angle; ϕ, heel (roll) angle; and p', nondimensional roll rate.

The remaining terms are mass and hydrodynamic coefficients for the ship. The right-hand side of each equation represents the forcing function generated by the ship's hydrodynamics. These forcing functions F_1 through F_4 are dependent on the ship's underwater shape, propeller operating characteristics, hull motions, and orientation. These are calculated using hydrodynamic coefficients typically obtained from scale model tests of the ship's hull.

With the inclusion of the propeller thrusts and yawing moment due to unequal port and starboard thrusts, a typical representation of hydrodynamic forces and moments used in the ship motion equations is given below:

$$F_1(u', v', r', \delta, \phi) = T'_{P\text{-Port}} + T'_{P\text{-Stbd}} - R'_T + X'_{\delta\delta u}\delta^2 u'$$
$$+ X'_{vv}v'^2 + X'_{v\delta}v'\delta + m'x'_G r'^2 + m'v'r' - m'z'_G p'r' + X'_{\delta\delta}\delta^2;$$

$$F_2(u', v', r', \delta, \phi) = Y'_\delta \delta + Y'_{\delta\delta\delta}\delta^3 + Y'_{\delta|v|}\delta|v'| + Y'_{\delta u}\delta u'$$
$$+ Y'_{\delta uu}\delta u'^2 + Y'_v v' + Y'_{v|v|}v'|v'| + Y'_{v|r|}v'|r'| + Y'_{vu}v'u'$$
$$+ (Y'_r - m'u')r' + Y'_{rrr}r'^3 + Y'_\phi \phi + Y'_{\phi|r|}\phi|r'|;$$

$$F_3(u', v', r', \delta, \phi) = N'_\delta \delta + N'_{\delta\delta\delta}\delta^3 + N'_{\delta|v|}\delta|v'| + N'_{\delta u}\delta u'$$
$$+ N'_v v' + N'_{v|v|}v'|v'| + N'_{v|r|}v'|r'| + N'_{vu}v'u'$$
$$+ (N'_r - m'x'_G u')r' + N'_{rrr}r'^3 + N'_\phi \phi + N'_{\phi u}\phi u'$$
$$+ N'_{|\phi|r}|\phi|r' + N'_{\phi|\delta|}\phi|\delta| + N'_T;$$

$$F_4(u', v', r', \delta, \phi) = K'_\delta \delta + K'_{\delta\delta\delta}\delta^3 + K'_{\delta|v|}\delta|v'| + K'_{\delta u}\delta u'$$
$$+ K'_{\delta uu}\delta u'^2 + K'_v v' + K'_{v|v|}v'|v'| + K'_{v|r|}v'|r'| + K'_{vu}v'u'$$
$$+ (K'_r + m'z'_G u')r' + K'_{rrr}r'^3 + K''_\phi \phi + K'_{\phi|r|}\phi|r'|$$
$$+ K'_{|\phi|r}|\phi|r' + K'_p p'.$$

IV. DECOUPLING AND SIMPLIFICATION

These equations are used only to illustrate the complexity of the multi-degrees-of-freedom ship motion problem and to focus on the need for decoupling and simplification prior to simulation and analysis. In practice, most ship simulations are done to investigate particular interest areas and use decoupling procedures to concentrate only on the variables of interest. For example, the simulation of a turning ship, in a seaway while changing velocity is almost hopelessly complex

Table I

Purpose of simulation	*Typical degrees-of-freedom*
Propulsion systems dynamic analysis	*Surge*
Propulsion system governing	*Surge, heave, pitch*
Propulsion system loads in turns	*Surge, sway, roll, yaw*
Maneuvering	*Surge, sway, roll, yaw*
Seakeeping	*Surge, heave, pitch, roll, yaw*

without simplifying assumptions. Fortunately, such simplification methods that apply as well to the machinery and control system models are providing relatively good guidelines for propulsion system designers via large scale simulations.

Examples of typical reductions in the degrees-of-freedom and the interest area of the resulting system of equations are given in Table I.

For propulsion plant dynamics and control studies, by far the most common simplification is to utilize only the surge equation while ignoring the other degrees-of-freedom. This constrains the ship to a straight ahead nonturning condition with no other ship motions in an undisturbed sea. Such a procedure concentrates on the dynamics and control of the propulsion system while permitting the power plant to see a dynamically varying propeller load under all conditions of acceleration or deceleration of the ship in one degree-of-freedom, i.e., surge. The control system is then developed to provide safe, efficient machinery operation under these transient conditions while assuring acceptable propulsion system loading, ship stopping, and acceleration performance.

Ship maneuvering studies require four degrees-of-freedom, e.g., surge, sway, yaw, and roll. During turning maneuvers, the ship's propeller(s) experiences forces greatly different than in straight ahead. As a consequence, the propulsion system loading changes significantly and the propulsion control system is designed to maintain acceptable system loads while maximizing the maneuvering performance.

Studies to determine ship seakeeping qualities and forces on the ship's structure introduce additional complexity to the ship calm water equations. The seaway is usually described using a wave energy spectrum representing the irregular sea to generate forces and moments on the ship's hull with the addition of pitch and heave dynamics. This procedure, involving superposition of regular waves, has limitations when used with the nonlinear ship motion equations [2].

V. THE PROPELLER THRUST AND TORQUE EQUATIONS

Decoupling leads to the two most important and fundamental equations in ship propulsion dynamics, namely, the coupled propeller thrust and torque equations. These coupled equations for the single screw ship are [3]:

$$m\dot{v} = T_p - R_T, \quad \text{(thrust equation)}$$

$$2\pi I\dot{n} = Q_d - Q_f - Q, \quad \text{(torque equation)}$$

where m and I are the ship mass and drive train moments of inertia, respectively, $\dot{v}$ and $\dot{n}$ are ship and propeller acceleration, respectively, T_p is propeller net thrust, R_T is ship resistance, Q_d, Q_f, and Q are the respective torques developed by the engine on the shafting, the friction torque, and propeller torque.

Simulation of these equations and the results obtained in terms of ship and machinery response and peak system loads are discussed in [3]. Of these two equations only the thrust (surge) equation originates in the ship's motion equations. The torque equation is a propulsion machinery system equation that becomes coupled to the thrust equation via the characteristics of a marine screw propeller which are determined by both v and n. The single degree-of-freedom, nonturning calm sea simulation based on these two equations is used extensively in the design of ship propulsion systems. These equations provide a dynamic model for the propeller and engine loading under a great variety of steady state or transient conditions. Additional design information is added with additional simulations developed to include turning or seaway effects on ship motions and propulsion plant loading. Thus, the simulation described by the coupled propeller thrust and torque equations can provide design information on ship acceleration or deceleration and worst-case transmission and engine loads caused by propulsion commands. These equations are nonlinear because of the nonlinear terms on the right-hand side, while the deceptively simple term Q_d involves the entire nonlinear dynamic response of the prime mover, shafting and propulsion control systems.

In propulsion dynamics and control simulations, the propulsion shafting is typically assumed to be totally rigid as described by the above torque equation. This is a good example of successful decoupling, where it is recognized that the exact shafting transient responses in terms of multimode oscillations do not bear significantly on either engine, control system, or ship dynamics. Shafting simulations of torsional and longitudinal responses to various excitations and determination of critical frequencies are very important and complex simulations in their own right, but are decoupled from the main propulsion plant simulation to simplify the already complex models.

It is useful to compare the time rates of change of the variables n and v in the coupled propeller thrust and torque equations. Solution of these equations shows that propeller speed response to changes in power level is many times faster than the ship response. Understanding the significant difference in the time responses of n and v aids greatly in the design of propulsion control systems. Simply stated, the enormous ship mass is a smoothing filter to sudden changes in propulsion power so that rapid changes in ship speed cannot be achieved with rapid changes in power.

Propeller loading caused by the seaway can cause cycling of the propeller, drive train, and engine(s). This cycling is greatly influenced by the design of the engine governing system, e.g., power or speed governing. It has been shown that a ship drive train acts as a first-order lag or low-pass filter to sea-imposed loads [4]. Engine cycling depends on the drive train inertia, frequency of disturbance, and governing system design. Under steady-state propulsion system conditions and with a calm sea, a speed-governed system can maintain a constant propeller speed, but is incapable of doing so with high propeller loading imposed by the sea or by high-speed turns.

VI. SHIP PROPULSION SYSTEMS

The major elements of a ship propulsion system are each discussed briefly below and ship reversing deserves particular attention. The ship's propeller by way of the reversing system can produce either positive or negative thrust, where negative thrust is used in backing and stopping maneuvers. Interestingly, a ship is the only vehicle that can use all its power in a stopping maneuver; a ship has no "brakes" as such. This use of power in reverse can result in remarkable stopping performance, on the order of several ship lengths, for naval ships with high power-to-displacement ratios. Conversely, giant ships such as the crude oil carriers can have stopping distances measured in miles. The reversing system may be incorporated into the engine (reversing steam turbine, reversing diesel), transmission (reversing reduction gear, electric drive), or thruster (controllable pitch propeller).

A. *ENGINE*

The engine(s) consisting of one or more prime movers may be a gas turbine, steam turbine, or diesel engine. In some ships, primarily Naval, combined plants, e.g., involving both diesels and gas turbines, are often used. Simulations of the engines and particularly gas turbines invariably require extensive manufacturer's data. Overall, a steam turbine plant is a more complex system than a gas turbine or diesel plant, particularly in its use of auxiliary systems integrally related to the main propulsion cycle. The modeling of a combined gas and steam turbine system is discussed in [5].

B. TRANSMISSION

The transmission and shafting couples a usually high- or medium-speed engine to the low-speed requirements of the thruster. This is an "impedance match" problem. Many direct-drive, low-speed diesel engines are also in use. The transmission is either a reduction gear or electric drive, i.e., generator-motor system. The mathematical models of electric drive systems are usually controller and lumped parameter differential equations. Reduction gears are often modeled as constant gear ratios although the dynamics of a reversing reduction gear requires two coupled nonlinear torque equations [6].

In an electric drive transmission, two coupled torque equations are required as shown in Fig. 2. These equations are coupled electrically through the generator (Q_g) and motor (Q_m) torques. The drive train inertia is split between the turbine-generator side (I_e) and motor-propeller side (I_m). With systems of this type, speed reduction and reversing can be accomplished electrically eliminating the reduction gear and using a fixed pitch propeller. Reversing reduction gear transmissions using friction elements are described by two nonlinear friction coupled torque equations similar to the

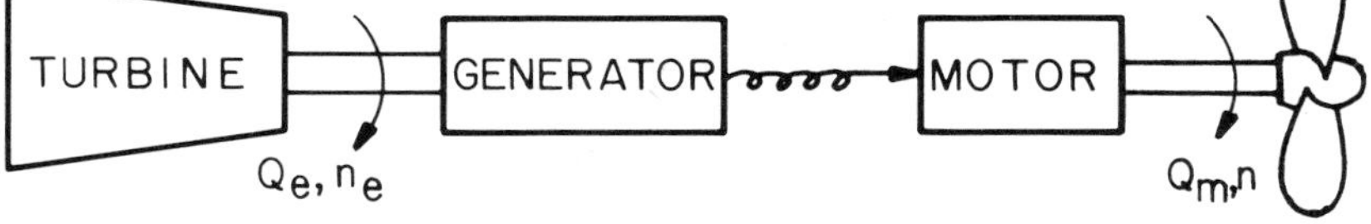

TURBINE GENERATOR MOTOR

Q_e, n_e Q_m, n

$$2\pi I_e \dot{n}_e = Q_e - Q_{fw} - Q_g \qquad 2\pi I_m \dot{n} = Q_m - Q_f - Q$$

Fig. 2. Electric drive ship transmission.

electric transmission equations. In such a system, one torque equation applies prior to and after the friction phase transient.

C. THRUSTER

Marine thrusters are screw propellers, waterjets, and in some cases air propellers as in air cushion craft. Thruster modeling is still almost entirely dependent on experimental data. Experience has shown that of all the major elements in a ship propulsion system, the modeling of the hydrodynamics and particularly the thruster is subject to the most uncertainty, generally due to the complex, unsteady flow phenomona.

D. AUXILIARY SYSTEMS

The ship's propulsion auxiliary systems include, for example, fuel, lubrication, cooling, air, and electrical systems. Such systems are usually not included as part of the propulsion dynamics simulations since they are effectively decoupled from engine-drive train dynamics. In steam plants certain systems such as feed pumps and forced draft blowers can be part of the main cycle propulsion dynamics.

E. CONTROL SYSTEMS

Control system models for the main propulsion system generally fall within two categories. The inner-loop controls are directly associated as an integral part of an engine, e.g., fuel control, stator vane, or turbine nozzle area controls on a gas turbine. The outer-loop controls connect various major machinery elements, e.g., pitch/RPM schedules or closed-loop propeller speed control and the control systems associated with multiple engine start-stop and clutching-declutching operations.

VII. PROPULSION SYSTEM SIMULATION

Developing the mathematical models comprising a propulsion simulation involves combining differential equations, experimental/empirical data fits from model tests or full-scale tests on machinery, control systems, hulls and thrusters, and algebraic, logic, or control equations into a nonlinear representation of the system response.

A gas turbine controllable pitch propeller propulsion system is used here as an example illustrating the simulation development of a portion (engine, drive train, and thruster) of a ship propulsion system shown in Fig. 3. This figure shows the various input, status feedback, and control signals together with the controlled machinery. The propulsion control system plays a dominant role in determining performance in terms of engine, transmission, thruster and ship dynamic response, and load levels. Typical modeling approaches are outlined below for the propeller, gas turbine, and control systems. The simulation example is developed in the surge degree-of-freedom.

A. HYDRODYNAMICS MODELING

To solve the coupled propeller thrust and torque equations requires models of the hydrodynamic behavior of the ship's hull and propeller. If additional degrees-of-freedom are desired, then considerable hydrodynamic data are needed as coefficients in the ship motion equations. Hydrodynamic data

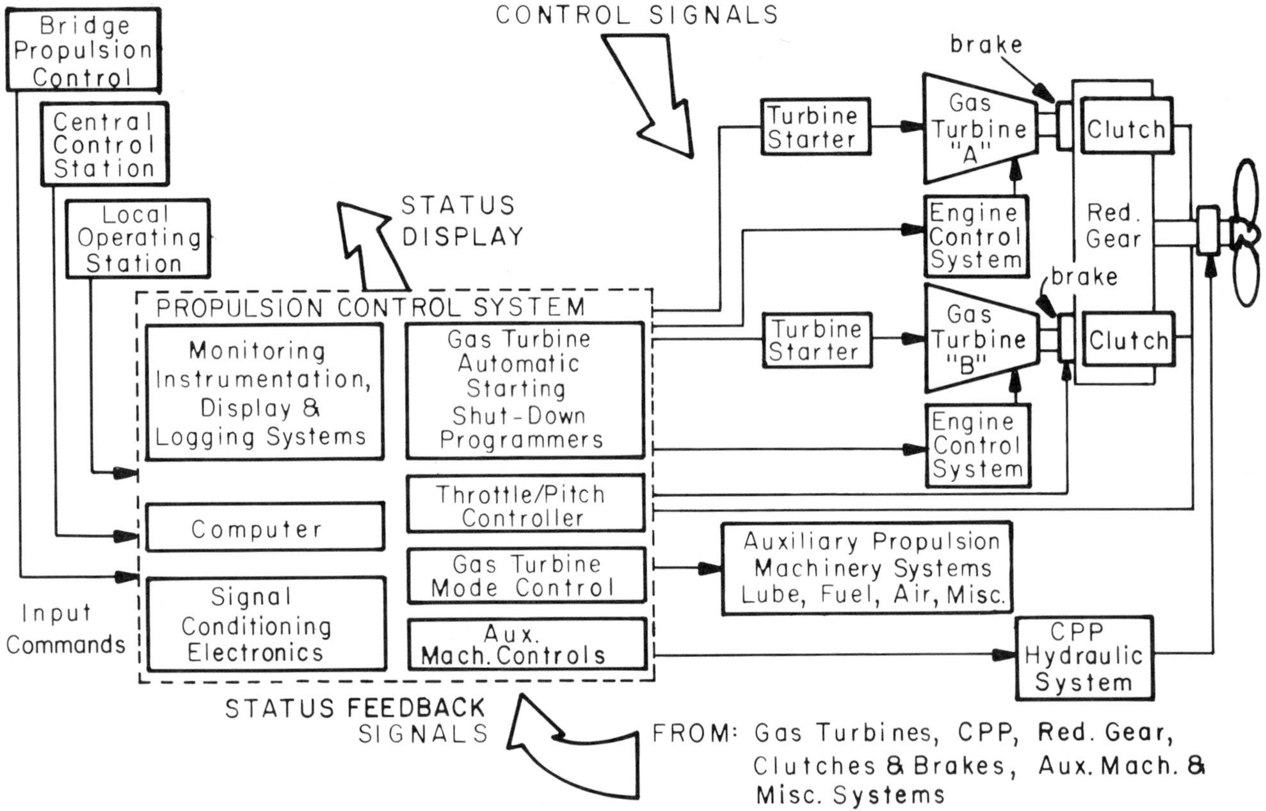

Fig. 3. Gas turbine ship propulsion system showing one shaft with two engines.

are largely experimental, derived from scale model and full-scale tests of propellers and hulls. Thus, in spite of expanded analytical computer techniques and continuing progress to develop performance predictions, based for example on finite-element methods, the model test basins of the world and full-scale ship trials still provide the hydrodynamic data for most studies.

Propeller modeling is important because propeller torque constitutes the propulsion system load and propeller thrust determines surge dynamics. Propeller thrust and torque for a given geometry and size propeller are functions of several variables, i.e., rotational speed (n), ship speed, and pitch ratio (P/D) for controllable pitch propellers. Depending on propeller speed, loading, and design, cavitation (equivalent to low-temperature boiling of a liquid at reduced pressure [7]) can occur. Cavitation decreases propeller performance and thus introduces still another variable in propeller modeling [3]. In this section cavitation is ignored.

Figure 4 shows experimental controllable pitch propeller thrust data for a particular propeller in terms of a modified advance coefficient μ. Each value of P/D represents a different pitch to diameter ratio, where the propeller pitch P is the distance the propeller would advance in one revolution without slip and D is the propeller diameter. Positive and negative pitch ratios are for ahead and astern operation, respectively. Each pitch ratio is, in effect, a different propeller. The direction of thrust (ahead or astern) is changed by changing the pitch of all blades via a hydraulically powered blade-turning mechanism in the propeller hub.

The modified advance coefficient (μ) is defined as

$$\mu = V_A/\left(V_A^2 + n^2D^2\right)^{1/2},$$

where V_A is the speed of advance of the propeller and n is propeller rotational speed. This definition is an improvement over the conventional advance coefficient (J), where $J = V_A/nD$. The propeller speed of advance is $V_A = V(1 - w)$, where the ship speed (V) is corrected by the wake factor $(1 - w)$ since the propeller operates in the wake of the hull. The propeller thrust (T) and torque (Q) are determined from the thrust and torque coefficients C_T and C_Q via the equations

$$T = C_T\rho D^2\left(V_A^2 + n^2D^2\right), \quad Q = C_Q\rho D^3\left(V_A^2 + n^2D^2\right),$$

where ρ is the water mass density. The propeller develops a net thrust $T_p = T(1 - tt_p)$ different from the propeller open-water thrust (T) by the thrust deduction factor $(1 - tt_p)$, which depends on both ship speed and propeller pitch ratio. Note that in this expression t is thrust deduction fraction and t_p is a thrust/pitch multiplier.

Use of the coefficient μ instead of J greatly simplifies the representation of propeller thrust and torque characteristics that can be described by two, single, closed figures, one for thrust and one for torque for each pitch ratio. This modified coefficient (μ) is well behaved compared with J, which becomes unbounded as n approaches zero. In the past, use of J as the independent variable resulted in a total of four curves each for both thrust and torque for each pitch ratio. Transient operation through $V_A = 0$ and $n = 0$ is typical, hence both J and 1/J representations were used in the earlier method with switching back and forth between curves.

With the introduction of the μ method (note the presence of both V_A and n in the denominator), dynamic simulation using such experimental data is greatly simplified. Propeller designers consider propeller operation principally in terms of J or μ, which are ratios of advance-to-rotational speeds. The two independent variables are thereby reduced by one with the assumption that propellers behave the same at either low or high translational and rotational speeds so long as J or μ are unchanged. For controllable pitch propellers, experimental data are typically available for a set of discrete propeller pitch ratios. Linear interpolation between lines of constant pitch ratio as well as nonlinear techniques such as cubic spline and Lagrange methods are used [8].

The propeller operation loci shown in Fig. 4 are simulated ship transients showing crashback (maximum astern from a maximum ahead condition) and crashahead (maximum ahead from a maximum astern condition) for a gas turbine, controllable pitch propeller driven ship. Note that the majority of the figure represents dynamic conditions. Only two steady-state conditions exist, i.e., two single points or a narrow range for all steady-state ahead and astern conditions. The experimental propeller data such as Fig. 4 are almost always obtained under "quasi-steady" conditions, yet simulations based on this data are used to generate "dynamic simulations." This limitation, discussed later, underlies many simulation approaches. Figure 5 shows the thrust coefficient data for a fixed pitch propeller plotted in terms of the modified advance coefficient as defined by

$$\nu = nD/\left(V_A^2 + n^2D^2\right)^{1/2},$$

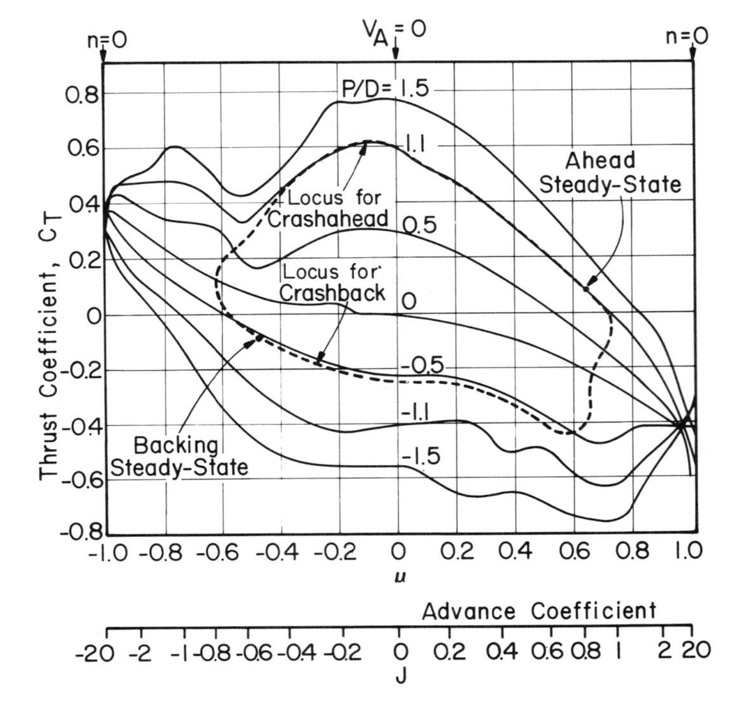

Fig. 4. Controllable pitch propeller thrust coefficient experimental data versus modified advance coefficient μ *for various pitch ratios showing typical transient loci.*

where the thrust and torque in terms of the thrust and torque coefficients is the same as previously defined. Either the μ or ν methods of representing propeller characteristics can be used for either fixed or controllable pitch propellers. For various reasons the authors prefer the μ method for CP propellers and the ν method for fixed pitch propellers.

Additional models are required for the wake and thrust deduction factors that are usually assumed to vary with ship speed and propeller pitch based on steady-state propulsion

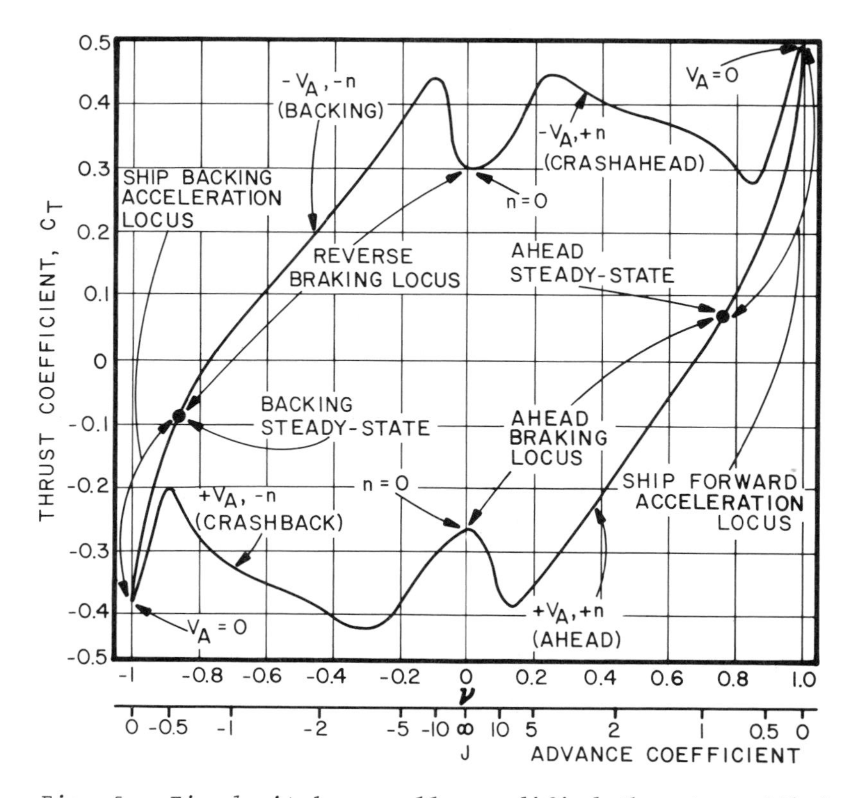

Fig. 5. Fixed pitch propeller modified thrust coefficient versus modified advance coefficient showing typical steady-state points and transient loci.

tests [3]. The use of wake and thrust deduction factors based on transient tests may be required to reproduce full-scale maneuvering results [9].

B. *GAS TURBINE ENGINE MODELING*

Gas turbine modeling utilizing the functional component approach is widely used in gas turbine simulations. In this method each individual component such as compressor, combustor, etc., is described by performance maps and equations representing mathematical models for major engine components. Such an approach has the great advantage of flexibility to adapt to

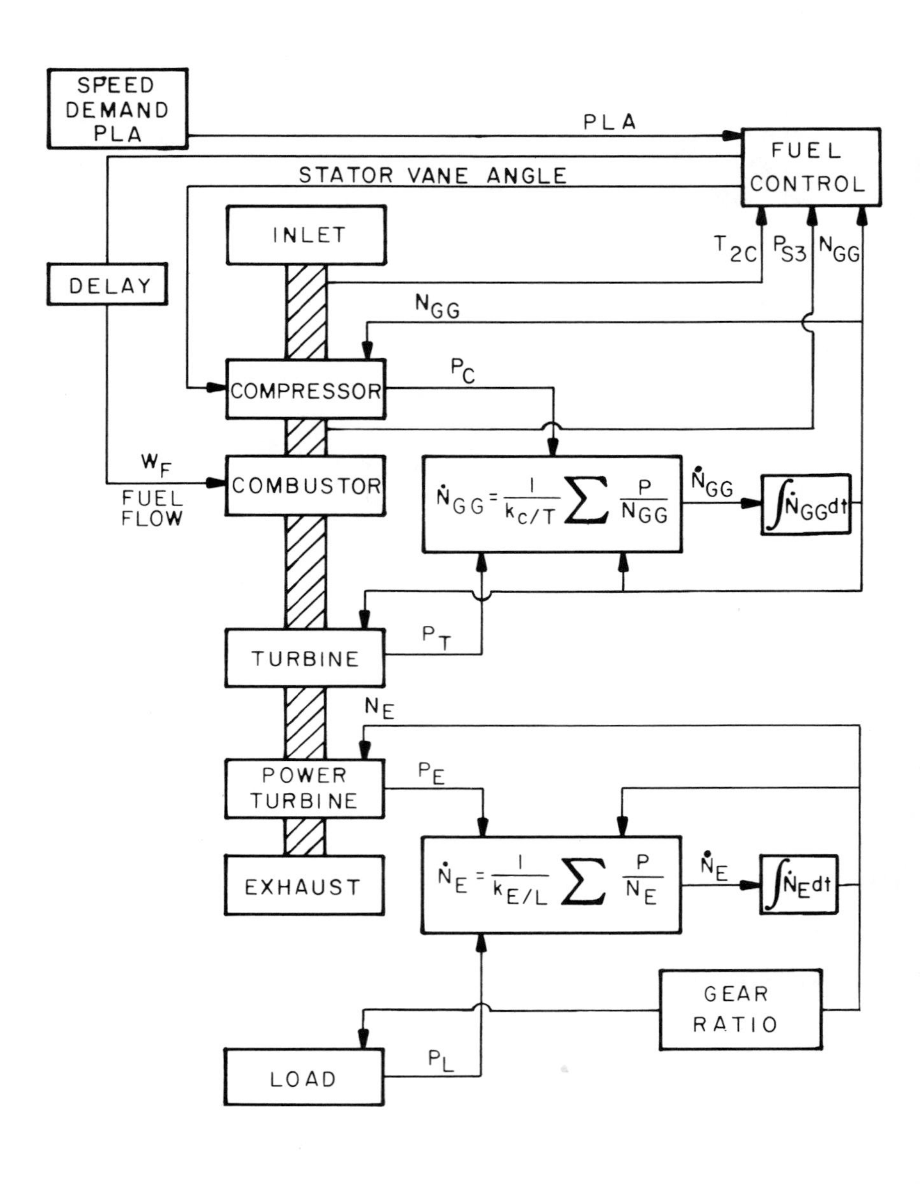

Fig. 6. Gas turbine engine dynamics simulation.

changes as the design evolves. A functional component block diagram for a ship propulsion gas turbine engine is shown in Fig. 6. This is a two-shaft "free-turbine" type engine that uses a jet engine as a gas generator coupled to a free turbine to convert the high-energy gas flow to rotational power [10]. In Naval applications, aircraft derivative gas turbines are used and such engines achieve about the highest power-to-weight ratios of any type of engine.

Control system design that affects the dynamic performance and stability of gas turbine engines usually progresses as part of the entire engine design. A nonlinear engine model is required to simulate the complete range of engine operation from idle to full power and to predict correctly the transient response. Linear theory is inadequate to model large transient excursions. Such nonlinear modeling in turn requires either analytical or experimental models of the functional components such as the compressor, combustor, turbine(s), control systems, etc. Such data as compressor, combustor, and turbine performance used as look-up tables or performance maps are developed experimentally under quasi-steady conditions. Lumped parameter approaches such as mass storage and delays at one or more points are used to simulate a continuous process. Such lags in a gas turbine simulation involve temperature changes following a fuel change and other lags involving pressure and flow. The block diagram of the gas turbine engine, Fig. 6, shows the major functional components of a free turbine, gas turbine engine. Its variable geometry consists of variable stator vanes between several stages of compressor blades. The initial iterative solution is arrived at by a solution of the power balances between the power developed by the high-pressure turbine and the power absorbed by the turbine-driven compressor. Similarly, the gas-coupled power turbine (the "free turbine") drives a load and, as in the compressor, a balance between free turbine power and power absorbed by the load determines the power turbine speed. Many equations are involved in this simulation for component and gas path performance, e.g., temperature rises in compressed air, compressor pressure ratio, mass flow and power, burner and turbine efficiency, etc. The principal dynamic engine response is determined by the speed of the rotors, which in turn depend on their inertia and the excess power causing acceleration. In turn, the speed of the compressor rotor, mass flow, and pressure are interrelated variables. Such a model assumes energy storage in the rotating components, but not in air compression. The compressor transient states are a succession of quasi-steady state points from the compressor characteristics determined instantaneously from the accelerating or decelerating rotor speed.

> "The net result of neglecting the pressure and temperature dynamics in the engine is a degradation in model high-frequency response. Temperature and pressure are states of the fluid within a control volume and thus represent energy storage. Ideally, to describe any system dynamically, the number of states should equal the number of energy storage devices in the system. Consideration of only rotor dynamics implies that the control volumes of compressor, combustor, and turbine do not store energy significant enough to affect transient performance. Model analysis of linear models of gas turbines having as many as 17 states reveals that such is indeed the case" [11].

A large number of variables are computed in a dynamic simulation model of a gas turbine engine. For a large complex engine, the authors compute up to 70 gas turbine parameters involving gas path temperatures, flows, pressures, energy levels, rotor speeds, torques, and control parameters during various engine transients. Such simulations, both steady state and dynamic, are complicated by the circular or iterative nature of the computations where a change in one parameter affects component performance so as to change the original parameter. This behavior leads to iterative solutions requiring convergence routines in several parameters, i.e., nested loops within loops. Often, transient simulation requires an iterative convergence of the simulation to a solution for each time step. In addition, to obtain a valid starting point for the simulation, an iterative procedure is often used to solve the coupled nonlinear differential equations.

A steady-state representation or macroscopic model for a gas turbine engine in terms of its output characteristics is the torque map shown in Fig. 7. Such a model describes steady state or slowly varying turbine performance; its limitations are discussed later in more detail.

C. *CONTROL SYSTEM MODELING*

One of the primary reasons for large-scale propulsion machinery system simulations is to design the total plant control system. The complex interrelationships between the characteristics and responses of machinery systems often requires a comprehensive control system design approach. Failure to consider the system as a whole may lead to

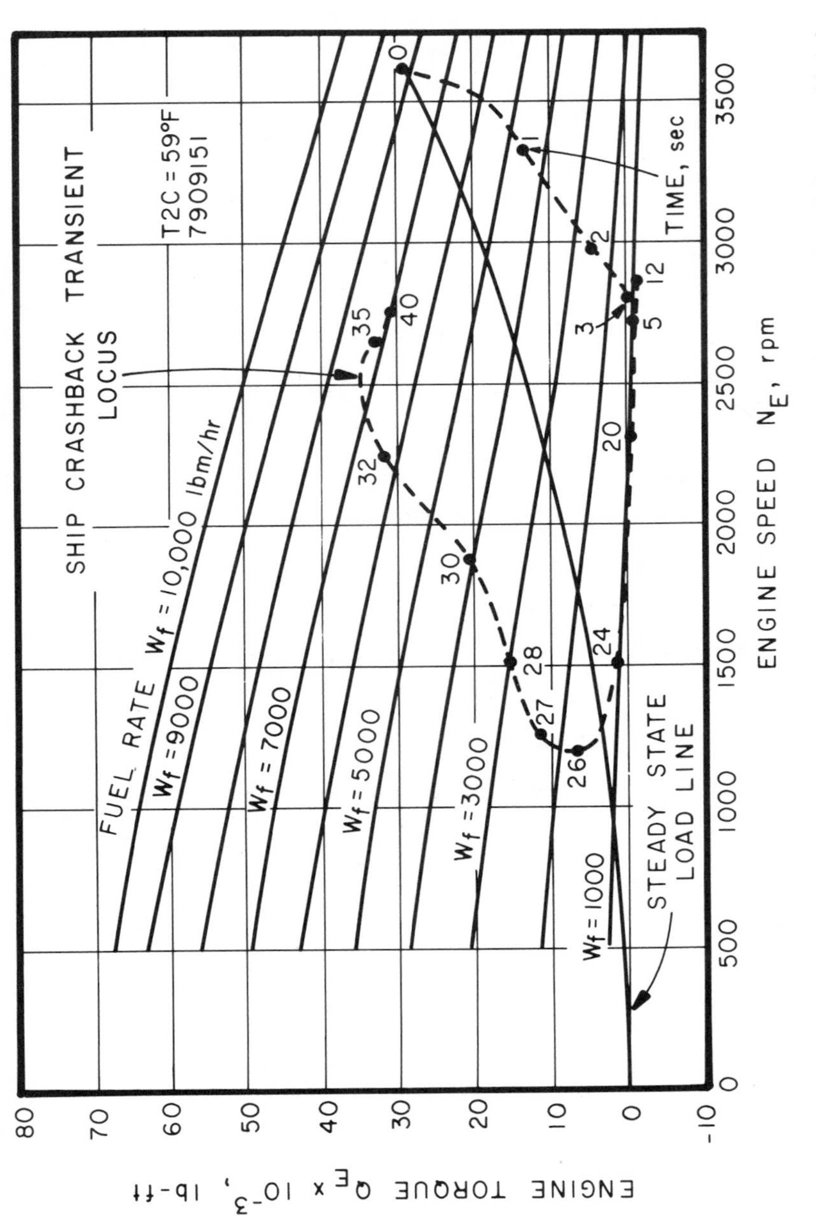

Fig. 7. Gas turbine engine torque map with dynamic torque-speed locus in ship crashback.

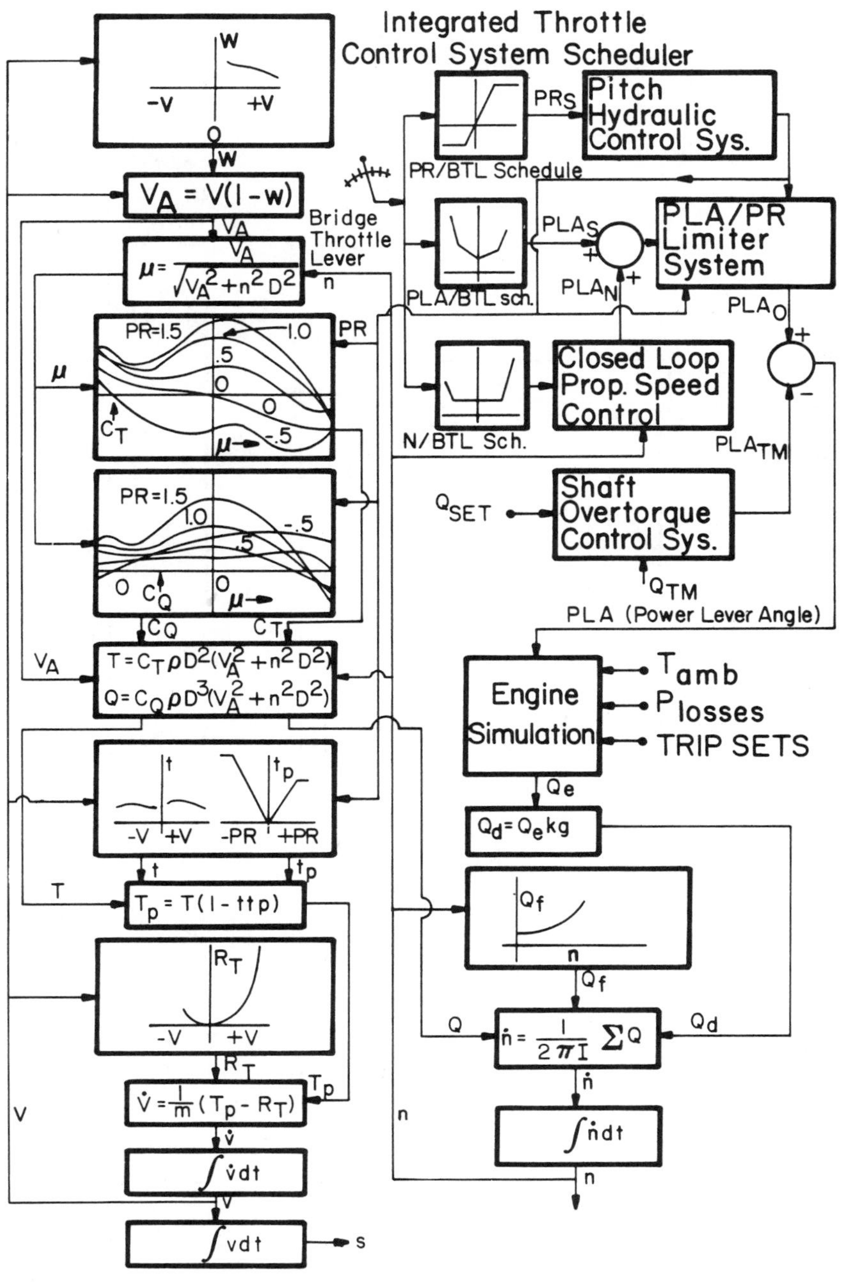

Fig. 8. Propulsion system simulation flow diagram.

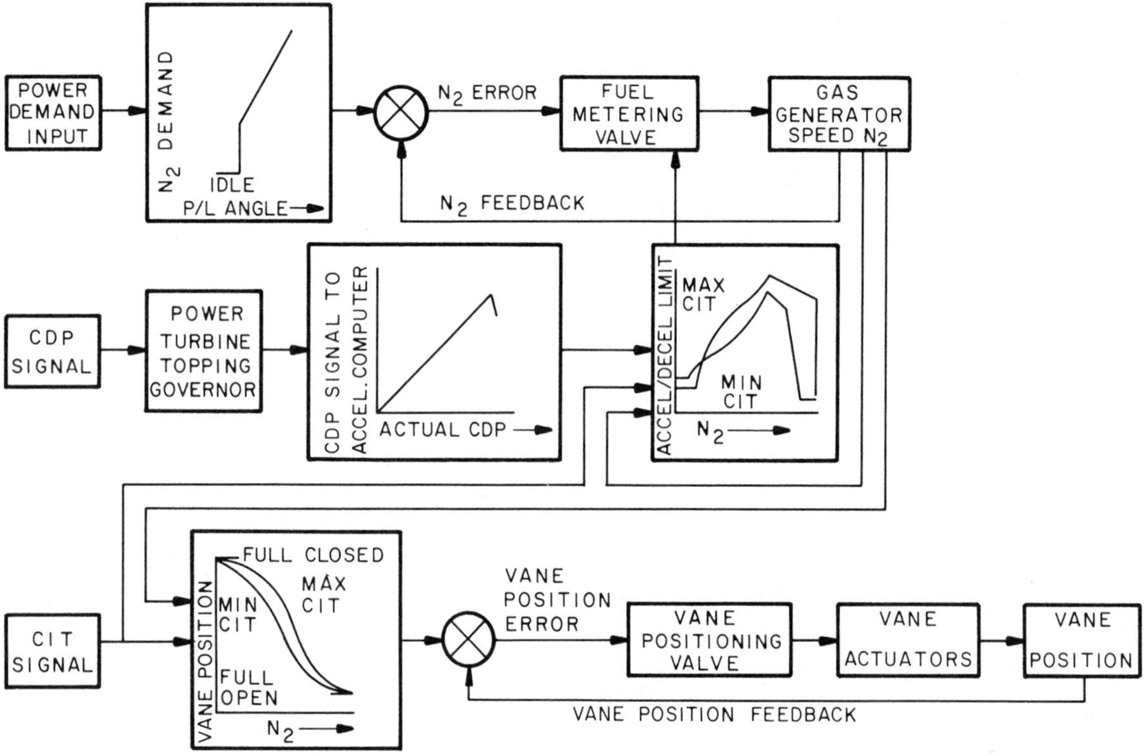

Fig. 9. Simplified fuel control block diagram of a typical gas turbine engine.

interactions between the control and machinery components leading to system cycling, wear, fatigue, unstable operation, or excessive transient loads.

Figure 8 illustrates the control system modeling requirements for a portion of a gas turbine ship control system. The system shown consists of the integrated throttle control system, propeller speed control system, and the shaft overtorque control system. For each system, the controller consists of basic control elements such as feedforward schedules, proportional and/or reset controls, and control range limiters. For example, the integrated throttle control system is responsible for scheduling, via analog or digital function generators, propeller pitch, propeller shaft speed, and gas turbine power level based on a single-command handle input. The propeller speed control system shown involves a reset control term whose input is the propeller speed error and whose output is limited by upper/lower range limits.

Figure 9 illustrates typical gas turbine fuel control systems [12]. These fuel control systems combine the inputs of power lever position, compressor discharge pressure, compressor inlet temperature, and gas generator speed to provide a regulated fuel flow for engine operation.

In addition to metering fuel for combustion, the fuel control may provide the following functions

(1) isochronous gas generator speed governing in response to power lever input during steady-state conditions;

(2) transient-rate fuel limiting;

(3) acceleration and deceleration fuel limiting as a function of compressor inlet temperature (CIT), compressor discharge pressure (CDP), and gas generator speed (N_2), which permits unlimited throttle manipulation without causing compressor stall or combustor blowout;

(4) a reduction of fuel flow under high CDP conditions;

(5) gas generator overspeed limiting; and

(6) a control of the variable vanes and inlet guide vanes as a function of N_2 and CIT.

The propulsion control system is typically in a design phase subject to continual evaluation greatly influenced by the simulation, therefore it tends to be the best-defined subsystem of the plant mathematical model. For some studies, where control system actions are minimal, the controller models may be greatly simplified, for example, by eliminating fast-acting elements and appropriate nonlinearities. Conversely, for other studies, control system actuator and sensor models must be highly detailed to assure adequate safety margins, e.g., during turbine overspeed control system design.

D. PROPULSION SYSTEM SIMULATION FLOWCHART

Figure 8 shows an overall computer flow diagram for the example gas turbine, controllable pitch propeller ship propulsion system. This diagram is typical of the many all digital ship simulations conducted by the authors and is generally applicable to various types of propulsion systems. The controllable pitch propeller is a general case and is more complex to model than a fixed pitch propeller. This flow diagram is applicable to any prime mover, e.g., gas turbine, diesel engine or steam turbine. However, marine steam propulsion systems typically embody a reversing turbine with a fixed pitch propeller. The flowchart illustrates the solution of the coupled propeller thrust and torque equations in one degree-of-freedom with ship speed (v) and propeller speed (n) as the outputs. Data in the simulation such as propeller characteristics, ship resistance, and wake and thrust deduction factors are entered as look-up tables with interpolation between curves for functions of more than one variable. The engine simulation is by far the most complex part of the propulsion dynamics simulation. For ease of illustration, some simplifications have been made in the diagram. In some cases propeller data with cavitation (as in high-speed gunboats) must be used, resulting in functions of three variables. Thus, each pitch ratio requires a family of curves, one curve for each cavitation number. Many additional auxiliary computations such as propeller blade spindle torque (moment about the blade's pitch change axis), shaft horsepower, propeller efficiency, etc., are not shown here. The gas turbine parameters often number about 70 with 40 or more additional parameters for the ship, drive train, and propulsion control systems. Several studies by the authors concerning combined cycle plant plants or ship turning simulations have involved as many as 250 computed parameters.

VIII. QUASI-STEADY VERSUS DYNAMIC BEHAVIOR

So-called "dynamic simulations" very often depend upon experimental data which are obtained under quasi-steady conditions. Quasi-steady has come to mean the behavior under dynamic, but nontransient conditions. Two good examples in ship propulsion simulations both involve pressure-flow phenomena; the gas turbine compressor for aerodynamic flow, and a ship's propeller for hydrodynamic flow. In both cases quasi-steady experimental data under all expected operating conditions are used as stored maps to predict behavior under transient conditions of the entire system. For example, as discussed earlier only a narrow region of steady-state points for a given propeller pitch exists in an entire propeller map, occurring for steady-ahead or steady-astern at various speeds. Yet, a complete range of propeller data is obtained by off-normal (or off-self propulsion) tests conducted under various fixed conditions of n and v. For a controllable pitch propeller, still another quasi-steady assumption involves the propeller pitch ratio. Such propellers are always tested at discrete fixed pitch ratios, but their dynamic performance is predicted during pitch changes by interpolation of data obtained from tests at the various fixed pitch ratios. Transient hydrodynamic testing is almost unknown. Even quasi-steady testing involves measurement problems and considerable data scatter. However, the more fundamental reason for the quasi-steady approach is a practical limitation. To obtain transient test results would require testing under an unreasonably large range of conditions and each transient is unique. This is typical of nonlinear systems. Fortunately, the quasi-steady approach in component modeling, while limited in accuracy and often the only feasible approach, can lead to quite acceptable results where the rates of change between quasi-steady states are small or where the system exhibits low sensitivity to such rates of change. Thus, ship propeller transients can never be simulated exactly when based on quasi-steady data and neglect of other nonlinearities such as cavitation, air drawing, wave effects, and transient hull-propeller hydrodynamic interactions [9].

A comparable situation exists with the modeling of a multistage axial flow gas turbine compressor. Here quasi-steady data are used to predict transient response of the entire engine and excursions of the compressor stall margin. The problem is extremely complex because of the distributed nature of the components that are typically modeled using lumped parameter techniques. For example, blade stall can occur circumferentially or axially almost anywhere within the compressor volume, but as a practical matter stall margin for the compressor as a whole is computed. Viewing the gas turbine engine as a "black box," its principal steady-state characteristics may be represented as the torque map shown in Fig. 7. The engine outputs are torque and speed, as related by nearly straight lines of negative slope, subject to the input of fuel rate (engine pressure ratio or gas generator speed are also used). Such a torque map is widely used as a model of an entire gas turbine engine in certain ship propulsion dynamics simulations. Its suitability depends on the purpose of the simulation. If changes in engine power are relatively slow or small in magnitude, such a torque map provides a good prediction of engine torque to drive the coupled propeller thrust and torque equations. Thus, for example, in ship turning simulations executed at nearly a constant power level, the torque map approach is usually perfectly adequate. Use of a torque map obviously assumes an instantaneous torque response to a fuel command, i.e., engine dynamics are neglected. A complete dynamic engine model is required for studies of engine governing or commands involving large power excursions. The modeling weaknesses in the dynamic engine model are still quasi-steady maps, lumped components, and delays together with many other simplifying assumptions. However, such a model is usually a good simulation of the nonlinear engine. If a torque map, which is a quasi-steady model of a gas turbine engine, is used for predicting large power changes, then fairly large errors in transient prediction of engine torque can result.

IX. NONLINEAR MODELS VERSUS LINEARIZATION

Simplifications often made in the mathematical modeling of complex systems are to linearize the model about a system operating point. This technique is extremely useful in considering the design of controllers whose purpose is to maintain a stable operating point. For example, the analysis of a turbogenerator speed control system may be accomplished by linearizing the system about the proper operating point and designing the controller to maintain the setpoint speed during electrical load changes. In this case, linearization is an adequate approach where the desired speed range is constrained to be within a few percent of the setpoint. This speed range represents a narrow vertical band of the torque map.

Generally, ship propulsion machinery models may not be linearized whenever the machinery response covers a wide range of propulsion operating conditions such as a crashback maneuver. In a maximum power crashback, for example, the prime mover can cover the entire range from idle to full power and operate considerably off its steady-state characteristics. Similarly, for a fixed pitch propeller, the entire steady-state ahead region may be very nearly represented by a single operating point, since $J = V_A/nD$ is nearly constant; yet during propulsion transients, the disparity in response times for v and n discussed earlier results in a considerable deviation from this steady-ahead region resulting in wide transient swings of J.

X. MECHANICS OF SIMULATION

The discussion of the mechanics of simulation is limited here to digital simulations. In developing the digital computer simulation, the primary question to be resolved is what languages of the many available should be used. Generally, with the exception of training simulators where real-time programming constraints may dictate the use of an assembly language, high-order languages such as FORTRAN have become universal [13]. The choice between a fundamental language such as FORTRAN and a specialized simulation language such as CSMP can be difficult [14]. Simulation in FORTRAN requires much more knowledge of the details, such as integration or iteration techniques, and how to write sometimes complex subprograms to handle the program input and output. This can lead to considerable increases in overall program development costs, yet to simpler, more efficient programs with lower execution costs. In addition, many excellent symbolic debugging software programs have evolved and form an integral part of the modern overall FORTRAN computer system. These debuggers result in improved program development procedures and enable more rapid isolation of problems discovered during model verification and validation phases. A simulation program such as CSMP enables the engineer to develop nonprocedural codes representing the machinery performance and greatly reduces involvement with the details cited above. However, CSMP acts primarily as a model translator and prepares the necessary FORTRAN subprograms for execution. In this role as translator, the code generated often is obscured by the CSMP naming convention and model code sorting procedures. Debugging a CSMP program can be difficult since CSMP does not contain a sophisticated symbolic debugger. To debug the FORTRAN code produced by the translator requires a FORTRAN understanding on par with that required to solve the problem completely in FORTRAN. Most of the simulations developed by the authors have been coded directly in FORTRAN.

The choice of integration, iteration, and interpolation methods to be employed are another major decision area. For integration a simple Euler technique will often prove quite adequate, especially in cases where the total dimensionality of the system is small. In other cases, more sophisticated techniques such as Runge-Kutta, Adams, or Kutta-Merson are recommended. Tradeoffs involve the time update increment and integration package sophistication. Simple techniques such as Euler often require small time increments while increases in the time increment may often be accomplished by utilizing a predictor-corrector technique. The variable time-step techniques such as Kutta-Merson are gaining favor since they dynamically adjust the time step in response to the nature of the dynamics using a small step during rapid transients and extending it as the system stabilizes at a new operating point.

Interpolation techniques can affect results more than is generally realized [8]. The difficulties in obtaining adequate data to define the characteristics of the system lead to the need for interpolation using, for example, multidimensional linear, Lagrange, and cubic spline techniques.

Iteration methods can be employed to solve implicit loops in the modeling and to generate a valid initial steady-state condition for the simulation. Some machinery models, such as gas turbines, require the solution to these implicit loops at each time step. The initial condition solution is often referred to as a time-zero balance procedure. In both cases, i.e., implicit loops and time-zero balances, the problem may be posed as the solution of a series of coupled nonlinear algebraic and logic equations since time may be thought of as constant. The solution can be achieved by the use of a Newton-Raphson technique where the number of iterations is dependent on the initial guess.

XI. SIMULATED SHIP CRASHBACK EXAMPLE

A brief discussion of the response of a ship with multiple gas turbines driving twin controllable pitch propellers is presented to illustrate some results achieved through simulation. Figure 10 presents the system response for a crashback maneuver from maximum ahead power to maximum astern power. For this maneuver also the dynamic engine torque locus computed from a dynamic engine model is shown on the torque map of Fig. 7. From the maximum ahead speed, the integrated throttle control handle is set to maximum astern. The gas turbine power lever angle (PLA) is ordered to an idling fuel and gas turbine fuel reduction begins reaching the idling value in less than 5 sec. Coincident with the ordered reduction in PLA, the pitch control system is ordered to the design astern pitch ratio and pitch reduction begins immediately at a

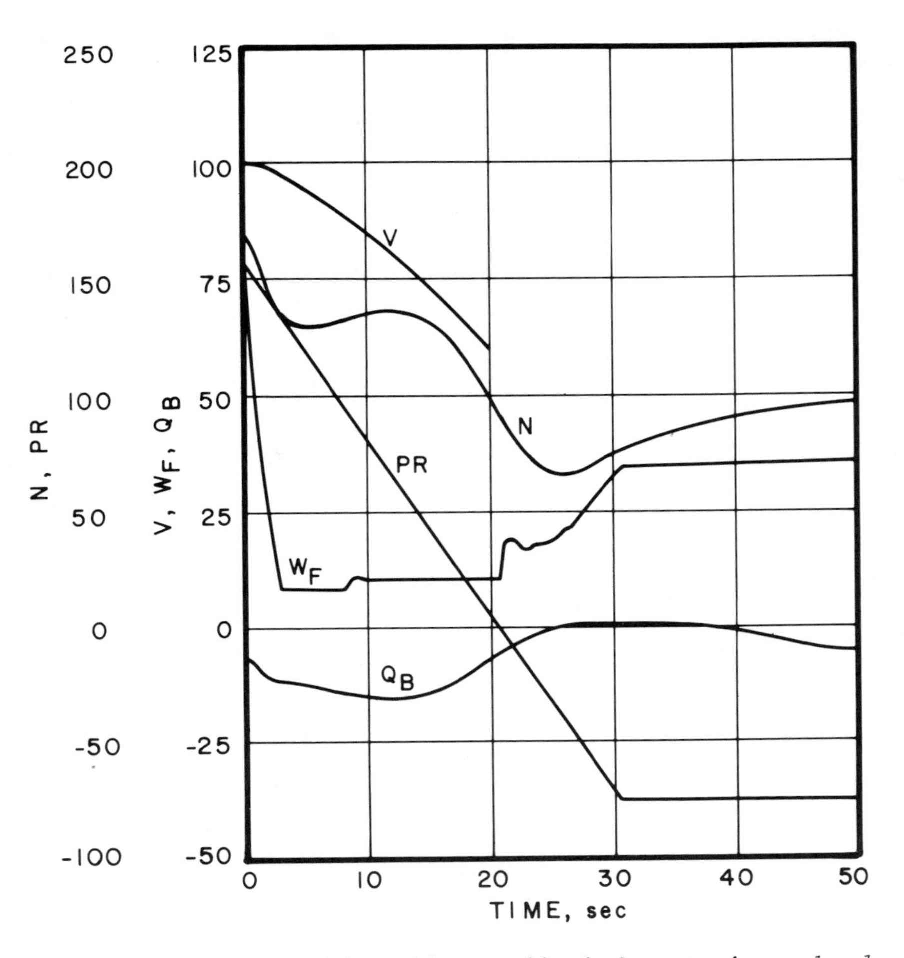

Fig. 10a. Gas turbine ship crashback from maximum ahead speed. Propeller speed, $N \times 1$ (rpm); propeller pitch ratio, $PR \times 10^{-2}$; ship speed, $V \times 1$ (%, arbitrary normalization); propeller blade torque, $Q_B \times 10^4$ (lb-ft); turbine fuel flow rate, $W_F \times 10^2$ (lbm/hr).

rate determined by the pitch control system. When the decreasing pitch reaches a predetermined value (typically zero), the gas turbine is commanded to its final astern power and fuel flow increases. During power reapplication, gas turbine power is controlled by the various maneuvering and turbine control systems. Propeller pitch ratio continues to decrease

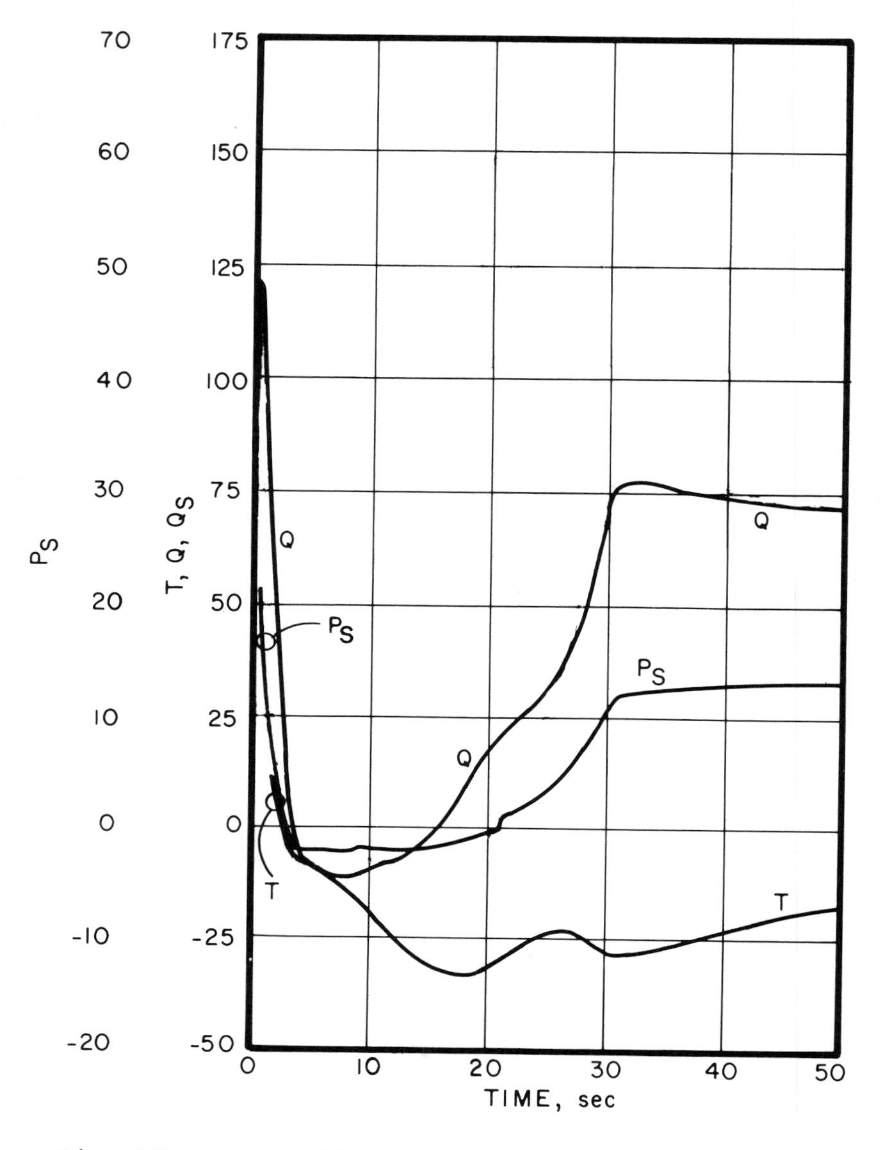

Fig. 10b. Gas turbine ship crashback from maximum ahead speed. Shaft power, $P_S \times 10^3$ (hp); propeller torque, $Q \times 10^4$ (lb-ft); propeller thrust, $T \times 10^4$ (lb).

until design astern pitch is achieved where it remains for the duration of the crashback maneuver. Gas turbine fuel flow, shaft torque, and shaft power increase reaching their final values with the ship moving astern. The sequence of events, causes, and effects during a typical crashback maneuver are as follows:

Fuel-to-idle phase: Following the command to idle the engines, a rapid fuel decrease occurs resulting in a rapid decrease in shaft torque and propeller speed. Propeller thrust (T) falls rapidly from its initial large positive value to some value near zero. This rapid thrust decrease is due primarily to the decreased N and secondarily to the decrease in propeller pitch ratio (PR). During the fuel-to-idle phase there are no peak values for N, Q, T, and propeller blade spindle torque.

Fuel-at-idle phase: During this phase, gas turbine and shaft torques reach a minimum value and may become slightly negative. Also, propeller deceleration becomes zero, then peaks and begins deceleration again. This peak in propeller speed or "water turbine effect" is characteristic of controllable pitch propeller propulsion plants during the early phase of decreasing pitch ratio. Large speed peaks are undesirable due to resulting loads. Propeller thrust can have one or more peaks, occurring at different times during the crashback maneuver, e.g., in this phase or the engine reacceleration phase. The thrust peaks typically occur in the region between a small positive ahead pitch ratio and the design astern propeller pitch.

Engine reacceleration phase: Engine reacceleration occurs as propeller pitch passes through zero. A scheduled fuel increase begins taking about 10 sec to reach the commanded power depending on propulsion control system actions. Propeller speed is decreasing rapidly due to the increasing astern propeller pitch. During this reacceleration phase turbine fuel rate is limited by the propulsion controls because of the reduced propeller pitch. Propeller speed begins to increase rapidly about 5 sec after the increase in engine power level begins. In addition, as turbine PLA approaches its commanded value, large system torques are possible and overtorque control protection is needed. For the engine reacceleration phase, the propulsion control system limits peak values of propeller shaft speed and torque. As previously noted, a peak in propeller thrust may also occur during this phase. This peak astern propeller thrust is potentially a serious stress problem in the ship propulsion system. Since the thrust is negative and can exceed the steady-state maximum ahead thrust, the thrust bearing must withstand high bidirectional transient loads.

XII. SUMMARY

The demand for new propulsion systems with increased performance, automation, and energy conservation are factors leading to greatly improved engineering methods in marine propulsion system development. These methods include large-scale simulations of marine propulsion systems to provide increased analysis and design capability for rapid solution with relative ease, of multi-degree-of-freedom, nonlinear systems of great complexity. These simulations are built upon fundamental principles drawn from many disciplines together with experimental data where analytical methods are lacking. This growth of large, comprehensive simulations has provided the means of identifying areas where new experimental approaches and data are needed, while posing questions and providing answers to heretofore impossible problems. This capability has further accelerated the growth of dynamics simulations. Marine propulsion dynamics simulations for ships, especially gas turbine, controllable pitch propeller plants and various combined plants are now extensively used worldwide in the design of ship propulsion systems. These simulations are required because of the complex interacting machinery systems heavily dependent on automation. With fast-responding gas turbine engines, multiple engines, combined plants, and controllable pitch propellers, possibilities exist for damaging torques, thrusts, and speeds in various maneuvering and machinery configuration transients. One of the greatest benefits in these simulations is the discovery of unexpected problems and interface mismatches between various systems, e.g., prime mover, thrust bearings, clutches, shafting, gears, propeller, and controls, early enough so they can be solved before expensive hardware modifications become necessary. Even with well-tested and sea-going propulsion plants, dynamics and control problems occur, but are amenable to solution by simulation. Thus, the dynamic simulation of a ship's power plant has become one of the best methods to answer immediate questions on many fleet problems particularly in the areas of dynamics and control. New approaches in control and monitoring systems can be developed based on the simulations and verified by testing-at-sea. Ship propulsion plant simulations can be developed and kept in continual improvement to keep pace with ship conditions, thus providing the fleet with immediate answers to systems performance and failures. Interactive man-machine systems with immediately available shore-based supporting simulations are reminiscent of the manned space program. This development is an exciting solution to the growing complexity and diagnostic burdens of a worldwide fleet.

REFERENCES

1. G. M. WEINBERG, "An Introduction to General Systems Thinking," Wiley, New York, 1975.

2. R. BHATTACHARYYA, "Dynamics of Marine Vehicles," Wiley, New York, 1978.

3. C. J. RUBIS, "Acceleration and Steady State Performance of a Gas Turbine Ship with Controllable Pitch Propeller," *Trans. Soc. Naval Arch. Marine Eng. 80* (1972).

4. C. J. RUBIS, "Gas Turbine Performance Under Varying Torque Propeller Loads," International Shipbuilding Progress, January 1974.

5. J. W. ABBOTT, J. G. McINTIRE, and C. J. RUBIS, "A Dynamic Analysis of a COGAS Propulsion Plant," *Naval Eng. J.* (1977).

6. C. J. RUBIS, "Braking and Reversing Ship Dynamics," *Naval Eng. J.* (1970).

7. J. P. COMSTOCK (Editor), "Principles of Naval Architecture," Society of Naval Architects and Marine Engineers, 1967.

8. T. R. HARPER and L. C. CARROLL, "The Effect of Controllable Pitch Propeller Data Interpolation Procedures on Propulsion Plant Simulations," Proceeding of the 1977 Summer Computer Simulation Conference, Chicago, Illinois, 1977.

9. C. J. RUBIS and T. R. HARPER, "Reversing Dynamics of a Gas Turbine Ship with Controllable Pitch Propeller," Proceedings of the Fifth Ship Control Systems Symposium, Annapolis, Maryland, 1978.

10. J. B. WOODWARD, "Marine Gas Turbines," Wiley, New York, 1975.

11. J. H. TONEY, "Comparative Analysis of Modeling Techniques for Ships with Gas Turbine and CRP Prime Movers," Proceedings of the Fifth Ship Control Systems Symposium, Annapolis, Maryland, 1978.

12. C. J. RUBIS and R. R. PETERSON, "Simulated Dynamics and Control of LM2500 Marine Gas Turbine Engine," Third Ship Control Systems Symposium, Bath, England, 1972.

13. D. D. McCRACKEN, "A Guide to FORTRAN IV Programming," Wiley, New York, 1972.

14. F. H. SPECKHART and W. L. GREEN, "A Guide to Using CSMP — The Continuous System Modeling Program," Prentice-Hall, Englewood Cliffs, New Jersey, 1976.

Toward A More Practical Control Theory for Distributed Parameter Systems

MARK J. BALAS

Electrical, Computer,
and Systems Engineering Department
Rensselaer Polytechnic Institute
Troy, New York

ISBN 0-12-012718-0

I. INTRODUCTION

The theme of this paper is the development of feedback control theory for *distributed parameter systems* (DPS) with practical constraints. By DPS, we mean those systems that are modeled by partial differential equations, as differentiated from *lumped parameter systems*, which are described by ordinary differential equations. Many of the ideas presented here will apply to differential equations with delays, as well, but these systems will not be emphasized. A general discussion of DPS control appeared in [1], which is an excellent reference for those unfamiliar with the basic issues.

The *practical constraints* that we propose must be imposed are the following:

(a) a finite (small) number of actuators and sensors with prescribed influence functions;

(b) some amount (no matter how small) of natural dissipation in the system; and

(c) a finite (small) dimensional controller.

Constraint (c) is essential if the controllers we generate for DPS are to be implemented with on-line computers of limited word length. This list seems to us to include the minimum constraints for a practical DPS control theory. A proposal for a practical theory of DPS was first introduced by Athans [2]. In addition to his versions of constraints (a)-(c), he added the further constraint that "the distributed nature of the system to be controlled should be retained until numerical results are required"; we have omitted this constraint here, not because we disagree, but because we think it is more a matter of style. We plan to show that much can be accomplished without always retaining the distributed nature of the problem quite to the bitter end; a principal difficulty, which becomes apparent, is to determine at what point it is most effective to replace the distributed problem with a lumped parameter approximation (i.e., a reduced-order model).

We hope, by developing a DPS control theory with a set of practical constraints like (a)-(c), that the results generated will appeal, not only to control theorists but to the audience of practicing control engineers with interests in a wide variety of systems applications.

The ideas presented here have had their origin in our work in active control of mechanically flexible structures (e.g., [3,4]) and, in particular, large aerospace structures (e.g., the survey [5]). The basic theory for these systems using finite dimensional controllers can be found in [6]; however, in this chapter, we want to extend this theory to a larger class of DPS that includes parabolic, as well as hyperbolic, systems. We shall not strive to produce new mathematical ideas here, although such may be available in the context of control theory (see the excellent survey [7]); rather, we shall use only some basic ideas of functional analysis, linear partial differential equations, and semigroup theory [8-11], to pose properly our control problems and investigate their fundamental structure.

Our results are in the same vein as those of [12] but with the added emphasis on an implementable finite-dimensional control. We should like to call attention to related work on control of parabolic (or diffusion-like) DPS with finite-dimensional controllers that has been reported recently [13-16].

We consider DPS described by the following:

$$\begin{cases} \partial v(t)/\partial t = Av(t) + Bf(t) + F_D(t), & v(0) = v_0, \\ y(t) = Cv(t), \end{cases} \tag{1}$$

where the system *state* $v(t)$ is in an infinite dimensional Hilbert space H with inner product $(\cdot, \cdot)$ and associated norm $\|\cdot\|$. (With some slight changes later, a more general Banach space can be substituted for the Hilbert space H). The operator A is an unbounded, linear, time-invariant, differential operator with domain $D(A)$, a subspace that is dense in H and contains all sufficiently smooth states which satisfy the boundary conditions of the DPS. The closed operator A generates a C_0 semigroup $U(t)$ on H; for more about semigroups, see [8-11].

The *control* is introduced via M inputs:

$$Bf(t) = \sum_{i=1}^{M} b_i f_i(t), \tag{2}$$

where the actuator influence functions b_i are in H. External disturbances and (possible) nonlinearities are represented by F_D, which may be a function of the state $v(t)$ as well as time t. Since it is possible to deal with these external disturbances by modifying the basic controller, we shall defer discussion of disturbances and nonlinearities until Section IX. Henceforth, we *assume* $F_D \equiv 0$ *in (1) unless otherwise stated.*

The system is *observed* via P sensors whose outputs $y_j(t)$ form the vector $y(t) = [y_1(t), \ldots, y_P(t)]^T$ and are given by

$$y_j(t) = (c_j, v(t)), \quad 1 \leq j \leq P, \tag{3}$$

where the sensor influence functions c_j are in H. The rank of the input operator B is M and that of the output operator C is P. In most DPS applications the control and observation must be done with a finite number of devices; hence, these restrictions on B and C are natural and coincide with constraint (a). Usually, the control devices will be *localized*, i.e., the influence functions b_i and c_j will be approximations of Dirac delta functions. However, *distributed* control devices, such as those using electromagnetic fields or electric charge diffusion, are not precluded in the preceding description, as long as b_i and c_j are in H.

The *weak formulation* of the DPS is given by

$$\begin{cases} v(t) = U(t)v_0 + \int_0^t U(t - \tau)B\, f(\tau)d\tau, \\ y(t) = C\, v(t). \end{cases} \tag{4}$$

Although it is convenient for us to talk about the "differential" formulation (1) of the DPS, usually we shall mean the weak formulation (4) was used in any proofs of theorems.

Such a formulation of a controlled DPS is not very restrictive. It does exclude the important class of DPS problems where observation and/or control can only be done at the boundary of the process rather than in the interior. However, many of these boundary-controlled DPS may be converted to equivalent interior-controlled DPS that fit the above formulation, e.g., [17]. Nevertheless, not all boundary-controlled problems may be treated this way; some give rise to unbounded operators B and C and must be handled differently, e.g., [13].

Although the preceding formulation may seem to exclude actuator-sensor dynamics, they may be easily incorporated by the addition of their states to the DPS state v(t). This may produce a hybrid system with lumped parameter, as well as distributed parameter, subsystems. Once the DPS control problem is understood, these hybrid systems may be handled by essentially the same methods; the theory does not change. Of course, from a practical standpoint, it is very important to include, and assess the effects of, actuator-sensor dynamics in the DPS control problem.

II. REDUCED-ORDER MODELS OF DISTRIBUTED PARAMETER SYSTEMS

Even though the DPS, as described by (1)-(3) is infinite-dimensional, it must be controlled by a finite-dimensional controller in order to satisfy constraint (c). The most natural way to obtain such a controller is to base it on a lumped parameter, i.e., finite dimensional, approximation of the DPS. Such an approximation or *reduced-order model* (ROM) is a projection (not necessarily orthogonal) of the DPS in H onto an appropriate *finite-dimensional* subspace H_N of H. The *ROM subspace* H_N has dimension N, and its projection is denoted by P_N; the *residual subspace* H_R associated with H_N completes the decomposition $H = H_N \oplus H_R$ and its projection is denoted by P_R. The total DPS state v is the sum of the *ROM state* $v_N \equiv P_N v$ and the *residual state* $v_R \equiv P_R v$:

$$v(t) = v_N(t) + v_R(t). \tag{5}$$

The choice of subspace H_N and the projection P_N completely specifies the residual subspace H_R and its corresponding projection P_R; this choice is usually dictated by the physical application and corresponding knowledge (or lack of knowledge) about the specific DPS involved. Therefore, we are willing to deal with the very general class of ROM methods already described.

A desirable subclass of ROM's occurs when the projection P_N (and, hence, also P_R) commutes with the operator A; these are called ROM's based on *reducing subspaces*. This occurs very naturally when a finite group of discrete eigenvalues $\lambda_1, \ldots, \lambda_N$ of A can be separated by a simple closed curve from the rest of the *spectrum* of A, denoted $\sigma(A)$, (i.e., the points λ in the complex plane for which the resolvent operator $R(\lambda, A) \equiv (\lambda - A)^{-1}$ fails to be a bounded operator on H); then, the *modal subspace*, i.e.,

$$H_N = \text{span}\,\{\phi_1, \ldots, \phi_N\},$$

where ϕ_k are the eigenfunctions corresponding to the eigenvalues λ_k, and its complement H_R form a pair of reducing subspaces for the operator A ([18], p. 178).

In many situations, the partial differential equations or the corresponding boundary conditions are too complicated to permit simple closed-form solutions for the eigendata. Consequently, an approximation of DPS, such as the Rayleigh-Ritz-Galerkin or finite element method, e.g., [19], must be used to deal with the problem numerically; this is one way to produce an ROM of the DPS, and some of the special properties for active control were pointed out in [31] and in particular for flexible structures in [20]. However, it is by no means the only way to generate such approximations. In the following, any pair of P_N, P_R projections on H, with P_N having a finite-dimensional range, is considered an acceptable model reduction of the DPS (1). The practical advantages of specific model reduction schemes for control of DPS will, we hope, become more apparent as we proceed. A rather complete bibliography of model reduction methods for lumped parameter systems (some of which can be extended to DPS) is given in [21].

The projection of the DPS (1)-(3) onto the subspaces H_N and H_R decomposes the system into the following:

$$\partial v_N(t)/\partial t = A_N v_N(t) + A_{NR} v_R(t) + B_N f(t), \tag{6}$$

$$\partial v_R(t)/\partial t = A_{RN} v_N(t) + A_R v_R(t) + B_R f(t), \tag{7}$$

$$y(t) = C_N v_N(t) + C_R v_R(t), \tag{8}$$

where $A_N = P_N A P_N$, $A_{NR} = P_N A P_R$, $B_N = P_N B$, $C_N = CP_N$, etc.; note that all these parameters, with the exception of A_R, are bounded operators because they involve projection onto the finite-dimensional subspace H_N. The terms $B_R f$ and $C_R v_R$ are called *control* and *observation spillover*, respectively; these definitions were introduced in [3]-[4]. The terms $A_{NR} v_R$ and $A_{RN} v_N$ are called the *modeling error terms* because they often represent errors that appear due to incorrectly modeled eigendata when approximate modal reducing subspaces are used for the ROM; *the modeling error terms are both zero when reducing subspaces are used*. It should also be noted that *the control (observation) spillover term is zero when the actuator (sensor) influence functions* $b_i (c_j)$ *are in the ROM subspace* H_N; this is a very special situation which we feel can be expected to occur *infrequently in practical systems*, especially in conjunction with reducing subspaces.

The ROM is obtained from (6)-(8) by ignoring the residuals $v_R(t)$. We call (A_N, B_N, C_N) parameters of the ROM of the DPS. The ROM is a finite-dimensional or lumped parameter approximation of the DPS, and the parameters may be identified with matrices by taking any basis for the subspace H_N. The controllability and observability of the ROM can be easily determined by checking the usual finite dimensional rank conditions; further simplification of these conditions occurs when modal or finite element subspaces are used (see [4] and [20]).

We make the following natural but very important assumption about the residual subsystem operator A_R: $(H_1) A_R$ generates a C_0 semigroup $U_R(t)$ with the following dissipative property:

$$\|U_R(t)\| \leq M_R \exp(-\sigma_R t), \quad t \geq 0, \tag{9}$$

where $M_R \geq 1$ and $\sigma_R \geq 0$. This assumption says that zero is not in the spectrum of A_R (i.e., all zero-frequency modes are included in the ROM), and the residual subsystem is *exponentially stable*. Such an assumption coincides with the practical constraint (b) in Section I.

In physical systems the natural energy dissipating mechanisms will tend to make (9) true although the stability margin σ_R may be quite small (as it is thought to be in most large aerospace structures). For dissipative hyperbolic systems [6] that represent mechanically flexible structures, (9) is true for the full system semigroup $U(t)$; consequently, it remains true for A_R when modal subspaces are used in the model reduction. For parabolic systems, the semigroup $U(t)$ is analytic; therefore, if the system operator A which generates $U(t)$ has compact resolvent, then only a finite number of discrete eigenvalues will lie to the left of a vertical line through $(-\sigma_R, 0)$ in the complex plane for any fixed, positive σ_R. Consequently, the modal subspace associated with this finite group of eigenvalues may be used as H_N in the model reduction, and (9) will hold on H_R (see [12]). Thus, the assumption (H_1)

is not unreasonable from a theoretical or a practical point of view and includes a wide variety of DPS applications. Further discussion of this will appear later in Section X on applications.

III. THE DISTRIBUTED PARAMETER SYSTEMS CONTROL PROBLEM

The basic control issues for DPS are very well described in [22], and we recommend this for the reader unfamiliar with them. Although we can propose performance indices to be minimized by an *optimal feedback control*, we feel that a far more basic issue is the use of a *finite-dimensional feedback controller to stabilize or improve the stability of the DPS*. This is the control problem we shall address here: *How much can we improve the stability of a DPS with a finite-dimensional controller*? The answer is not so surprising: The finite dimensional controller can redistribute the overall DPS stability to improve the stability of the most critical finite dimensional subsystem. As we shall see, sometimes (chiefly, with parabolic DPS) this redistribution can be done with little or no reduction in the overall system stability; however, many systems (usually of hyperbolic type) must be treated more carefully in order to guarantee an acceptable overall stability while enhancing the critical subsystem. Nevertheless, the controller synthesis and stability analysis proceed along the same lines for all DPS satisfying the assumptions we have set forth in Sections I and II.

Consequently, the *basic DPS Control problem* is whether the stability of the system (1) can be improved by a finite-dimensional controller of the form

$$\begin{cases} f(t) = H_{11}y(t) + H_{12}z(t), \\ \dot{z}(t) = H_{21}y(t) + H_{22}z(t), \end{cases} \tag{10}$$

where $\dim z = N < \infty$. The DPS control problem includes the selection of the controller dimension N and appropriate controller parameter matrices H_{ij}, as well as the selection and location of the M control actuators and P control sensors; this constitutes the *controller synthesis*.

The control algorithm (10) will be implemented by an on-line digital computer. Consequently, the controller dimension N will be related to the memory capacity and the memory access speed of this computer. To be quite precise, the algorithm (10) should be written in discrete, rather than continuous, time since this is the way the digital computer will actually operate. This can be done but it obscures the basic structure of the DPS control problem. Hence, we shall leave everything in a continuous-time formulation but recognize that, in practice, we must rework the problem in discrete-time (for example, see [39] and [53]). The amplitude quantization of the control commands from a digital computer should also be considered as a practical issue; it will certainly cause an increase in the spillover terms in (6)-(8) and, hence, will lead to further difficulties in implementation of DPS control.

IV. INFINITE-DIMENSIONAL DISTRIBUTED PARAMETER SYSTEMS CONTROLLERS

In this section we present some results for infinite dimensional DPS controllers which are analogous to the finite-dimensional state-space controllers for lumped parameter systems, e.g., [23]. Unlike their finite-dimensional counterparts, these controllers cannot be implemented with practical computers and devices in general. Nevertheless, such results give further insight into the DPS control problem.

The first result gives conditions under which the full state $v(t)$ of the DPS can be recovered asymptotically from the finite number of available measurements $y(t)$ by an *infinite dimensional state* estimator (Kalman filter or Luenberger observer).

Theorem IV.1. If (A^*, C^*) is exponentially stabilizable, then there is a bounded operator K mapping R^P into D(A) such that the estimated state $\hat{v}(t)$ generated by the *state-estimator:*

$$\begin{cases} \partial\hat{v}(t)/\partial t = A\hat{v}(t) + Bf(t) + K[y(t) - \hat{y}(t)], \\ \hat{v}(0) = 0, \\ \hat{y}(t) = C\hat{v}(t), \end{cases} \tag{11}$$

converges in norm to the actual state $v(t)$ at an exponential rate (determined by K).

The second result gives conditions under which stability of the DPS may be achieved using the state-estimator (11):

Theorem IV.2. In addition to the hypothesis of theorem IV.1 if (A, B) is also exponentially stabilizable, then there is a bounded operator G from D(A) into R^M such that the control law

$$f(t) = G\hat{v}(t), \tag{12}$$

where $\hat{v}(t)$ is generated by (11), produces an exponentially stable closed-loop system consisting of (1) and (11)-(12).

The proofs for these results are given in Appendix I; except for some infinite-dimensional technicalities, they are the same as those for the finite dimensional case.

Recall that A^* is the adjoint operator associated with A and defined on $D(A^*)$ dense in H; see e.g., [9] or [18]. Also, recall that (A, B) *exponentially stabilizable* means there is a bounded operator G such that the semigroup $U_C(t)$ generated by $\tilde{A}_C = A + BG$ is *exponentially stable*, i.e.,

$$\|U_C(t)\| \leq M_C \exp(-\sigma_C t), \quad t \geq 0, \tag{13}$$

for constants $\sigma_C > 0$ and $M_C \geq 1$. Further discussion of stability and stabilizability for infinite-dimensional systems appears in the excellent surveys [7] and [24]. Note that for finite dimensional systems, (A, B, C) controllable and observable would be sufficient to satisfy the hypotheses of theorems IV.1-IV.2; however, in infinite dimensions this is not the case when controllability and observability are taken in the approximate (and most reasonable) sense of [25] or [9] Chapter 4.

Furthermore, it is known [26] that (A, B) cannot be exponentially stabilizable when A is energy-conserving and B is compact (as it is in (2)) unless A already generates an exponentially stable semigroup. In other words, unless we include a practical constraint, (b) of Section I, on the energy dissipation in the uncontrolled DPS (1), we may not be able to satisfy the hypothesis of theorems IV.1-IV.2. The following result gives conditions under which these hypothesis are satisfied.

Theorem IV.3. If assumption (H1) in Section II is satisfied for some pair of reducing subspaces H_N and H_R where dim $H_N = N < \infty$, then (A, B) and (A^*, C^*) are both exponentially stabilizable and a stabilizing infinite-dimensional controller (11)-(12) exists when (A_N, B_N, C_N) is controllable and observable.

The proof is given in Appendix II. This result says that, for some model reductions (i.e., those based on modal subspaces), it is possible to guarantee the existence of the infinite dimensional controller under the practical constraints (H1) and a controllable-observable ROM.

Of course, this may be an academic point since the controller produced by (11)-(12) is not generally implementable. The controller is given by

$$f(t) = \int_0^t G\hat{U}(t - \tau)Ky(\tau)d\tau, \tag{14}$$

where $\hat{U}(t)$ is the semigroup generated by $L = A + BG - KC$; the convolution (14) cannot usually be implemented with finite-dimensional devices even though it has a finite number of inputs $y(t)$ and a finite number of outputs $f(t)$. The approximate implementation of (14) or equivalently (11)-(12) will be considered in Section V.

The infinite-dimensional state estimator (11) was developed for the heat equation in [27] and used in a feedback control formulation (11)-(12) for the heat equation in [15]. The stability and convergence results of [15] and [27] follow directly from our results in this section.

The relationship of stabilizability and approximate controllability and observability for general DPS with contractive semigroups was studied in [25] and [28]. However, when practical constraints are imposed on the DPS, it is not surprising that it is only the controllability and observability of the finite-dimensional (modal) ROM that determines stabilizability.

V. FINITE-DIMENSIONAL CONTROLLER SYNTHESIS FOR DISTRIBUTED PARAMETER SYSTEMS

In this section we develop two basic procedures for synthesizing *finite-dimensional controllers* for the DPS (1):

(1) *direct model reduction*, i.e., perform a model reduction on the DPS (1) as in Section II and synthesize the controller directly from this ROM;

(2) *indirect model reduction*, i.e., perform a model reduction on the infinite-dimensional controller (11)-(12) in Section IV to obtain a finite-dimensional approximation.

The direct procedure is quite straightforward and is the most natural one to use from a practical standpoint. It requires nothing but ROM information for the controller synthesis and can be carried out even though the conditions of Section IV for existence of an infinite dimensional controller are not verified. The indirect procedure requires the existence of an infinite-dimensional controller and some knowledge of the gain operators G and K. When this knowledge is available, it seems reasonable to take advantage of it; the finite-dimensional approximation of the infinite-dimensional controller may perform better! This is the motivation for the constraint in [2] to retain the distributed parameter nature of the problem as long as possible.

Clearly, there are technical drawbacks to the indirect procedure while the direct procedure can always be performed. The advantages of the indirect procedure will only be apparent in the analysis of the closed-loop system with such finite-dimensional controllers. At the end of this section, we shall present conditions under which the two procedures yield the same controller.

A. *FINITE-DIMENSIONAL CONTROLLER SYNTHESIS: DIRECT MODEL REDUCTION*

The ROM (A_N, B_N, C_N) is chosen to represent the most critical part of the DPS. Once this ROM is selected as in Section II and its controllability and observability are verified, the controller synthesis follows standard lumped parameter techniques, e.g., [23]. The controller is synthesized as though the finite-dimensional ROM were the full system, i.e., the control and observation spillover and modeling error terms are all assumed to be zero and the residual subsystem is ignored.

The *feedback controller* based on the ROM has the following format:

$$\begin{cases} f(t) = G_N \hat{v}_N(t), \\ \partial \hat{v}_N(t)/\partial t = A_N \hat{v}_N(t) + B_N f(t) + K_N[y(t) - \hat{y}(t)], \\ \hat{y}(t) = C_N \hat{v}_N(t), \quad \hat{v}_N(0) = 0, \end{cases} \tag{15}$$

where (A_N, B_N, C_N) are the same as in (6) and (7). This is a *linear control law* with feedback from a *state estimator* or observer acting on only the available sensor outputs. By taking any basis in the ROM subspace H_N, the controller in (15) can be identified with a finite-dimensional or lumped parameter system. The gains G_N: $H_N \to R^M$ and K_N: $R^P \to H_N$ are bounded (finite rank).

Let

$$\hat{v}_N(t) = v_N(t) + e_N(t), \tag{16}$$

and it is easy to see (from (15) that $\hat{v}_N$ and e_N are in H_N. Define

$$F_N = \begin{bmatrix} A_N + B_N G_N & B_N G_N \\ 0 & A_N - K_N C_N \end{bmatrix}. \tag{17}$$

If a stability margin of $\sigma_N > 0$ is desired, the controller gains G_N and K_N can be determined by standard finite-dimensional pole placement or regulator techniques to cause the (finite rank) operator F_N in (17) to have its spectrum to the left of a vertical line through $(-\sigma_N, 0)$; consequently, the semigroup $U_N(t)$ generated by F_N has the property:

$$\|U_N(t)\| \leq M_N \exp(-\sigma_N t), \quad t \geq 0 \tag{18}$$

for some $M_N \geq 1$ and the desired $\sigma_N > 0$. Thus, the stability of the ROM subsystem of the DPS can be greatly improved by the feedback controller (15) and the synthesis of such a controller can be easily carried out by well-known techniques. However, *the stability of the actual DPS in closed-loop with this controller remains in question.*

The special case where the exact modes of the DPS (1) are known and *modal* reducing subspaces are used is called *modal control*. Such controllers for DPS have been considered extensively in the literature, but they are not the most useful designs in practice where the modes are unknown or very poorly known (see V.B or [31]).

B. FINITE-DIMENSIONAL CONTROLLER SYNTHESIS: INDIRECT MODEL REDUCTION

To perform a model reduction on the finite-dimensional controller (11)-(12), we rewrite it:

$$\begin{cases} f(t) = G\hat{v}(t), \\ \partial\hat{v}(t)/\partial t = L\hat{v}(t) + Ky(t), \\ \hat{v}(0) = 0, \end{cases} \tag{19}$$

where L = A + BG - KC has the same domain as that of A. Proceeding as in Section II, we obtain the pair of subspaces H_N and H_R and let $\hat{v}_N = P_N\hat{v}$ and $\hat{v}_R = P_R\hat{v}$. Then, (19) projected onto H_N becomes

$$\begin{cases} f(t) = \overline{G}_N\hat{v}_N(t) + \overline{G}_R\hat{v}_R(t), \\ \partial\hat{v}_N(t)/\partial t = L_N\hat{v}_N(t) + L_{NR}\hat{v}_R(t) + \overline{K}_N y(t), \\ \hat{v}_N(0) = 0, \end{cases} \tag{20}$$

where $L_N = P_N L P_N$, etc., and $\overline{G}_N = GP_N$, $\overline{G}_R = GP_R$, and $\overline{K}_N = P_N K$.

But, (20) cannot be implemented without knowledge of $\hat{v}_R(t)$, which is generated by an infinite-dimensional system. Consequently, the following finite-dimensional controller is the natural choice for the approximation of (20):

$$\begin{cases} f(t) = G_N\tilde{v}_N(t), \\ \partial\tilde{v}_N(t)/\partial t = L_N\tilde{v}_N(t) + \overline{K}_N y(t) \\ \qquad = A_N\tilde{v}_N(t) + B_N f(t) + \overline{K}_N[y(t) - C_N\tilde{v}_N(t)], \end{cases} \tag{21}$$

$$\tilde{v}_N(0) = 0,$$

Thus, (21) is the *finite-dimensional controller synthesized by indirect model reduction*; it is simply (20) with the residuals $\hat{v}_R(t)$ ignored. The parameters and gains (A_N, B_N, C_N, $\overline{G}_N$, $\overline{K}_N$) are obtained by projecting the infinite-dimensional data (A, B, C, G, K) onto H_N; of course, these parameters may be identified with their matrices in any appropriate basis of H_N. The only difference between (21) and (15) is that in (21) we are not free to choose the gains G_N and K_N — they are determined by the G and K of the infinite-dimensional controller.

Let

$$\tilde{v}_N(t) = v_N(t) + \tilde{e}_N(t) = \hat{v}_N(t) + q_N(t), \tag{22}$$

and it is easy to see from (20)-(21), that

$$\begin{cases} \partial q_N(t)/\partial t = L_N q_N(t) - L_{NR}\hat{v}_R(t), \\ \quad q_N(0) = 0. \end{cases} \tag{23}$$

Even under the hypothesis of theorem IV.3 (where modal subspaces are used), $L_{NR} = -\bar{K}_N C_R$, which is not usually zero. Consequently, in general,

$$\tilde{v}_N(t) \neq P_N\hat{v}(t). \tag{24}$$

Furthermore, even though A + BG and A - KC are exponentially stable, $A_N + B_N\bar{G}_N$ and $A_N - \bar{K}_N C_N$ may not yield the stability margin $\sigma_N > 0$ in (18) when F_N is replaced by

$$\bar{F}_N = \begin{bmatrix} A_N + B_N\bar{G}_N & B_N\bar{G}_N \\ 0 & A_N - \bar{K}_N C_N \end{bmatrix} \tag{25}$$

However, the following result gives sufficient conditions under which *the direct and indirect procedures yield the same finite-dimensional controller.*

Theorem V.1. If the hypothesis of theorem IV.3 is satisfied for some pair of reducing subspaces and these subspaces are used for the direct and indirect model reduction, then controllers (15) and (21) are the same.

The proof follows from the fact that $\bar{G}_N$ and $\bar{K}_N$ in (25) can be chosen equal to G_N and K_N in (17). Consequently, *for modal subspaces the direct and indirect procedures yield the same controller* and there is no advantage to keeping the distributed parameter nature of the problem as long as possible. Nonetheless, when the exact modes are unknown, there may be substantial advantage in one of the procedures over the other.

To be more specific, suppose we use the *Galerkin method* to reduce the infinite-dimensional controller (11)-(12), i.e., given the subspace H_N, choose $\tilde{v}_N(t)$ in H_N such that

$$\tilde{P}_N(E) = 0, \tag{26}$$

where P_N is the Galerkin (or orthogonal) projection on H_N and $E(t)$ is the *equation error* defined by

$$\begin{cases} \partial\tilde{v}_N(t)/\partial t = L\tilde{v}_N(t) + Ky(t) + E(t), \\ \tilde{v}_N(0) = P_N\hat{v}(0) = 0, \end{cases} \tag{27}$$

where L and K are as in (19). Since $\tilde{P}_N\tilde{v}_N = \tilde{v}_N$, we have, from (27) the *Galerkin finite-dimensional controller*:

$$\begin{cases} f(t) = \tilde{G}_N\tilde{v}_N(t) \\ \partial\tilde{v}_N(t)/\partial t = L_N\tilde{v}_N(t) + \tilde{K}_N y(t) \\ \qquad = \tilde{A}_N\tilde{v}_N(t) + \tilde{B}_N f(t) + \tilde{K}_N[(y(t) - \tilde{C}_N\tilde{v}_N(t)] \\ \tilde{v}_N(0) = 0 \end{cases} \tag{28}$$

where $\tilde{A}_N = \tilde{P}_N A\tilde{P}_N$, etc. In this case, the parameters $(\tilde{A}_N, \tilde{B}_N, \tilde{C}_N)$ and the gains $(\tilde{G}_N, \tilde{K}_N)$ are obtained by *orthogonal projection onto* H_N. Therefore, on nonreducing subspaces, the indirect procedure using the *Galerkin method* will yield a different controller from that obtained by the direct procedure. However, if the Galerkin method is used for the original model reduction of Section II, then the direct procedure can yield (28) with the choice of gains $G_N = \tilde{G}_N$ and $K_N = \tilde{K}_N$ [31].

VI. CLOSED-LOOP STABILITY ANALYSIS: REGULAR PERTURBATIONS

The finite-dimensional controllers obtained in Section V are based on model reduction (direct or indirect) of the original DPS (1). The stability of such controllers in closed-loop with the actual DPS is the concern of this (and the next two) sections. Such stability cannot be taken for granted, especially in highly oscillatory DPS. This was pointed out by examples in [4] and [30], where unstable residual modes were caused by finite-dimensional modal controllers.

Since the form of the closed-loop stability problem is the same whether a direct or indirect procedure has been used to obtain the controller, we shall concentrate here on a controller of the form (15) in closed-loop with the DPS (1). The same analysis can be carried out using (28). The ROM (A_N, B_N, C_N) is assumed to be controllable and observable and the controller gains G_N and K_N are assumed to satisfy (18). The assumption (H1) is imposed as well.

Let the *closed-loop system state* be

$$\omega = [v_N \; e_N \; v_R]^T \text{ in } H_N \times H_N \times H_R, \tag{29}$$

where

$$||\omega||^2 = ||v_N||^2 + ||e_N||^2 + ||v_R||^2. \tag{30}$$

From (6)-(8) and (15)-(17), obtain

$$\partial\omega(t)/\partial t = A_c\omega(t), \tag{31}$$

where

$$A_c = \begin{bmatrix} A_{11} & A_{12} \\ A_{21} & A_{22} \end{bmatrix}$$

and $A_{11} = F_N$, $A_{22} = A_R$, $A_{21} = [B_R G_N + A_{RN}, B_R G_N]$, and

$$A_{12} = \begin{bmatrix} A_{NR} \\ K_N C_R - A_{NR} \end{bmatrix}.$$

We define control and observation *spillover bounds* β_N and Γ_N:

$$\|A_{21}\| < \beta_N, \quad \|A_{12}\| \leq \Gamma_N. \tag{32}$$

These coefficients exist because A_{12} and A_{21} have finite rank and, hence, are bounded for each N; they give a measure of the effects of spillover and modeling error in the closed-loop system. In general, the terms A_{12} and A_{21} may not be uniformly bounded for all N; such uniform boundedness is not required in what follows.

We note that A_c is a bounded perturbation of the semigroup generator:

$$\begin{bmatrix} A_{11} & 0 \\ 0 & A_{22} \end{bmatrix}$$

and, therefore, A_c generates a C_0 semigroup $U_c(t)$ by [8], theorem X.9. It is the stability of this semigroup in the norm generated by (30) that we want to determine. Note that norms equivalent to (30) may be used, but we must be careful since, unlike the finite-dimensional situation, in infinite dimensions not every norm is equivalent.

The following theorem deals with a special case of interest:

Theorem VI.1. If either

(a) $A_{12} = 0$ (in particular, $A_{NR} = 0$ and $C_R = 0$) or

(b) $A_{21} = 0$ (in particular, $A_{RN} = 0$ and $B_R = 0$), then

$$\|U_c(t)\| \leq M_0 \exp(-\sigma_0 t), \tag{33}$$

where $\sigma_0 = \min(\sigma_N, \sigma_R)$ and, when (a) is true,

$$M_0 = M_N M_R (1 + E + E^2)^{1/2}$$

$$E = \beta_N (|\sigma_N - \sigma_R|)^{-1}$$

with (M_N, σ_N) from (18) and (M_R, σ_R) from (9). When (b) is true, (33) holds with Γ_N replacing β_N in E.

In other words, if the model error and either control or observation spillover were zero, then (31) is exponentially stable. This idealized situation would occur, for example, when reducing subspaces H_N and H_R were used and either the actuator or sensor influence functions were contained entirely in the ROM subspace H_N. Theorems VI.1 generalizes the results reported in [14] and [16] for a larger class of DPS which includes parabolic systems. Note that (33) gives *a bound on the effect of control spillover* (and one modeling error term) *alone*.

Even though model error might be zero in certain DPS (e.g., when modal subspaces were used), in general, spillover would not be zero. Correspondingly, one might include the actuator and sensor influence functions in H_N and use the indirect procedure of Section V with the Galerkin method; see also [31]. This will eliminate the spillover terms quite nicely, but it increases the modeling error terms simultaneously. Thus, we must seek stability conditions when A_{12} and A_{21} are not zero.

A sufficient condition for stability is given by the following:

Theorem VI.2. The closed-loop system satisfies:

$$\|U_c(t)\| \leq M_0 \exp(-\sigma_c t), \tag{34}$$

where

$$\sigma_c = \sigma_0 - M_0 \Gamma_N$$

with σ_0 and M_0 given in theorem VI.1, (β_N, Γ_N) in (32), (M_N, σ_N) in (18), and (M_R, σ_R) in (9). The closed-loop system is exponentially stable when

$$\Gamma_N < \sigma_0 / M_0. \tag{35}$$

The "dual" result with β_N and Γ_N interchanged is also true.

This result helps to answer the closed-loop stability question and makes it possible to estimate the stability margin σ_c with data obtained from the ROM and bounds on the residual subsystem and the spillover. It seems to us that this is the most practical form in which to answer the question as well. *In actual DPS, the residual data and the spillover are likely to be poorly known, but engineers may be able to specify* (or approximate) *their bounds* β_N and Γ_N. Also, in many applications $M_R = 1$ since energy norms are used.

The proofs of theorems VI.1 and VI.2 follow directly from our general stability lemma.

Lemma VI.3. Consider

$$\omega_t = A_c \omega = \begin{bmatrix} A_{11} & A_{12} \\ A_{21} & A_{22} \end{bmatrix} \omega, \tag{36}$$

where A_{ij} are bounded for $i \neq j$ and A_{ii} generates the C_0 semigroup $U_i(t)$, with growth property:

$$\|U_i(t)\| \leq K_i \exp(-\sigma_i t), \quad t \geq 0 \tag{37}$$

for $i = 1, 2$. Assume $\sigma_1 \neq \sigma_2$. Then A_c generates the C_0 semigroup $U_c(t)$ with growth property:

$$\|U_c(t)\| \leq K_c \exp(-\sigma_c t), \quad t \geq 0, \tag{38}$$

where

$$\sigma_c = \sigma_0 - K_c\|A_{21}\|,$$

$$\sigma_0 = \min(\sigma_1, \sigma_2),$$

$$K_c = K_1K_2(1 + \gamma + \gamma^2)^{1/2}$$

and

$$\gamma = \|A_{12}\|/|\sigma_1 - \sigma_2|.$$

The dual result with A_{12} and A_{21} interchanged in (38) is also true. The proofs of this lemma and theorems VI.1 and VI.2 appear in Appendix III.

We offer the following result on "dissipative perturbations" as an indication of closed-loop stability under special conditions:

Theorem VI.4. If $A_c = A_0 + \Delta A$ in (31) with ΔA bounded, A_0 the generator of an exponentially stable semigroup $U_0(t)$ with stability margin $\sigma_0 > 0$ and $\text{Re}(\omega, \Delta A\omega) \leq 0$ for all ω in $D(A_0)$, then A_c generates an exponentially stable semigroup with stability margin $\sigma_c \geq \sigma_0$.

The proof of this result follows directly from [32], theorem II.6. It does not seem as useful a result as theorem VI.2, but it can be used with special control laws that are entirely "dissipative," i.e., ΔA satisfies above. Such a control law is *direct velocity feedback* (see [5]).

We classify the theorems in this section as "regular perturbations" because we prove stability results about the closed-loop system (31) by separating

$$A_C = A_0 + \Delta A,$$

where A_0 is known to be stable (by design) and ΔA, a regular perturbation of A_0, is bounded and contains the important stability-reducing spillover and modeling error terms. The stability bounds we have obtained here are lower bounds for the overall stability margin σ_C in terms of a known (designed) stability margin σ_0 and the (possibly) deleterious effects of spillover and modeling error:

$$\sigma_C \geq \sigma_0 - h(\Delta A, A_0),$$

where h is an appropriate nonnegative function of ΔA and A_0.

These estimates are necessarily conservative since norms and perturbation methods are used to obtain them. Of course, we should like to have "sharper" estimates i.e., $h(\cdot, \cdot)$ smaller; however, the sharpness of any estimate will be directly related to how well we know the residual subsystem data (A_R, B_R, C_R), *when this is poorly known we cannot expect to do better than a conservative estimate*. In the later sections, we shall present some further approaches to stability analysis and make use of further knowledge (when available) of (A_R, B_R, C_R) in the controller synthesis and stability estimates.

It should also be noted that the stability estimates presented above give the control system designer a better idea of the tradeoffs to be made in DPS control. This knowledge can be used to redesign the controller, i.e., reduce controller gains and/or increase the controller dimension to control more "modes" when necessary and to indicate when some form of spill-over compensation, such as a sensor prefilter [5], should be added to improve the overall closed-loop stability.

VII. CLOSED-LOOP STABILITY ANALYSIS: SINGULAR PERTURBATIONS

In many cases the DPS (1) may be modeled by a *singularly perturbed system*:

$$\begin{cases} E(\epsilon)\dfrac{\partial v(t)}{\partial t} = Av(t) + Bf(t), \quad v(0) = v_0 \\ y(t) = Cv(t) \end{cases} \tag{39}$$

where the only difference between (39) and (1) is the bounded, linear operator $E(\epsilon)$: $H \to H$, which is a function of the singular perturbations parameter $\epsilon \geq 0$. The inverse $E^{-1}(\epsilon)$ is also a bounded linear operator on H when $\epsilon > 0$ and finite; however, $E(0)$ is singular (i.e., not one to one). The literature on singular perturbations and control of lumped parameter systems has been well surveyed in [33-34].

The small parameter ϵ will have various physical interpretations depending on the application. In [35], we have used the formulation (39) on large flexible aerospace structures with ϵ representing a time-scale separation and, alternatively, a frequency-scale separation. Many other formulations are possible. The singular perturbations method has been used extensively in aerospace applications, e.g., [36].

In [37], we have attempted to extend many of the singular perturbations ideas for lumped parameter systems to linear DPS. These ideas will be summarized here; the details can be found in [37].

A. *REDUCED-ORDER MODELING: SINGULAR PERTURBATIONS*

For certain choices of subspaces H_N and H_R as in Section II, it is possible to obtain, from (39), the following *singularly perturbed formulation of the DPS*:

$$\begin{cases} \partial v_N(t)/\partial t = A_N v_N(t) + A_{NR} v_R(t) + B_N f(t), \\ \epsilon(\partial v_R(t)/\partial t) = A_{RN} v_N(t) + A_R v_R(t) + B_R f(t), \\ y(t) = C_N v_N(t) + C_R v_R(t), \end{cases} \tag{40}$$

where $A_N = P_N A P_N$, etc. and all parameters except A_R are (finite rank) bounded operators since H_N is finite dimensional. We *assume* that (40) satisfies (H1); consequently, A_R^{-1} is a bounded operator. The *singular perturbation ROM* is obtained from (40) by formally setting $\epsilon = 0$:

$$\begin{cases} \partial v_N(t)/\partial t = \overline{A}_N v_N(t) + \overline{B}_N f(t), \\ y(t) = \overline{C}_N v_N(t) + \overline{D} f(t), \end{cases} \tag{41}$$

where $\overline{A}_N = A_N - A_{NR} A_R^{-1} A_{RN}$, $\overline{B}_N = B_N - A_{NR} A_R^{-1} B_R$, $\overline{C}_N = C_N - C_R A_R^{-1} A_{RN}$, and $\overline{D} = -C_R A_R^{-1} B_R$.

This is somewhat different from the ROMs produce in Section II because the parameters $(\overline{A}_N, \overline{B}_N, \overline{C}_N, \overline{D})$ are the usual ROM parameters (A_N, B_N, C_N) modified by "static" $(\epsilon = 0)$ correction terms involving residual subsystem information (A_R, B_R, C_R) and the modeling error terms (A_{RN}, A_{NR}). Even in the special case of reducing subspaces $(A_{NR} = 0$ and $A_{RN} = 0)$, although $\overline{A}_N = A_N$, $\overline{B}_N = B_N$, and $\overline{C}_N = C_N$, the term $\overline{D} = -C_R A_R^{-1} B_R$ is not usually zero; hence, *this ROM differs from those in Section II*. Therefore, it takes a more precise knowledge of the residual subsystem to obtain (41). We *assume* that the ROM $(\overline{A}_N, \overline{B}_N, \overline{C}_N)$ is a controllable-observable finite dimensional system; this is easy to verify.

B. REDUCED-ORDER CONTROLLER SYNTHESIS: SINGULAR PERTURBATIONS

The *reduced-order controller* is based on the ROM (41):

$$\left\{\begin{aligned} f(t) &= \overline{G}_N \hat{v}_N(t), \\ \partial \hat{v}_N(t)/\partial t &= \overline{A}_N \hat{v}_N(t) + \overline{B}_N f(t) + \overline{K}_N [y(t) - \hat{y}(t)], \\ \hat{y}(t) &= \overline{C}_N \hat{v}_N(t) + \overline{D} f(t), \quad \hat{v}_N(0) = 0, \end{aligned}\right. \tag{42}$$

where the gains $\overline{G}_N$ and $\overline{K}_N$ are designed to obtain closed-loop stability with the ROM (41), i.e., when $\epsilon = 0$; this can be done since $(\overline{A}_N, \overline{B}_N, \overline{C}_N)$ is controllable-observable.

Note that (42) differs from the controller (15) obtained in Section V by the direct model-reduction procedure; the difference is even present when reducing subspaces are used because a term $\overline{D}f(t)$ is added to $\hat{y}(t)$. This difference can be seen more clearly by rewriting (42) in the following form:

$$\left\{\begin{aligned} f(t) &= \overline{G}_N \hat{v}_N(t), \\ \partial \hat{v}_N(t)/\partial t &= A_N \hat{v}_N(t) + B_N f(t) + \overline{K}_N [y(t) - C_N \hat{v}_N(t)] \\ &\quad + N_1 \hat{v}_N(t) + N_2 f(t), \end{aligned}\right. \tag{43}$$

where

$$N_1 = (\overline{K}_N C_R - A_{NR}) A_R^{-1} A_{RN},$$

$$N_2 = (\overline{K}_N C_R - A_{NR}) A_R^{-1} B_R.$$

It is the additional knowledge of the residual subsystem contained in N_1 and N_2 that is required to obtain the singular perturbations controller (42); the gains $\overline{G}_N$ and $\overline{K}_N$ are different from those in (15), but this is less important conceptually. Even when reducing subspaces are used in the model reduction procedures, we have $N_1 = 0$ but $N_2 = \overline{K}_N C_R A_R^{-1} B_R$, which is not usually zero.

C. *CLOSED-LOOP STABILITY: SINGULAR PERTURBATIONS*

The controller (42) has been designed for stable closed-loop operation with the ROM (41); consequently, when $\epsilon = 0$, we have stable operation with the DPS (40). However, the fundamental question is whether the same controller (42) will maintain stable operation with the actual DPS (40) when $\epsilon > 0$. We expect that the singular perturbations correction terms that appear in (41) and (42) will improve the controller performance by making it a little more "aware" of the residual subsystem. The results below show that this is the case when ϵ is small.

The following is a distributed parameter version of the well-known lumped parameter *Klimushchev-Krasovskii lemma:*

Theorem VII.1. Let $\omega(t) = [\omega_1(t)\omega_2(t)]^T$ in H with $\omega_i(t)$ in H_i, where $\dim H_1 < \infty$. Consider

$$\begin{cases} \partial\omega_1(t)/\partial t = H_{11}\omega_1(t) + H_{12}\omega_2(t), \\ \epsilon(\partial\omega_2(t)/\partial t) = H_{21}\omega_1(t) + H_{22}\omega_2(t). \end{cases} \tag{44}$$

If H_{22} generates an exponentially stable semigroup on H_2 and the finite-rank operator $\overline{H} = H_{11} - H_{12}H_{22}^{-1}H_{21}$ is stable (i.e., all eigenvalues in the open left-hand plane), then there exists an $\epsilon_0 > 0$ such that for all $0 < \epsilon \leq \epsilon_0$, (44) is exponentially stable, i.e., the operator

$$\begin{bmatrix} H_{11} & H_{12} \\ \frac{H_{21}}{\epsilon} & \frac{H_{22}}{\epsilon} \end{bmatrix}$$

generates an exponentially stable semigroup on H. Furthermore, the upper bound ϵ_0 can be calculated from the norms of H_{ij}.

Now consider the closed-loop system consisting of (40) and (42); this can be written in the form (44) where $e_N = \hat{v}_N - v_N$ and

$$\omega_1 = \begin{bmatrix} v_N \\ e_N \end{bmatrix}, \qquad \omega_2 = v_R.$$

The parameters in (44) become:

$$H_{11} = \begin{bmatrix} A_N + B_N\overline{G}_N & B_N\overline{G}_N \\ Q_N & A_N - \overline{K}_N C_N + Q_N \end{bmatrix},$$

$$Q_N = (\overline{K}_N C_R - A_{NR})A_R^{-1}(A_{RN} + B_R\overline{G}_N),$$

$$H_{12} = \begin{bmatrix} A_{NR} \\ \overline{K}_N C_R - A_{NR} \end{bmatrix},$$

and

$$H_{21} = [A_{RN} + B_R\overline{G}_N \quad B_R\overline{G}_N],$$

$$H_{22} = A_R.$$

Since A_R satisfies (H1) and

$$\overline{H} = H_{11} - H_{12}H_{22}^{-1}H_{21} = \begin{bmatrix} \overline{A}_N + \overline{B}_N\overline{G}_N & \overline{B}_N\overline{G}_N \\ 0 & \overline{A}_N - \overline{K}_N\overline{C}_N \end{bmatrix}$$

is stable by design, the hypothesis of the theorem VII.1 is satisfied and we obtain the following:

Theorem VII.2. There exists ϵ_0 such that, for all $0 < \epsilon \leq \epsilon_0$, the closed-loop system consisting of the DPS (40) and the finite-dimensional controller (42) is exponentially stable. Furthermore, the upper bound ϵ_0 can be calculated from the ROM and bounds on the residual subsystem parameters. The proofs of theorems VII.1 and VII.2 appear in [37].

Consequently, when the DPS can be formulated in the singular perturbations format (39) or (40), it is possible to obtain a finite-dimensional controller that will stabilize the infinite-dimensional closed-loop system for small ϵ. *This result depends on more knowledge of the residual subsystem than was presumed in the previous sections.*

Of course, when such knowledge is available, it should be used. Further discussion of the *nontrivial* problem of obtaining singularly perturbed formulations of system models is contained in [38].

VIII. RELATED TOPICS: STABILIZING SUBSPACES AND IMBEDDING

In Sections VI and VII we have examined closed-loop stability form both regular and singular perturbations viewpoints. In this section, we shall briefly describe two alternatives for stability analysis: stabilizing subspaces and imbedding.

A. *STABILIZING SUBSPACES*

This method is a variation on the basic approach in Section VI. We shall say (A, B) in (1) has a pair of stabilizing *subspaces* H_N and H_R, if, in addition to the requirements of Section II, they also satisfy

$\tilde{A}_C = A + BG$ exponentially stable with a desired stability margin σ_C [see (13)] for some $G:H \to R^M$, such that

$$G = GP_N = G_N. \tag{45}$$

Note that (45) says that $G_R = GP_R = 0$, i.e., $\tilde{A}_C$ is stabilized by a gain that is restricted to the finite-dimensional subspace H_N. This situation occurs, for example, when modal reducing subspaces are used in "parabolic" problems, e.g., [12]-[16]. It means that all but a finite number of the modes of the system have the desired stability margin and those that do not may be stabilized by the gains (45). Furthermore, using (45), obtain

$$A_c = \begin{bmatrix} A_N + B_N G_N & A_{NR} \\ A_{RN} + B_R G_N & A_R \end{bmatrix}. \tag{46}$$

We use the finite-dimensional controller (15) based on the stabilizing subspace ROM (A_N, B_N, C_N) of (1) where $A_N = P_N A P_N$, $B_N = P_N B$, and $C_N = CP_N$. Assume this ROM is controllable and observable (in the finite-dimensional sense). Let $e_N = \hat{v}_N - v_N$ and obtain (from (1) and (15):

$$f(t) = G\hat{v}_N(t) = Gv_N(t) + G_N e_N(t), \tag{47a}$$

$$\partial e_N(t)/\partial t = (A_N - K_N C_N) e_N(t) + (K_N C_R - A_{NR}) P_R v(t). \tag{47b}$$

Note that we may omit P_R from (47b) since it is already incorporated into C_R and A_{NR}.

The closed-loop system (1) and (15) can be written [from (45)-(47)]:

$$\begin{cases} \partial v(t)/\partial t = \tilde{A}_c v(t) + BG_N e_N(t), \\ \partial e_N(t)/\partial t = (K_N C_R - A_{NR}) v(t) + (A_N - K_N C_N) e_N(t). \end{cases} \tag{48}$$

This may be written in the form (36) with $A_{11} = \tilde{A}_c$, $A_{12} = BG_N$, $A_{21} = K_N C_R - A_{NR}$, and $A_{22} = A_N - K_N C_N$. The following result follows immediately from Lemma VI.3:

Theorem VIII.1. If H_N and H_R are stabilizing subspaces for (A, B) and the ROM of (1) associated with these subspaces is controllable and observable (in the finite-dimensional sense), then the gain K_N may be chosen in (15) so that this finite-dimensional controller produces an exponentially stable closed-loop system (48) when $\|K_N C_R - A_{NR}\|$ is sufficiently small.

Note that, from (46) and (48), the above result is a variation of theorem VI.2. We feel that the stabilizing subspace concept makes these results much more transparent. In addition, stabilizing subspaces can be used to tie the direct and indirect methods of controller synthesis together in discrete and continuous-time [39], [53].

B. IMBEDDING

The basic controller (15) is obtained by ignoring the residual subsystem. When additional information about the residuals is available, we want to make use of it. This was done in Section VII via singular perturbations that altered the basic controller (15) by adding "static" correction terms, derived from the residual subsystem, to produce (42). A different approach, which we shall describe here, is to *imbed* the controller in the full order DPS problem rather than designing it by ignoring residuals. We mean to treat the finite-dimensional controller as a reduced-order controller of the infinite-dimensional system (1); previously, we have synthesized the controller by making it a full-order design on the ROM. This discussion will follow [40] where all the details can be found.

Let H_N and H_R be any pair of model reduction subspaces satisfying the conditions of Section II.

For all $v_R(t)$ in the domain of A_R, we define $e(t)$ by

$$\hat{v}_N(t) = v_N(t) + T_R v_R(t) + e(t), \tag{49}$$

where T_R is a (bounded) linear operator from H_R into H_N. The original controller (15) is altered to produce the following *modified controller* of the same dimension:

$$\begin{cases} f(t) = G_N\hat{v}_N(t) + H_N y(t), & (50a) \\ \partial\hat{v}_N(t)/\partial t = A_N\hat{v}_N(t) + B_N f(t) + K_N[y(t) - \hat{y}(t)] & \\ \qquad + L_1\hat{v}_N(t) + L_2 f(t), & (50b) \\ y(t) = C_N\hat{v}_N(t), \quad \hat{v}_N(0) = 0. & (50c) \end{cases}$$

We define the gain operators L_1 and L_2 by

$$\begin{cases} L_1 = T_R A_{RN}, \\ L_2 = T_R B_R, \end{cases} \tag{51}$$

where T_R satisfies the *nonlinear equation*:

$$T_R = h(T_R) = FT_R A_R^{-1} + T_R A_{RN} T_R A_R^{-1} + (K_N C_R - A_{NR}) A_R^{-1} \tag{52}$$

and

$$F = A_N - K_N C_N. \tag{53}$$

Using either the implicit function theorem or the contraction mapping theorem (see [41]), conditions are obtained in [40] to guarantee a solution T_R of (52) and the following convergent (successive approximation) algorithm is presented:

$$\begin{cases} T_R^{k+1} = h\left(T_R^k\right), \\ T_R^0 = (K_N C_R - A_{NR}) A_R^{-1}. \end{cases} \tag{54}$$

From (49)-(53), we obtain

$$\partial e(t)/\partial t = (A_N - K_N C_N + T_R A_{RN}) e(t) \tag{55}$$

and

$$f(t) = Gv(t) + G_N e(t), \tag{56}$$

where we define the (bounded) gain operator G from H into R^M by

$$\begin{cases} GP_N = G_N + H_N C_N, \\ GP_R = G_N T_R + H_N C_R, \end{cases} \tag{57}$$

with P_N, P_R the projections defined in Section II. The closed-loop system consisting of (1) with the modified controller (50) can now be rewritten, using (55) and (56), as

$$\begin{aligned} \partial v(t)/\partial t &= (A + BG)v(t) + BG_N e(t) \\ \partial e(t)/\partial t &= (A_N - K_N C_N + T_R A_{RN}) e(t). \end{aligned} \tag{58}$$

The stability of this system can be analyzed via lemma (31) to produce the following result:

Theorem VIII.2. The closed-loop system (58), which uses the imbedded controller (50), is exponentially stable if

(a) (A, B) is exponentially stabilizable by G;

(b) (A_N, C_N) is observable in the finite-dimensional sense; and

(c) $\|A_{RN}\|$ is sufficiently small.

In [40], the suboptimality of the finite-dimensional controller (50) is assessed. This is done by comparison with the DPS optimal quadratic regulator and an upper bound on the suboptimality of such optimal control laws generated by the reduced-order controller (50) is produced. Such results are the DPS versions of those of Bongiorno and Youla in [42]. Of course, to produce these improved results requires that the residual subsystem data be known sufficiently accurately to solve (51)-(52) for the controller modifications. If these data are not trustworthy, it is better to stick with the original controller (15).

IX. DISTURBANCES AND NONLINEARITIES

In the previous sections we have omitted discussion of the term $F_D(t)$ in (1). This term can represent either external disturbances or system nonlinearities; we shall treat each of these possibilities here briefly.

A. *DISTURBANCE ACCOMMODATION*

We shall assume that external disturbances can enter the DPS (1) in the following way:

$$F_D(x, t) = \Gamma_0 f_D(t) = \sum_{\ell=1}^{M_D} \gamma_\ell(x) f_\ell^D(t). \tag{59}$$

In other words, there are a finite number (M_D) of disturbances with influence functions $\gamma_\ell(x)$ in H and corresponding amplitudes $f_\ell^D(t)$. The form (59) can represent most practical situations. In general, the disturbance influence functions, but not amplitudes, would be known to the control system designer, i.e., we know where the disturbances enter the system but not much else.

When a model reduction from Section II is performed on (1), the disturbances appear in (6)-(7):

$$\partial v_N(t)/\partial t = A_N v_N(t) + A_{NR} v_R(t) + B_N f(t) + F_N(t), \tag{60}$$

$$\partial v_R(t)/\partial t = A_{RN} v_N(t) + A_R v_R(t) + B_R f(t) + F_R(t), \tag{61}$$

$$y(t) = C_N v_N(t) + C_R v_R(t), \tag{62}$$

where

$$F_N(t) = \Gamma_N f_D(t) = \sum_{\ell=1}^{M_D} (P_N \gamma_\ell) f_D(t) \tag{63}$$

and

$$F_R(t) = \Gamma_R f_D(t) = \sum_{\ell=1}^{M_D} (P_R \gamma_\ell) f_D(t). \tag{64}$$

Although *it is not essential* to do so, henceforth, we shall *assume*:

$$F_R(t) = 0 \text{ (i.e., } \Gamma_R = 0). \tag{65}$$

This is not unreasonable because the model reduction would be chosen to retain the most "excitable modes" or, in other terms, to retain the major disturbance effects in the reduced-order model.

Stochastic disturbance amplitudes (as well as sensor noise) can be handled directly by applying the *separation theorem* of finite-dimensional control [23] to the ROM (60) and (62) with $v_R(t)$ set to zero. The form of the basic controller (15) does not change when this is done; however, the controller gains K_N are calculated from the noise and disturbance covariances (and the state estimator is called a Kalman filter). Therefore, the controller synthesis is not essentially different from Section V (although the closed-loop stability analysis of Section VI must be modified somewhat). We shall not consider stochastic disturbances further.

It often happens in practical problems that, instead of purely stochastic models of the disturbances, models of *disturbances with known waveform but unknown amplitude* may be used more effectively. For example, a narrow-band disturbance may be approximated quite well by a linear combination of sines and cosines at a small number of discrete frequencies; the frequencies are known but the amplitudes are not. Such disturbances may be generated by the following *waveform system*:

$$\begin{cases} f_D(t) = \theta_D z_D(t), \\ \dot{z}_D(t) = F_D z_D(t) + E_D v(t), \end{cases} \tag{66}$$

where the dimension of z_D is $N_D < \infty$ and θ_D, F_D, E_D are appropriate matrices (or bounded operators). Note that for the above-mentioned narrow-band disturbance model F_D would have only diagonal blocks of the form

$$\begin{bmatrix} 0 & 1 \\ -\omega_k^2 & 0 \end{bmatrix},$$

where ω_k are the frequencies included, and $E_D = 0$. Other types of disturbances, such as unknown biases, can be realistically modeled this way, too; see the survey [43] for details. For the rest of this subsection, we shall consider how to modify the basic controller (15) to accommodate disturbances generated by a waveform system (66) in the DPS (1); this is based on [44]-[45]. In the same spirit as (65) we shall assume in (66) that

$$E_R = E_D P_R = 0. \tag{67}$$

We begin by modifying the state estimator of (15) to include the waveform generator:

$$\begin{cases} \partial \hat{v}_N(t)/\partial t = A_N \hat{v}_N(t) + B_N f(t) + K_N[y(t) - \hat{y}(t)] \\ \qquad\qquad + \Gamma_N \theta_D \hat{z}_D(t), \\ \dot{\hat{z}}_D(t) = F_D \hat{z}_D(t) + E_N \hat{v}_N(t) + K_D[y(t) - \hat{y}(t)], \\ \hat{y}(t) = C_N \hat{v}_N(t), \quad \hat{v}_N(0) = 0, \quad \hat{z}_D(0) - 0. \end{cases} \tag{68}$$

where $E_N \equiv E_D P_N$. Note that the dimension of this state estimator is $N + N_D$. This system will attempt to estimate the disturbance amplitudes z(t) as well as the ROM states $v_N(t)$

from the sensor outputs $y(t)$. In the absence of the residual subsystem [i.e., $v_R(t) = 0$], the estimator gains K_N and K_D can always be chosen to yield *asymptotically covergent estimates* (at any desired covergence rate) *when* $(\tilde{A}_N, \tilde{C}_N)$ *is observable* in the finite-dimensional sense, where

$$\tilde{A}_N = \begin{bmatrix} A_N & \Gamma_N \theta_D \\ E_N & F_D \end{bmatrix} \quad \text{and} \quad \tilde{C}_N = [C_N \quad 0].$$

Now the controller has the capability to suppress (or accommodate) the disturbances by using a two-part *disturbance absorbing control law*:

$$\begin{cases} f(t) = G_N \hat{v}_N(t) - H_D \hat{f}_D(t), \\ \hat{f}_D(t) = \theta_D \hat{z}_D(t), \end{cases} \tag{69}$$

where G_N is *chosen as though no disturbances were present* and H_D is chosen so that

$$B_N H_D \approx \Gamma_N. \tag{70}$$

The best situation occurs when (70) is an exact equality, then the disturbance is completely absorbed. The ability of the control system to make (70) hold is related to how well the disturbance influence can be approximated by the actuator influence, i.e., *(70) is exact when*, for each $1 \leq \ell \leq M_D$, *we have*

$$P_N \gamma_\ell \text{ is in span } \{P_N b_1, \ldots, P_N b_M\}. \tag{71}$$

Consequently, the modified controller (68)-(69) is able to accommodate disturbances of known waveform but unknown amplitude. However, a price must be paid and that is an increase in the controller dimension by the number N_D. The size of this increase is a direct function of how well the known waveform of the disturbance can be represented by the waveform generator (66). Luckily, in many applications, N_D is small and the modified controller is quite satisfactory.

Of course, the closed-loop stability of this modified controller with the actual DPS (1) remains in question. The residual subsystem can alter this stability and, hence, the disturbance accommodating capability as well. Nonetheless, this analysis proceeds in a straightforward way as in Section VI and we refer the reader to [45] for details.

B. COMPENSATION FOR SYSTEM NONLINEARITIES

In this subsection, we shall assume that the disturbance term $F_D(t)$ in (1) is actually due to ignored system nonlinearities, i.e.,

$$F_D(t) = h[v(t), t]. \tag{72}$$

The nonlinear function h is *assumed to be continuous in the time argument and Lipschitz in the state*, i.e., for all v_1 and v_2 in a neighborhood N(r) of the origin

$$\|h(v_1, t) - h(v_2, t)\| \leq \mu\|v_1 - v_2\|, \tag{73}$$

where the constant $\mu > 0$. Furthermore, we *assume* for convenience:

$$h(0, t) = 0. \tag{74}$$

It is especially important to use the weak formulation instead of (1) when nonlinearities are present:

$$\begin{cases} v(t) = U(t)v_0 + \displaystyle\int_0^t U(t - \tau)[Bf(\tau) + h(v(\tau), \tau)]d\tau, \\ y(t) = Cv(t). \end{cases} \tag{75}$$

This makes the idea of a solution to the nonlinear system (1) sensible; it is the solution of (75) when v_0 is in N(r).

At the start, we may simply use the basic linear controller (15) in closed-loop with the *nonlinear* DPS (75). The stability of this system was analyzed in [46]. However, it is natural to expect that a *nonlinear* controller might handle the situation better; the problem is how to incorporate the nonlinearities into (15). The following approach seems reasonable: We leave the control law in (15) *linear* but modify the state estimator to be a *nonlinear state estimator*:

$$\begin{cases} \partial\hat{v}_N(t)/\partial t = A_N\hat{v}_N(t) + B_N f(t) + K_N[y(t) - \hat{y}(t)] \\ \qquad\qquad + g_N[\hat{v}_N(t), t] + L_1\hat{v}_N(t) + L_2 f(t), \\ \hat{y}(t) = C_N\hat{v}_N(t), \quad \hat{v}_N(0) = 0, \end{cases} \tag{76}$$

where L_1 and L_2 come from (51) and the *aggregated nonlinearity* is given by

$$g_N(z, t) = h_N(Lz, t) + T_R h_R(Lz, t) \tag{77}$$

with $h_N(\omega, t) = P_N h(\omega, t)$ and $h_R(\omega, t) = P_R h(\omega, t)$ and the (bounded) linear operator L from H_N into H defined by

$$L = T^{\#}, \text{ the } \textit{pseudo-inverse} \text{ of } T, \tag{78}$$

where $T = [I_N \quad T_R]$ and T_R is obtained from (52). This is the state estimator that was used in the imbedding method of Section VIII except that the aggregated nonlinearity has been added.

The stability analysis of the DPS (1) in closed-loop with this nonlinear controller is carried out in [47]. It is important for this analysis that the *control law remains linear even though the state estimator (76) is nonlinear*. These results for DPS may be compared with the finite-dimensional ones for nonlinear observers in [48]-[50]. It should be noted that *in finite dimensions* it is possible to choose T = L = I (the identity on H), i.e., the full nonlinearity $h(v, t)$ may be incorporated into (76); however, this is not possible for a DPS when (76) is finite dimensional.

X. APPLICATION: CONTROL OF MECHANICALLY FLEXIBLE STRUCTURES

Feedback control to suppress vibrations in mechanically flexible structures is relevant to many current engineering problems. Control of flexible structures, in general, was addressed in [3], [4], [6], [20], and [46] and the new application area, control of large aerospace structures, was surveyed in [5]. The ideas and results presented here for DPS have application to (and in many cases were motivated by) control of flexible structures. In this section, the flexible structure problem will be examined from the viewpoint established in the previous sections.

We consider the class of flexible systems that can be described by a generalized wave equation; this represents the idealization of many mechanically flexible structures. The generalized wave equation is given by

$$(\partial^2 u/\partial t^2) + D_0(\partial u/\partial t) + A_0 u = F, \tag{79}$$

which relates the vector of displacements $u(x, t)$ of a body Ω, a bounded open set with smooth boundary $\partial\Omega$ in n-dimensional Euclidean space R^n, to the applied control forces $F_0(x, t)$. The operators A_0 and D_0 are time-invariant, symmetric differential operators with compact resolvent and lower semibounded spectrum. The domain $D(A_0)$ of A_0 is dense in the Hilbert space $H_0 = L^2(\Omega)$ with $(\cdot, \cdot)_0$ denoting the usual inner product and $||\cdot||_0$ denoting the associated norm. The natural damping in the structure is modeled by the term $D_0(\partial u/\partial t)$, where D_0 has domain dense in H_0 and containing $D(A_0)$. The control forces

$$F_0(x, t) = B_0 f(t) = \sum_{i=1}^{M} b_i^0(x) f_i(t) \tag{80}$$

are provided by M actuators with influence functions $b_i^0(x)$. The displacements are measured by P-averaging sensors

$$y(t) = C_0 u(x, t), \tag{81}$$

where $y_j(t) = \int_\Omega c_j^0(x) u(x, t) dx$ with $j = 1, 2, \ldots, P$. The actuator and sensor functions $b_i^0(x)$, $c_j^0(x)$ are in H_0 and normalized to have unit integral. When the support of $b_i(x)$ is in a small neighborhood of a point x_i, we say it is a *point actuator* and, similarly, we define a *point sensor*. The point actuator and point sensor situation is of special interest here. This class of distributed parameter systems includes interior and boundary control of vibrating strings, membranes, thin beams, and thin plates. Although only displacement sensors are considered in (81), this is not a restriction, more general sensors may be modeled, as well as actuator and sensor dynamics.

It is well known ([18], p. 277) that the spectrum of A_0 contains only isolated eigenvalues λ_k with corresponding eigenfunctions ϕ_k such that

$$\lambda_1 \leq \lambda_2 \leq \cdots$$

and $A_0 \phi_k = \lambda_k \phi_k$. We shall *assume* that λ_1 is positive. The resonant *mode frequencies* ω_k of the structure are given by $\omega_k = (\lambda_k)^{1/2}$ and the corresponding eigenfunctions ϕ_k are the *mode shapes.* Thus A_0 satisfies

$$(A_0 u, u)_0 \geq a||u||^2, \quad a > 0 \tag{82}$$

and has a square root $A_0^{1/2}$. We shall assume that $D(D_0) \supseteq D\left(A_0^{1/2}\right)$. Every vector $u \in H_0$ has a unique representation

$$u(x) = \sum_{k=1}^{\infty} u_k \phi_k(x), \tag{83}$$

where $u_k = \int_\Omega u\phi_k \, dx$ and we define the orthogonal projections P_N^0, P_R^0 by

$$\begin{cases} P_N^0 u = \sum_{k=1}^{N} u_k \phi_k, \\ P_R^0 u = \sum_{k=N+1}^{\infty} u_k \phi_k. \end{cases} \tag{84}$$

Let V be the domain of A_0 and W be the domain of $A_0^{1/2}$. A new operator A is defined in H by

(a) $D(A) = V \times W \equiv H_1$

(b) $$A\begin{bmatrix} u \\ w \end{bmatrix} = \begin{bmatrix} w \\ -D_0 w - A_0 u \end{bmatrix}, \quad \text{for } u \in V,\ w \in W \tag{85}$$

The *energy inner product* $(\cdot, \cdot)$ is defined on H_1 by

$$\left(\begin{bmatrix} u_1 \\ w_1 \end{bmatrix}, \begin{bmatrix} u_2 \\ w_2 \end{bmatrix}\right) \equiv \left(A_0^{1/2} u_1, A_0^{1/2} u_2\right)_0 + (w_1, w_2)_0 \tag{86}$$

for $u_1, u_2 \in V$ and $w_1, w_2 \in W$, and the Hilbert space H is defined as the closure of H_1 in this energy inner product. The associated *energy norm* is denoted by $||\cdot||$ and is a measure of the total potential and kinetic energy in $(u, \partial u/\partial t)$ where u is a solution of (79). Let $v = [u^T, \partial u^T/\partial t]^T$ be in H and write (79)-(81) as

$$\begin{cases} \partial v/\partial t = Av + Bf, \; v_0 \in H, \\ \quad y = Cv, \end{cases} \tag{87}$$

where $B = \begin{bmatrix} 0 \\ B_0 \end{bmatrix}$ and $C = [C_0 \quad 0]$; this is in the form (1).

$$v = \sum_{k=1}^{\infty} \begin{bmatrix} u_k \\ \dot{u}_k \end{bmatrix} \phi_k \tag{88}$$

and $\|v\|^2 = \Sigma_{k=1}^{\infty} \left[\lambda_k u_k^2 + \dot{u}_k^2\right]$. Also, A generates a C_0 semigroup U(t) by [10], p. II-59.

The form of the damping operator D_0 substantially affects the behavior of the semigroup U(t). Here we shall study two extreme cases:

Case I: $D_0 = 2\alpha_0 I$ with $\alpha_0 > 0$

Case II: $D_0 = 2\xi_0 A_0^{1/2}$ with $\xi_0 > 0$

where

$$A_0 \phi_k = \omega_k^2 \phi_k \tag{89a}$$

$$A_0^{1/2} \phi_k = \omega_k \phi_k. \tag{89b}$$

The first case may be thought of as linear viscous damping in the structure. It affects all vibration modes equally and hence is a good guess for the damping operator when there is some doubt about the form (as there is in large aerospace structures). Of course, the physics of a given situation may dictate another form, but from a conservative viewpoint viscous damping is reasonable when further insight is unavailable. The second case is a kind of structural damping that produces a constant damping coefficient ξ_0 in all modes, i.e., the damping increases linearly with the mode frequency. The true form of the structural damping operator may lie somewhere in between these two extremes and need not even be a true differential operator. Consequently, we shall look at these two cases to see how they affect the control problem.

For case I, the semigroup $U(t)$ satisfies (see [24], proposition III.5):

$$\|U(t)\| < M \exp(-\sigma t), \quad t \geq 0, \tag{90}$$

where

$$\sigma = \frac{\alpha_0 a}{a + \alpha_0\left[\alpha_0^2 + \left(\alpha_0^2 + a\right)^{1/2}\right]} \tag{91}$$

and a is given in (82). In the *special case* $\alpha_0 < a$, i.e., the damping is smaller than the square of the lowest mode frequency, we have in (90) from [51]:

$$\begin{aligned} \sigma &= \alpha_0, \\ M &= [1 + (\alpha_0/a)][1 - (\alpha_0/a)]^{-1}. \end{aligned} \tag{92}$$

Another bound of this type is available in [46] without the restriction that $\alpha_0 < a$; however, the coefficient α_0 is not so simply related to σ and M. Nevertheless, in all versions of this type of damping, σ is directly related to α_0 and the semigroup $U(t)$ is exponentially stable.

Recalling (84), we define the projections P_N and P_R on H by

$$\begin{cases} P_N v = \begin{bmatrix} P_N^0 & u \\ P_N^0 & w \end{bmatrix}, \\ P_R v = \begin{bmatrix} P_R^0 & u \\ P_R^0 & w \end{bmatrix}, \end{cases} \tag{93}$$

where $v = \begin{bmatrix} u \\ w \end{bmatrix}$. These projections are orthogonal in the energy inner product (86). Consequently, $H_N = P_N(H)$ and $H_R = P_R(H) = H_N^{\perp}$ form a pair of reducing (modal) subspaces for (87) and partition this system into the form (6)-(8) with $A_{RN} = 0$ and $A_{NR} = 0$.

We note that, from [29] theorem X.9, H_N and H_R are reducing subspaces for $U(t)$ since they are for A. Consequently, by [18, p. 173], P_N and P_R commute with $U(t)$. Let $U_R(t) = P_R U(t) P_R$ and we have that $U_R(t)$ is the unique C_0 semigroup generated by $A_R = P_R A P_R$ because P_R commutes with $U(t)$. Therefore, we have, since $\|P_R\| = 1$ due to orthogonality, that

$$\|U_R(t)\| = \|P_R U(t)\| \leq \|P_R(t)\| \ \|U(t)\|$$
$$\leq \|U(t)\| \leq M \exp(-\sigma t), \quad t \geq 0. \tag{94}$$

So, the damping in case I satisfies our basic hypothesis (H1) in Section II. Furthermore, σ is fixed (and determined by α_0) no matter how large the ROM subspace dimension N is taken to be.

In case II, it is known from [52] that the semigroup $U(t)$ is *analytic* (see [8] or [10] for definition), as well as exponentially stable. Therefore, not only does the stability property (90) hold, but also the stability increases in the residual subspaces, i.e., hypothesis (H1) holds with

$$\|U_R(t)\| \leq M \exp(-\sigma_R t), \quad t \geq 0 \tag{95}$$

with σ_R increasing as the dimension N of H_N increases. This is consistent with the physical interpretation of this type of damping, the damping increases linearly with the mode frequency. Theoretically, it is possible to make σ_R large (without changing M) by retaining a large number of modes in the

ROM subspace H_N. Of course, in practice, such a large finite-dimensional controller may be impossible to implement on-line.

In both cases I and II, the ROM hs the parameters (A_N, B_N, C_N), which may be identified with the modal matrices:

$$\left(\begin{bmatrix} 0 & I_N \\ -\Lambda_N & -D_N \end{bmatrix}, \begin{bmatrix} 0 \\ B_N^0 \end{bmatrix}, \begin{bmatrix} C_N^0 & 0 \end{bmatrix} \right)$$

respectively, with $\Lambda_N = \text{diag}\left[\omega_1^2, \ldots, \omega_N^2\right]$, $B_N^0 = [(b_i, \phi_k)]^T$, and $C_N^0 = [(c_j, \phi_k)]$. In case I, $D_N = 2\alpha_0 I_N$, and in case II, $D_N = 2\xi_0 \text{ diag}[\omega_1, \ldots, \omega_N]$. When the damping is small (i.e., either α_0 or ξ_0 is small), the controllability and observability of (A_N, B_N, C_N) may be easily assessed using the modal conditions of [3] or [4]. These conditions are related to the interaction of the actuator-sensor influence functions and the retained mode shapes. These conditions can be used to help the designer locate the actuators and sensors on the structure. The minimum number of actuators or sensors necessary is given by the maximum multiplicity of the retained mode frequencies.

Once the controllability and observability of the modal ROM is determined, the modal controller (15) can be synthesized directly. It makes no difference whether we use the direct or indirect method of Section V because the actual mode shapes are being used in the ROM (theorem V.1). A basic assumption made in synthesizing such a modal controller is that the N-retained modes will adequately represent the critical vibrational behavior of the whole flexible structure with respect to the control task; this assumption can often be satisfied in practice and it is the basis for most structure controllers.

Having synthesized an implementable modal controller to tailor adequately the vibrational response of N-critical modes in the flexible structure, we must not think that we are done. The closed-loop behavior of our controller is affected by the *residual modes* in H_R for which we have not designed, as well as the *retained modes* in H_N for which we have designed. These interactions occur through the control and observation spillover terms defined in Section II and can drastically alter the desired performance of the controlled structure; in fact, examples are presented in [4] and [30] where the closed-loop system becomes unstable. Consequently, the analysis of the closed-loop behavior presented in Section VI is especially pertinent. The principal results of theorems VI.1-VI.2 give and indication of how much spillover can be tolerated in a stable closed-loop system; for further details, consult [6]. In addition, the singular perturbations results of Section VII may be applied to improve the stability of the controller by modifying it to the form (43). Of course, this requires a singularly perturbed formulation (40) of the structure (79)-(81); several such formulations were discussed in [35].

The methods and results of Sections VIII and IX may be applied, as well, to structures. Naturally, in most structural applications, although we may know the modes exist (i.e., the compact resolvent property of A_0 is easy to verify), we may not know their exact form. Hence, Galerkin or finite-element techniques have been used by structural engineers to approximate these mode shapes and frequencies. The use of such methods for control is the subject of [31] and [20]; with some minor modification, the basic ideas of modal control for structures can be used with this approximate modal data instead of the true modes.

XI. SUMMARY AND CONCLUSIONS

In an attempt to develop further the prescription for a practical distributed parameter control theory set out in [2], we have presented the above formulation and results. Much of this is a summary of our previous work, although we feel our present viewpoint is expressed most clearly here and we have tried to restate and unify our past work from this viewpoint.

The principal concern of this chapter has been with finite-dimensional controllers for distributed parameter systems (DPS). Such controllers are practical in the sense that they can be implemented with a small number of actuators and sensors and one, or more, small on-line computers. We feel that the methods suggested here for controller synthesis make maximum use of well-known lumped parameter control system knowledge and design experience. Furthermore, the emphasis on closed-loop analysis cannot be overstressed when infinite-dimensional systems are under consideration. Many of the perturbation techniques for stability analysis can be successfully applied to large-scale, as well as distributed parameter, systems.

We hope that we have fulfilled our promise of a *more* practical DPS theory, but we suggest that this is only a beginning. The process of improving DPS theory was set in motion in [2] and it is essential that it continue, in order to provide viable approaches to the wide variety of current and future engineering systems which exhibit a distributed parameter nature.

APPENDIX I: PROOFS OF THEOREMS IV.1 AND IV.2

Define $e(t) = \hat{v}(t) - v(t)$, which is in $D(A)$; hence, from (1) and (11):

$$\partial e/\partial t = (A - KC)e, \quad e(0) = -v_0. \tag{A.I.1}$$

Since (A^*, C^*) is exponentially stabilizable, there is a bounded operator $-K^*$ such that $A^* - C^*K^*$ generates the exponentially stable semigroup $W^*(t)$. But, from [29] theorem X.8, $W^*(t)$ is the adjoint of $W(t)$ which is generated by $A - KC$ and, therefore, $\|W^*(t)\| = \|W(t)\|$. Consequently, $W(t)$ is exponentially stable and from (A.I.1):

$$\|e(t)\| \le \|W(t)\| \; \|v_0\| \le M_2 \exp(-\sigma_2 t)\|v_0\| \tag{A.I.2}$$

for some $M_2 \ge 1$ and $\sigma_2 > 0$. This proves theorem IV.1.

In addition, if (A, B) is exponentially stabilizable, then $A + BG$ generates the semigroup $U_c(t)$ which satisfies (13). From (1) and (12),

$$\partial v/\partial t = (A + BG)v + BGe, \tag{A.I.3}$$

where $e(t)$ comes from (A.I.1). Since (A.I.3) means

$$v(t) = U_c(t)v_0 + \int_0^t U_c(t - \tau)BGe(\tau)d, \tag{A.I.4}$$

we have, from (A.I.1), the following:

$$\|v(t)\| \le M_c \exp(-\sigma_c t)\Big[\|v_0\| + M_2([\exp(\Delta t) - 1]/\Delta)\|BG\| \; \|v_0\|\Big], \tag{A.I.5}$$

where $\Delta = \sigma_c - \sigma_2$. Let $\sigma = \min(\sigma_c, \sigma_2)$. So (A.I.2) and (A.I.5) become (for $t \ge 0$):

$$\|v(t)\| \le M_c\Big[1 + (M_2/|\Delta|)\|BG\|\Big]\|v_0\| \exp(-\sigma t),$$

$$\|e(t)\| \le M_2\|v_0\| \exp(-\sigma t). \tag{A.I.6}$$

Hence, the closed-loop system (1) and (11)-(12), which is equivalent to (A.I.1) and (A.I.3), is exponentially stable with rate σ.

APPENDIX II: PROOF OF THEOREM IV.3

Suppose (H1) is satisfied for some pair of reducing subspaces H_N and H_R. Since (A_N, B_N) is controllable and dim $H_N = N < \infty$, there exists abounded operator G_N: $H_N \to R^M$ such that $A_N + B_N G_N$ generates the semigroup $U_N(t)$ with the property (see [23]):

$$||U_N(t)|| \leq M_N \exp(-\sigma_N t), \quad t \geq 0. \tag{A.II.1}$$

Also, by (H1), A_R generates the semigroup $U_R(t)$ with the property

$$||U_R(t)|| \leq M_R \exp(-\sigma_R t), \quad t > 0. \tag{A.II.2}$$

We can assume G_N was chosen so that $\sigma_N \neq \sigma_R$.

Define the bounded operator $G = G_N P_N$: $H \to R^M$ and consider that $A_c = A + BG$ generates the semigroup $U_c(t)$ since BG is a bounded perturbation of the semigroup generator A (see theorem X.9, [8]). Consider the projections P_N and P_R onto H_N and H_R, respectively. These produce (since H_N and H_R reduce A)

$$\begin{aligned} P_N(A + BG)P_N &= A_N + B_N G_N, \\ P_N(A + BG)P_R &= 0, \\ P_R(A + BG)P_N &= B_R G_N, \\ P_R(A + BG)P_R &= A_R. \end{aligned} \tag{A.II.3}$$

Consequently,

$$\dot{\omega} = A_c \omega \tag{A.II.4}$$

decomposes into

$$\begin{aligned} \dot{\omega}_N &= (A_N + B_N G_N)\omega_N, \\ \dot{\omega}_R &= B_R G_N \omega_N + A_R \omega_R, \end{aligned} \tag{A.II.5}$$

i.e.,

$$A_c = \begin{bmatrix} A_N + B_N G_N & 0 \\ B_R G_N & A_R \end{bmatrix}.$$

Therefore,

$$\|\omega_N(t)\| \leq M_N \exp(-\sigma_N t)\|\omega_N(0)\|$$

$$\|\omega_R(t)\| \leq M_R \exp(-\sigma_R t)\Big[\|\omega_R(0)\| \qquad \text{(A.II.6)}$$

$$+ \|B_R G_N\| M_N([\exp(\Delta t) - 1]/\Delta)\|\omega_N(0)\|\Big].$$

where $\Delta = \sigma_N - \sigma_R \neq 0$.

Let $M = M_N M_R$ and $\sigma = \min(\sigma_N, \sigma_R)$. Since $\|\omega_N\| = \|P_N \omega\| \leq \|P_N\| \; \|\omega\|$, and $M_N, M_R \geq 1$, (A.II.6) becomes

$$\|\omega_N(t)\| \leq M \exp(-\sigma t)\|P_N\| \; \|\omega(0)\|$$

$$\|\omega_R(t)\| \leq M \exp(-\sigma t)\Big(\|P_R\| \qquad \text{(A.II.7)}$$

$$+ \Big[\|B_R G_N\|/|\Delta|\Big]\|P_N\|\Big)\|\omega(0)\|.$$

Consequently,

$$\|\omega(t)\| \leq \|\omega_N(t)\| + \|\omega_R(t)\| \leq M \exp(-\sigma t)\|\omega(0)\|,$$

and $U_c(t)$ is exponentially stable. This proves (A, B) is exponentially stabilizable and a similar argument proves (A^*, C^*) is also. #

APPENDIX III: PROOFS OF LEMMA VI.3 AND THEOREMS VI.1 and VI.2

Proof of Lemma VI.3. Since we can write

$$A_0 = \begin{bmatrix} A_{11} & 0 \\ 0 & A_{22} \end{bmatrix} + \begin{bmatrix} 0 & A_{12} \\ 0 & 0 \end{bmatrix}$$

which is the sum of a semigroup generator and a bounded perturbation, A_0 generates a C_0 semigroup $U_0(t)$ by [8] theorem X.9. Furthermore,

$$A_c = A_0 + \Delta A \tag{A.III.1}$$

where

$$\Delta A = \begin{bmatrix} 0 & 0 \\ A_{21} & 0 \end{bmatrix}$$

also generates a C_0 semigroup $U_c(t)$ for the same reason.

If we write (36) in weak form, we obtain

$$\omega(t) = U_0(t)\omega_0 + \int_0^t U_0(t - \tau)\Delta A\omega(\tau)\,d\tau. \tag{A.III.2}$$

Taking norms of this,

$$\|\omega(t)\| \le K_0 \exp(-\sigma_0 t)\left[\|\omega_0\| + \int_0^t \exp(\sigma_0\tau)\|\Delta A\| \; \|\omega(\tau)\|\,d\tau\right].$$

Let $z(t) \equiv \exp(\sigma_0 t)\|\omega(t)\|$ and this becomes

$$z(t) \le K_0\|\omega_0\| + \int_0^t K_0\|\Delta A\|z(\tau)\,d\tau. \tag{A.III.3}$$

Now using the Gronwall Inequality [8] on the above, we have

$$z(t) \le K_0\|\omega_0\| \exp(K_0\|\Delta A\|t)$$

or equivalently,

$$\|\omega(t)\| \le K_0\|\omega_0\| \exp(-\sigma_c t), \tag{A.III.4}$$

where $\sigma_c = \sigma_0 - K_0\|\Delta\|$. It remains to determine K_0 and σ_0 in

$$\|U_0(t)\| \le K_0 \exp(-\sigma_0 t). \tag{A.III.5}$$

Consider the weak form of $\partial\omega/\partial t = A_0\omega$, which is given by

$$\begin{cases} \omega_1(t) = U_1(t)\omega_1(0) + \int_0^t U_1(t - \tau)A_{12}\omega_2(\tau)\,d\tau, \\ \omega_2(t) = U_2(t)\omega_2(0). \end{cases} \tag{A.III.6}$$

Taking norms of this, we have

$$\|\omega_2(t)\| \le K_2 \exp(-\sigma t)\|\omega_2(0)\| \tag{A.III.7}$$

and, from (A.III.7),

$$\begin{aligned}\|\omega_1(t)\| &\le K_1 \exp(-\sigma_1 t)\Big[\|\omega_1(0)\| \\ &\quad + \int_0^t \exp(\sigma_1 \tau)\|A_{12}\| \ \|\omega_2(\tau)\|d\tau\Big] \\ &\le K_1 \exp(-\sigma_1 t)\Big[\|\omega_1(0)\| \\ &\quad + \frac{\exp[(\sigma_1 - \sigma_2)\tau] - 1}{\sigma_1 - \sigma_2} \|A_{12}\|K_2\|\omega_2(0)\|\Big].\end{aligned}$$

Case I: $\sigma_1 < \sigma_2$

$$\|\omega_1(t)\| \le K_1 \exp(-\sigma_1 t)\left[\|\omega_1(0)\| + \frac{K_2\|A_{12}\|}{\sigma_2 - \sigma_1} \|\omega_2(0)\|\right].$$

Case II: $\sigma_1 > \sigma_2$

$$\begin{aligned}\|\omega_1(t)\| &\le K_1 \exp(-\sigma_2 t)\Big[\|\omega_1(0)\| \\ &\quad + \frac{1 - \exp[-(\sigma_1 - \sigma_2)]}{\sigma_1 - \sigma_2} \|A_{12}\|K_2\|\omega_2(0)\|\Big] \\ &\le K_1 \exp(-\sigma_2 t)\left[\|\omega_1(0)\| + \frac{K_2\|A_{12}\|}{\sigma_1 - \sigma_2} \|\omega_2(0)\|\right].\end{aligned}$$

Combining cases I and II, we have

$$\begin{aligned}\|\omega_1(t)\| &\le K_1 \exp(-\sigma_0 t)\left[\|\omega_1(0)\| + \frac{K_2\|A_{12}\|}{|\sigma_1 - \sigma_2|}\|\omega_2(0)\|\right] \\ &\le K_1K_2 \exp(-\sigma_0 t)\Big[\|\omega_1(0)\| \\ &\quad + \frac{\|A_{12}\|}{|\sigma_1 - \sigma_2|} \|\omega_2(0)\|\Big]\end{aligned} \tag{A.III.8}$$

because $K_2 \ge 1$.

Consider (since $K_1 \geq 1$)

$$\|\omega(t)\| = \left[\|\omega_1(t)\|^2 + \|\omega_2(t)\|^2\right]^{1/2} \leq K_1K_2 \exp(-\sigma_0 t)$$

$$\times \left[\left(\|\omega_1(0)\| + \gamma\|\omega_2(0)\|\right)^2 + \|\omega_2(0)\|^2\right]^{1/2}$$

$$\leq K_1K_2 \exp(-\sigma_0 t)(1 + \gamma + \gamma^2)^{1/2} \|\omega(0)\|. \qquad \text{(A.III.9)}$$

This latter follows from the inequality [16]:

$$(a + \gamma b)^2 + b^2 \leq (1 + \gamma + \gamma^2)(a^2 + b^2)$$

by taking $a = \|\omega_1(0)\|$ and $b = \|\omega_2(0)\|$.

From (A.III.9) we have $\sigma_0 = \min(\sigma_1, \sigma_2)$ and $K_0 = K_1K_2(1 + \gamma + \gamma^2)^{1/2}$ in (A.III.5). This proves (38) with $K_c = K_0$ and $\sigma_c = \sigma_0 - K_0\|\Delta A\|$. The dual result with A_{12} and A_{21} interchanged can be proved in a similar way. #

Proof of Theorem VI.1. Let $A_{21} = 0$. Then $\Delta A = 0$ in lemma VI.3 and $\sigma_c = \sigma_0$ in (38). This yields (33). The dual result is obtained similarly. #

Proof of Theorem VI.2. Use lemma VI.3 with $K_1 = M_N$ and $K_2 = M_R$ and $\|A_{12}\| \leq \Gamma_N$, $\|A_{21}\| \leq \beta_N$. So we have (34) with M_0 defined in theorem VI.1. The dual result is obtained similarly. #

ACKNOWLEDGMENTS

This work was partially supported by the National Science Foundation under Grant No. ECS-80-16173 and by the National Aeronautics and Space Administration under Contract No. NAS9-16053 and Grant No. NAG-1-171. Any opinions findings and conclusions or recommendations are those of the author and do not necessarily reflect the views of NSF or NASA.

REFERENCES

1. P. K. C. WANG, "Control of Distributed Parameter," in "Advances in Control Systems," Vol. 1 (C. T. Leondes, ed.), Academic Press, New York, 1964.

2. M. ATHANS, "Toward a Practical Theory of Distributed Parameter Systems," *IEEE Trans. Autom. Control AC-15*, 245-247 (1970).

3. M. BALAS, "Modal Control of Certain Flexible Dynamic Systems," *SIAM J. Control Opt. 16*, 450-462 (1978).

4. M. BALAS, "Feedback Control of Flexible Systems," *IEEE Trans. Autom. Control AC-23*, 673-679 (1978).

5. M. BALAS, "Some Trends in Large Space Structure Control: Fondest Hopes, Wildest Dreams," *IEEE Trans. Autom. Control*, June 1982 (to appear).

6. M. BALAS, "Feedback Control of Dissipative Hyperbolic Distributed Parameter Systems Using Finite Dimensional Controllers," *J. Math. Anal. Appl.* (to appear).

7. D. RUSSELL, "Controllability and Stabilizability Theory for Linear Partial Differential Equations: Recent Progress and Open Questions," *SIAM Rev. 20*, 639-739 (1978).

8. R. CURTAIN and A. PRITCHARD, "Functional Analysis in Modern Applied Mathematics," Academic Press, New York, 1977.

9. A. BALAKRISHNAN, "Applied Functional Analysis," Springer-Verlag, New York, 1976.

10. J. GOLDSTEIN, "Semigroups of Operators and Applications," Tulane University, New Orleans, Louisiana, Mathematics Lecture Notes, 1979.

11. J. WALKER, "Dynamical Systems and Evolution Equations: Theory and Applications," Plenum, New York, 1980.

12. R. TRIGGIANI, "On the Stabilizability Problem in Banach Space," *J. Math. Anal. Appl. 52*, 383-403 (1975).

13. R. CURTAIN, "Finite-Dimensional Compensator Design for Parabolic DPS with Point Sensors and Boundary Inputs," Groningen University, Groningen, The Netherlands, Report R.U.G., 1979.

14. R. CURTAIN and J. SCHUMACHER, "Finite-Dimensional Compensator Design for Parabolic Systems," presented at Mathematical Systems Theory Meeting, University of Warwick, Coventry, England, 1980.

15. N. FUJII, "Feedback Stabilization of DPS by a Functional Observer," *SIAM J. Control Opt. 18*, 108-120 (1980).

16. M. BALAS, "Feedback Control of Linear Diffusion Processes," *Int. J. Control 29*, 523-533 (1979).

17. H. FATTORINI, "Boundary Control Systems," *SIAM J. Control 6*, 349-385 (1968).

18. T. KATO, "Perturbation Theory for Linear Operators," Springer-Verlag, New York, 1966.

19. G. STRANG and G. FIX, "An Analysis of the Finite Element Method," Prentice-Hall, Englewood Cliffs, New Jersey, 1973.

20. M. BALAS, "Finite Element Models and Feedback Control of Flexible Aerospace Structures," *Proc. Joint Autom. Control Conf.*, San Francisco, California, 1980.

21. R. GENESIO and M. MILANESE, "A Note on the Derivation and Use of Reduced-Order Models," *IEEE Trans. Autom. Control AC-21*, 118-122 (1976).

22. D. RUSSELL, Book Review of *Infinite-Dimensional Linear Systems Theory* by R. Curtain and A. Pritchard, *Bull. Am. Math. Soc. 3*, 724-728 (1980).

23. H. KWAKERNAAK and R. SIVAN, "Linear Optimal Control Systems," Wiley, New York, 1972.

24. A. PRITCHARD and J. ZABCZYK, "Stability and Stabilizability of Infinite-Dimensional Systems," *SIAM Rev. 23*, 25-52 (1981).

25. M. SLEMROD, "A Note on Complete Controllability and Stabilizability for Linear Control Systems in Hilbert Space," *SIAM J. Control 12*, 500-508 (1974).

26. J. S. GIBSON, "A Note on Stabilization of Infinite-Dimensional Linear Oscillators by Compact Linear Feedback," *SIAM J. Control Opt. 18*, 311-316 (1980).

27. Y. SAKAWA and T. MATSUSHITA, "Feedback Stabilization of a Class of Distributed Systems and Construction of a State Estimator," *IEEE Trans. Autom. Control AC-20*, 748-753 (1975).

28. C. BENCHIMOL, "A Note on Weak Stabilizability of Contraction Semigroups," *SIAM J. Control Opt. 16*, 373-379 (1978).

29. P. FUHRMANN, "Linear Systems and Operators in Hilbert Space," McGraw-Hill, New York, 1981.

30. M. BALAS and S. GINTER, "Attitude Stabilization of Large Flexible Spacecraft," *J. Guidance Contr. 4*, 561-564 (1981).

31. M. BALAS, "The Galerkin Method and Feedback Control of Linear Distributed Parameter Systems," *J. Math. Anal. Appl.* (to appear); presented at Opt. Days Conference, Montreal, Quebec, Canada, 1981.

32. J. WALKER, "On the Application of Liapunov's Direct Method to Linear Dynamical Systems," *J. Math. Anal. Appl. 53*, 187-220 (1976).

33. P. KOKOTOVIĆ, R. O'MALLEY, and P. SANNUTI, "Singular Perturbations and Order Reduction in Control Theory — An Overview," *Automatica 12*, 123-132 (1976).

34. P. KOKOTOVIĆ, J. ALLEMONG, J. WINKELMAN, and J. CHOW, "Singular Perturbation and Iterative Separation of Time-Scales," *Automatica 16*, 23-33 (1980).

35. M. BALAS, "Closed-Loop Stability of Large Space Structures via Singular and Regular Perturbation Techniques," *Proc. IEEE Control Decision Conf.*, Albuquerque, New Mexico, 1980.

36. M. ARDEMA, "Singular Perturbations in Flight Mechanics," NASA Report TM-X-62, 380, August 1974.

37. M. BALAS, "Reduced-Order Feedback Control of DPS via Singular Perturbation Methods," *J. Math. Anal. Appl.* (to appear); presented at Princeton Conf. Inform. Sci. Systems, Princeton, New Jersey, 1980.

38. P. KOKOTOVIĆ, "Subsystems, Time-Scales, and Multi-Modeling," IFAC Symp. Large-Scale Systems, Toulouse, France, 1980.

39. M. BALAS, "Stabilizing Subspaces and Linear DPS: Discrete and Continuous Time Control," Proc. of 15th Asilomar Conf. on Circuits, Systems, and Computers, Pacific Grove, California, 1981.

40. M. BALAS, "Suboptimality and Stability of Linear DPS with Finite-Dimensional Controllers," Proc. of 3rd AIAA Symp. on Dynamics and Control of Large Flexible Spacecraft, Blacksburg, Virginia, 1981; to appear in revised form in *J. Opt. Appl.*

41. R. MARTIN, JR., "Nonlinear Operators and Differential Equations in Banach Spaces," Wiley, New York, 1976.

42. J. BONGIORNO, JR. and D. YOULA, "On Observers in Multivariable Control Systems," *Int. J. Control 8*, 221-243 (1968); and "Discussion of above Reference," *12*, 183-190 (1970).

43. C. D. JOHNSON, "Theory of Disturbance-Accommodating Controllers," in "Control and Dynamic Systems: Advances in Theory and Application," Vol. 12 (C. T. Leondes, ed.), Academic Press, New York, 1976.

44. M. BALAS, "Disturbance-Accommodating Control of DPS: An Introduction," *J. Interdiscip. Modeling*, Jan. 1980.

45. M. BALAS, "Disturbance Accommodating Control of Certain DPS," *Int. J. Policy Inform. 4*, 43-50 (1980).

46. M. BALAS, "Control of Flexible Structures in the Presence of Certain Nonlinear Disturbances," *Proc. Joint Autom. Control Conf.*, Denver, Colorado, 1979.

47. M. BALAS, "Nonlinear State Estimation and Feedback Control of Nonlinear and Bilinear DPS," *ASME J. Dyn. Systems, Meas., Control 102*, 78-83 (1981).

48. F. THAU, "Observing the State of Nonlinear Dynamic Systems," *Int. J. Contr. 17*, 471-479 (1973).

49. S. KOU, D. ELLIOT, and T. TARN, "Exponential Observers for Nonlinear Dynamic Systems," *Information and Contr. 29*, 204-216 (1975).

50. M. VIDYASAGAR, "On the Stabilization of Nonlinear Systems Using State Detection," *IEEE Trans. Autom. Contr. AC-25*, 504-509 (1980).

51. J. GOLDSTEIN and S. ROSENCRANS, "Energy Decay and Partition for Dissipative Wave Equations," *J. Differential Eqs.* (to appear).

52. G. CHEN and D. RUSSELL, "A Mathematical Model for Linear Elastic Systems with Structural Damping," *Quarterly Appl. Math.* (to appear).

53. M. BALAS, "Discrete-Time Control of DPS," Proc. of Intl. Symp. on Engr. Sci. and Mechanics, Natl. Cheng Kung University, Taihan, Taiwan, R.O.C.

INDEX

E

F

G

H